UNIT

1

INCREASED BY THE SCALE OF TENS. | DIVIDED, ACCORDING TO SCALE OF TENS.

&c. 10000 1000 100 10 1 | .1 .01 .001 .0001 .00001 &c.

1

INCREASED BY VARYING SCALES. | DIVIDED, ACCORDING TO VARYING SCALES.

		£	s.	d.	far.
		1	1	1	1
T.	cwt.	qr.	lb.	oz.	dr.
1	1	1	1	1	1

&c., for all denominate numbers.

$\frac{1}{4}$ $\frac{6}{9}$ $\frac{8}{17}$ $\frac{29}{47}$ } Simple.

$\frac{1}{6}$ *lb.* $\frac{1}{8}$ *oz.* $\frac{1}{7}$ *cwt.* } Denominate.

PROPORTION OF 1 TO ALL NUMBERS.
and of the numbers to each other.

APPLICATIONS.

For explanation of the Diagram, see "GRAMMAR OF ARITHMETIC."

ARITHMETICAL DIAGRAM.

THE

UNIVERSITY ARITHMETIC,

EMBRACING THE

SCIENCE OF NUMBERS,

AND THEIR NUMEROUS APPLICATIONS

BY

CHARLES DAVIES, LL. D.,

AUTHOR OF FIRST LESSONS IN ARITHMETIC; ARITHMETIC; ELEMENTARY ALGEBRA: ELEMENTARY GEOMETRY; ELEMENTS OF DRAWING AND MENSURATION; ELEMENTS OF SURVEYING; ELEMENTS OF ANALYTICAL GEOMETRY; DESCRIPTIVE GEOMETRY; SHADES, SHADOWS, AND PERSPECTIVE; AND DIFFERENTIAL AND INTEGRAL CALCULUS

REVISED AND IMPROVED EDITION.

NEW YORK:

PUBLISHED BY A. S. BARNES & CO.,

No 51 JOHN STREET.

1850.

Stereotyped by
RICHARD C. VALENTINE,
New York.

F. C. GUTIERREZ,
PRINTER,
Cor. John and Dutch-streets, N. Y.

PREFACE.

SCIENCE, in its popular signification, means knowledge reduced to order; that is, knowledge so classified and arranged, as to be easily remembered, readily referred to, and advantageously applied.

ARITHMETIC is the science of numbers. It lies at the foundation of the exact and mixed sciences, and a knowledge of it is an important element either of a liberal or practical education. While Arithmetic is a science in all that concerns the properties of numbers, it is yet an art in all that relates to their practical application. It is the first subject in a well-arranged course of instruction to which the reasoning powers of the mind are applied, and is the guide-book of the mechanic and man of business. It is the first fountain at which the young votary of knowledge drinks the pure waters of intellectual truth.

It has seemed to the author of the first importance that this subject should be well treated in our Elementary Text Books. In the hope of contributing something to so desirable an end, he has prepared a series of arithmetical works, embracing three books, entitled First Lessons in Arithmetic; Arithmetic; and University Arithmetic—the latter of which is the present volume

The First Lessons in Arithmetic are designed for beginners. The subjects treated are divided into separat lessons, each lesson embracing one combination of numbers, or one set of combinations.

The Arithmetic is designed for the use of schools and

academies, and contains all that is usually taught in a course of academical instruction.

The University Arithmetic is intended to answer another object. In it, the entire subject is treated as a science. The scholar is supposed to be familiar with the operations in the four ground rules, which are now taught to small children either orally or from elementary treatises. This being premised, the language of figures, which are the representatives of numbers, is carefully taught, and the different significations of which the figures are susceptible, depending on the manner in which they are written, are fully explained. It is shown, for example, that the simple numbers in which the value of the unit increases from right to left according to the scale of tens, and the Denominate or Compound numbers in which it increases according to a different scale, belong in fact to the same class of numbers, and that both may be treated under a common set of rules. Hence, the rules for Notation, Addition, Subtraction, Multiplication, and Division, have been so constructed as to apply equally to all numbers. This arrangement, which the author has not seen elsewhere, is deemed an essential improvement in the science of Arithmetic.

In developing the properties of numbers, from their elementary to their highest combinations, great labor has been bestowed in classification and arrangement. It has been a leading object to present the entire subject of arithmetic as forming a series of dependent and connected propositions: so that the pupil, while acquiring useful and practical knowledge, may at the same time be introduced to those beautiful methods of exact reasoning, which science alone can teach.

Great care has also been taken to demonstrate fully al'

the rules and to explain the reason of every process from the most simple to the most difficult. It has been thought that the Teachers of the country would like to possess a work of this kind, and that it might be studied advantageously as a text book in our advanced schools and academies. To adapt it to such a use, a large number of practical examples has been added, many of which have been selected from an English work by Keith.

In the preparation of the work, another object has been kept constantly in view, viz., to adapt it to the business wants of the country. For this purpose much pains have been bestowed in the preparation of the articles on Weights and Measures, foreign and domestic; on Banking, Bank Discount, Interest, Coins and Currency, and Exchanges.

Although by law the hundred weight is estimated at 100 pounds, and consequently the quarter at 25 pounds, in the United States, yet the old hundred of 112 pounds is still much used; and in all our intercourse with Great Britain, goods and wares are so estimated. Hence, it was thought best in this arithmetic, intended for general instruction, to retain the old standard.

In fine, it has been the aim of the author to publish both a scientific and practical treatise on the subject of Arithmetic, and one which shall in some measure correspond to the higher qualifications of teachers and the improved methods of communicating instruction.

Several excellent works, of an elementary character, having recently been published on Book-keeping, it has seemed best to omit, in the present edition, the article on that subject, and to supply its place by matter of a practical character.

FISHKILL LANDING, JANUARY, 1850.

DAVIES'
COURSE OF MATHEMATICS.

DAVIES' FIRST LESSONS IN ARITHMETIC—For beginners.

DAVIES' ARITHMETIC.—Designed for the use of Academies and Schools.

KEY TO DAVIES' ARITHMETIC

DAVIES' UNIVERSITY ARITHMETIC—Embracing the Science of Numbers, and their numerous applications.

KEY TO DAVIES' UNIVERSITY ARITHMETIC

DAVIES' ELEMENTARY ALGEBRA—Being an Introduction to the Science, and forming a connecting link between ARITHMETIC and ALGEBRA.

KEY TO DAVIES' ELEMENTARY ALGEBRA.

DAVIES' ELEMENTARY GEOMETRY.—This work embraces the elementary principles of Geometry. The reasoning is plain and concise, but at the same time strictly rigorous.

DAVIES' ELEMENTS OF DRAWING AND MENSURATION—Applied to the Mechanic Arts.

DAVIES' BOURDON'S ALGEBRA—Including Sturms' Theorem,—Being an Abridgment of the work of M. Bourdon, with the addition of practical examples.

DAVIES' LEGENDRE'S GEOMETRY AND TRIGONOMETRY.—Being an Abridgment of the work of M. Legendre, with the addition of a Treatise on MENSURATION OF PLANES AND SOLIDS, and a Table of LOGARITHMS and LOGARITHMIC SINES.

DAVIES' SURVEYING—With a description and plates of the THEODOLITE, COMPASS, PLANE-TABLE, and LEVEL: also, Maps of the TOPOGRAPHICAL SIGNS adopted by the Engineer Department—an explanation of the method of surveying the Public Lands, and an Elementary Treatise on NAVIGATION.

DAVIES' ANALYTICAL GEOMETRY—Embracing the EQUATIONS OF THE POINT AND STRAIGHT LINE—of the CONIC SECTIONS—of the LINE AND PLANE IN SPACE—also, the discussion of the GENERAL EQUATION of the second degree, and of SURFACES of the second order.

DAVIES' DESCRIPTIVE GEOMETRY,—With its application to SPHERICAL PROJECTIONS.

DAVIES' SHADOWS AND LINEAR PERSPECTIVE.

DAVIES' DIFFERENTIAL AND INTEGRAL CALCULUS.

CONTENTS.

FIRST FIVE RULES.

VULGAR FRACTIONS.

DECIMAL FRACTIONS.

RATIO AND PROPORTION OF NUMBERS.

PRACTICE—TARE AND TRET.

PERCENTAGE—INTEREST.

ARITHMETIC.

NOTATION AND NUMERATION.

1. SCIENCE in its popular sense, is knowledge reduced to order: that is, knowledge so classified and arranged, as to be easily remembered, readily referred to, and advantageously applied. In a strictly technical sense, it refers to the laws which connect the facts and principles of any subject of knowledge with each other.

2. ARITHMETIC is both a science and an art. It is a science in all that concerns the properties, laws and proportions of numbers; and an art in all that relates to their uses and applications.

3. NUMBERS are expressions for one or more things of the same kind: thus, the words *one, two, three, four, five, six, seven, eight, nine, ten, eleven, twelve, thirteen,* &c., are called numbers.

4. The *unit* of a number is one of the equal things which the number expresses. Thus, if the number express six apples, one apple is the unit; if it express five pounds of tea, one pound of tea is the unit; if ten feet of length, one foot is the unit; if four hours of time, one hour is the unit.

5. In common language numbers are expressed by words: in the language of arithmetic they are generally expressed by figures. In our language there are twenty-six different

---

QUEST.—1. What is Science? 2. What is Arithmetic? When is it a science and when an art? 3. What are numbers? Give an example. 4. What is the unit of a number? What is the unit of six apples? Of five pounds of tea? Of ten feet in length? Of four hours of time? 5. How are numbers expressed in common language? How are they expressed in the language of arithmetic? How many characters are there in our language?

characters called *letters:* in the language of arithmetic there are but *ten* characters which represent numbers; they are called *figures.* They are

naught,	one,	two,	three,	four,	five,	six,	seven,	eight,	nine.
0	1	2	3	4	5	6	7	8	9

The character 0 is used to denote the absence of a thing. As, if we wish to express by figures that there are no apples in a basket, we write, the number of apples in the basket is 0. The nine other figures are called *significant figures,* or *digits.*

6. Besides the figures which represent numbers, there are certain other characters used, called signs, which indicate the operations to be performed on numbers. They are the following:

The sign + is called *plus,* and when placed between two numbers, indicates that they are to be added together: thus, 3 + 2 shows that 3 and 2 are to be added, and is read, 3 plus 2.

The sign — is called *minus,* and when placed between two numbers, indicates that the one on the right is to be taken from the one on the left: thus, 4 — 3 shows that 3 is to be taken from 4, and is read, 4 minus 3.

The sign = is called the sign of equality, and when placed between two numbers, indicates that they are equal to each other: thus, 2 + 3 = 5 shows that 2 added to 3 gives a sum equal to 5, and is read, 2 plus 3 equals 5.

The sign × is called the sign of multiplication, and when placed between two numbers, indicates that they are to be multiplied together: thus, 12 × 3 shows that 12 is to be multiplied by 3, and is read, 12 multiplied by 3.

The sign ÷ is called the sign of division, and when placed between two numbers, indicates that the one on the left is to

---

QUEST.—What are the characters called? In arithmetic, how many characters are there which represent numbers? What are they called? Name them. What is the 0 used for? What are the other nine figures called? 6. What signs are used to indicate the operations to be performed on numbers? Name each, and explain its use.

be divided by the one on the right: thus, $8 \div 4$ shows that 8 is to be divided by 4: and is read, 8 divided by 4.

The parenthesis is used to indicate that the sum of two or more separate numbers is to be multiplied by a single number: thus, $(3 + 5) \times 6$ shows that the sum of 3 and 5 is to be multiplied by 6.

7. We have now learned the alphabet of the arithmetical language, and understand that

A single thing, or a unit of a number, may be expressed by 1,					
two things of the same kind, or two units,			"	"	by 2,
three	"	"	or three units	" "	by 3,
four	"	"	or four units	" "	by 4,
five	"	"	or five units	" "	by 5,
six	"	"	or six units	" "	by 6,
seven	"	"	or seven units	" "	by 7,
eight	"	"	or eight units	" "	by 8,
nine	"	"	or nine units	" "	by 9.

The units of the numbers expressed above are called *simple units*, or units of the *first order*.

8. The next step, in the arithmetical language, is to write the 0 on the right of the 1; thus, 10. This sign is the arithmetical expression for the word *ten*. The character 1 still expresses a single thing, viz., *one ten*. This ten, however, is *ten times as great* as the simple unit, and is called a unit of the *second order*.

9. We next write two 0's on the right of the 1; thus, 100. This is the arithmetical expression for one hundred, that is, for *ten tens*. Here, again, the 1 expresses but a single thing, viz., *one* hundred; but this *one* hundred is equal to ten units of the second order, or to one hundred units of the first order. In a similar manner we may form as many

---

QUEST.—7. What character stands for four things? What for eight? What are the units of such numbers called? 8. What is the next step in the language of figures? What does 1 still express? What is the single thing called? What is it equal to? 9. What is the next step? What does 1 still express? To how many units of the second order is it equal? To how many of the first?

orders of units as we please: thus, a single unit of the first order is expressed by - - - - - - 1,
a unit of the second order by 1 and a 0; thus, 10,
a unit of the third order by 1 and two 0's; thus, 100,
a unit of the fourth order by 1 and three 0's; thus, 1000,
a unit of the fifth order by 1 and four 0's; thus, 10000,
a unit of the sixth order by 1 and five 0's; thus, 100000,
and so on for the units of higher orders.

When units simply are named, *units of the first order are always meant.*

10. We see, from the language of figures, that units of the first order always occupy the place on the right; units of the second order the second place from the right; units of the third order, the third place; and so on for places still to the left.

We also see that ten units of the first order make one of the second; ten of the second, one of the third; ten of the third, one of the fourth; and so on for the higher orders. Hence, the language expresses that, *When figures are written by the side of each other, ten units of any one place make one unit of the place next to the left.*

11. For the purpose of reading figures, they are often separated into periods of three figures each. The units of the first order are read, simply, *units;* those of the second order are generally read, *tens;* those of the third, *hundreds;* those of the fourth, *thousands*, &c., according to the following

---

QUEST.—How is a single unit of the first order expressed? How do you express one unit of the second order? One of the third? One of the fourth? One of the fifth? 10. What places do units of different orders occupy? When figures are written by the side of each other, how many units of one order make one unit of the place next to the left? 11. How are figures separated for the purpose of reading? How are units of the first order read? Those of the second? Those of the third? Those of the fourth, &c.?

NUMERATION TABLE.*

Hundreds of Quintillions Tens of Quintillions Quintillions	7th period, or period of Quintillions.
Hundreds of Quadrillions Tens of Quadrillions Quadrillions	6th period, or period of Quadrillions.
Hundreds of Trillions Tens of Trillions Trillions	5th period, or period of Trillions.
Hundreds of Billions Tens of Billions Billions	4th period, or period of Billions.
Hundreds of Millions Tens of Millions Millions	3d period, or period of Millions.
Hundreds of Thousands Tens of Thousands Thousands	2d period, or period of Thousands.
Hundreds Tens Units	1st period, or period of Units.

The words at the head of the numeration table, *units, tens, hundreds*, &c., are equally applicable to all numbers, and must be committed to memory. The table may be continued to any extent. The higher periods take the names of Sextillions, Septillions, Octillions, Nonillions, Decillions, Undecillions, Duodecillions.

12. Expressing or writing numbers in figures is called NOTATION Reading the signification of the figures correctly, when written, is called NUMERATION.

EXAMPLES IN READING FIGURES.

1. In how many ways may the figures 658 be read?

1st. The common way, six hundred and fifty-eight.

2d. We may read, six hundreds, five tens, and eight units.

3d. We may read, sixty-five tens and eight units.

* NOTE.—This table is formed according to the French method of numeration. The English method gives six places to thousands, &c.

QUEST.—Are the words at the head of the table applicable to all numbers? May the table be continued? After what method is the table formed? What is the difference between it and the old English method? 12. What is notation? What is numeration? In how many ways may the figures 658 be read?

2. How may the figures 8046 be read?

1st. Eight thousand and forty-six. 2d. Eight thousand, no hundreds, four tens, and six units. 3d. Eighty hundreds and forty-six, or eighty hundreds, four tens, and six units. 4th. Eight hundred and four tens, and six units.

3. Give all the readings of the number 49704.

4. Give all the readings of the number 740692.

5. Give all the readings of the number 99800416.

6. Give all the readings of the number 80741047.

NOTE.—The pupil should be much exercised in these readings. He should remark that the lowest order of units used in any reading, whether it be units, tens, hundreds, &c., &c., gives the name or denomination to the part or whole of the number used in the reading.

We are now able to express any number whatever in the language of figures.

EXAMPLES.

1. Write, in figures, six units of the first order. *Ans.* 6.

2. Write, in figures, eight units of the second order. *Ans.* 80.

3. Write, in figures, nine units of the third order. *Ans.* 900.

4. Write, in figures, seven units of the fifth order. *Ans.* 70000.

5. Write, in figures, nine units of the first order, three of the third, and none of the second. *Ans.* 309.

6. Write, in figures, eight units of the eighth order, six of the fifth, seven of the seventh, five of the sixth, none of the fourth, none of the third, one of the second, and one of the first, and read the number. *Ans.* 87560011.

7. Write, in figures, six quintillions, four hundred and fifty-one billions, sixty-five millions, forty-seven ten thousands, and one hundred and four.

8. Write, in figures, nine hundred and ninety-nine octillions, sixty-five millions, eight hundred and forty-one billions, four trillions, and eleven nonillions.

QUEST.—How may the figures 8046 be read? Also, 49704? 740692? What gives the name or denomination to the number?

9. Write, in figures, sixty-five decillions, eight hundred quadrillions, seven hundred and fifty billions, seven hundred and fifty-one trillions, nine hundred and seventy-five thousand, three hundred and ten.

OF THE DENOMINATION OF NUMBERS.

13. A SIMPLE NUMBER is one which expresses a collection of units of the same kind, without expressing the particular value of the unit. Thus, 6 and 25 are simple numbers.

14. A DENOMINATE NUMBER expresses the kind of unit which is considered. For example, 6 dollars is a denominate number, the *unit* 1 dollar being denominated or named.

15. When two numbers have the same unit, they are said to be of the same denomination: and when two numbers have different units, they are said to be of different denominations. For example, 10 dollars and 12 dollars are of the same denomination; but 8 dollars and 20 cents express numbers of different denominations, the unit of 8 dollars being 1 dollar, and of 20 cents, 1 cent. *The kind of unit always indicates the denomination.*

In simple numbers, the unit in the place of units is different from the unit of the second order in the place of tens, and this last is different from that of the third order in the place of hundreds, and so on for places still to the left. These units, as we have seen, have different names or denominations, viz., simple units, or units of the first order; tens, or units of the second order; hundreds, or units of the third order, &c., and considered in this relation to each other, may be regarded as denominate numbers.

The following tables show the various kinds of denominate

QUEST.—13. What is a simple number? 14. What is a denominate number? 15. When are two numbers said to be of the same denomination? When of different denominations? What indicates the denomination? In simple numbers, how are the units of the different places? How do they compare in value with each other?

numbers in general use, and also the relative values of their different units.

OF FEDERAL MONEY.

16. Federal money is the currency of the United States. Its denominations, or names, are Eagles, Dollars, Dimes, Cents, and Mills.

The coins of the United States are of gold, silver, and copper, and are of the following denominations.

1. Gold: eagle, half-eagle, quarter-eagle, dollar.

2. Silver: dollar, half-dollar, quarter-dollar, dime, half-dime.

3. Copper: cent, half-cent.

If a given quantity of gold or silver be divided into 24 equal parts, each part is called a *carat*. If any number of carats be mixed with so many equal carats of a less valuable metal, that there be 24 carats in the mixture, then the compound is said to be as many carats fine as it contains carats of the more precious metal, and to contain as much alloy as it contains carats of the baser.

For example, if 20 carats of gold be mixed with 4 of silver, the mixture is called gold of 20 carats fine, and 4 parts alloy.

17. The standard, or degree of purity, of the gold coin, is fixed by Congress. Nine hundred equal parts of pure gold, are mixed with 100 parts of alloy, of copper and silver, (of which not more than one half must be silver,) thus forming 1000 parts, equal to each other in weight. The silver coins contain 900 parts of pure silver, and 100 parts of pure copper. The copper coins are of pure copper.

The eagle contains 258 grains of standard gold; the dollar 412½ grains of standard silver; the cent 168 grains of copper.

QUEST.—16. What is Federal Money? What are its denominations? Of what are the coins of the United States made? What are their denominations? What is a carat? What do you understand by 'carats fine?' What would be 20 carats fine of gold? 17. What is the standard of gold coin in the United States? What of silver? What of copper? What is the weight of the eagle? What of the dollar? What of the cent?

TABLE.

Mills.	Cents.	Dimes.	Dollars.	Eagle.
m.	*cts.*	*d.*	$	*E.*
10	= 1			
100	10	= 1		
1000	100	10	= 1	
10000	1000	100	10	= 1

This table is read, ten mills make one cent, ten cents one dime, ten dimes one dollar, ten dollars one eagle. In this table, ten units of each denomination make one unit of the denomination next higher, the same as in simple numbers.

In expressing Federal Money in the language of figures, the dollars are separated from the cents and mills by a comma: thus, 36,645 is read, 36 dollars, 64 cents, 5 mills; but may also be read, 36 dollars, 6 dimes, 4 cents, 5 mills; 375,043 is read 375 dollars, 4 cents, 3 mills.

ENGLISH CURRENCY.

18. The relative proportion between gold and silver in the English coins, according to the mint regulations, both for the old and new coinage, is as follows: in the *old coinage*, a pound of gold is worth 15.2096 times a pound of silver. In the new coinage, a pound of gold is worth 14.2878 times a pound of silver.

A standard gold coin is composed of 22 parts of pure gold and 2 parts of copper.

A standard silver coin is composed of 224 parts of pure silver and 18 parts of copper.

In the copper coin 24 pence make one pound avoirdupois.

QUEST.—Repeat the table of Federal money. How many units of each denomination make one of the next higher? In expressing Federal money in figures, how are the dollars separated from the cents? What place do the mills occupy, counting from the comma? 18. What is the relative proportion between gold and silver in the old and new coinage of English money? What is the standard of the English gold? Of the silver? What is the weight of the English penny?

TABLE.

Farthings.	Halfpence.	Pence.	Fourpences.	Sixpences.	Shillings.	Half-crowns.	Crowns.	7 shilling pieces.	Half-sovereigns.	Half-guineas.	Sovereigns.	Guineas.
far.		*d.*			*s.*						£	
2	=1											
4	2	=1										
16	8	4	=1									
24	12	6	$1\frac{1}{2}$	=1								
48	24	12	3	2	=1							
120	60	30	$7\frac{1}{2}$	5	$2\frac{1}{2}$	=1						
240	120	60	15	10	5	2	=1					
336	168	84	21	14	7	$2\frac{4}{5}$	$1\frac{2}{5}$	=1				
480	240	120	30	20	10	4	2	$1\frac{3}{7}$	=1			
504	252	126	$31\frac{1}{2}$	21	$10\frac{1}{2}$	$4\frac{1}{5}$	$2\frac{1}{10}$	$1\frac{1}{2}$	$1\frac{1}{20}$	=1		
960	480	240	60	40	20	8	4	$2\frac{5}{7}$	2	$1\frac{19}{21}$	=1	
1008	504	252	63	42	21	$8\frac{2}{5}$	$4\frac{1}{5}$	3	$2\frac{1}{10}$	2	$1\frac{1}{20}$	=1

AVOIRDUPOIS WEIGHT.

19. The standard avoirdupois pound of the United States, as determined by Mr. Hassler, is the weight of 27.7015 cubic inches of distilled water weighed in air.

By this weight are weighed all coarse articles, such as hay, grain, chandlers' wares, and all the metals, excepting gold and silver.

In this weight the words *gross* and *net* are used. Gross is the weight of the goods, with the boxes, casks, or bags in which they are contained. Net is the weight of the goods only; or what remains after deducting from the gross weight the weight of the boxes, casks, or bags.

An hundred weight, in its general sense, means 112 pounds, as appears in the table. But by the laws of the United States it is fixed at 100 pounds, (See Preface.)

Quest.—Repeat the table of English money. 19. What is the standard avoirdupois pound of the United States? For what is this weight used? What is the meaning of the terms *gross* and *net*? What is a hundred weight? How are goods now generally bought and sold?

TABLE.

Drams.	Ounces.	Pounds.	Quarters.	Hundreds.	Tons.
dr.	*oz.*	*lb.*	*qr.*	*cwt.*	*T.*
16	= 1				
256	16	= 1			
7168	448	28	= 1		
28672	1792	112	4	= 1	
573440	35840	2240	80	20	= 1

TROY WEIGHT.

20. By this weight are weighed gold, silver, jewels, and some liquids.

The standard troy pound of the United States, as determined by Mr. Hassler, is the weight of 22.794377 cubic inches of distilled water weighed in air. Hence, the pound is less than the pound avoirdupois.

TABLE.

Grains.	Pennyweights.	Ounces.	Pounds.
gr.	*pwt.*	*oz.*	*lb.*
24	= 1		
480	20	= 1	
5760	240	12	= 1

COMPARISON WITH AVOIRDUPOIS WEIGHT.

7000 troy grains = 1 lb. avoirdupois.
175 troy pounds = 144 lbs. "
175 troy ounces = 192 oz. "
$437\frac{1}{2}$ troy grains = 1 oz. "

QUEST.—Repeat the table of avoirdupois weight. 20. What articles are weighed by troy weight? What is the standard troy pound of the United States? Is it greater or less than the avoirdupois pound? Repeat the table of troy weight. How does it compare with avoirdupois weight?

APOTHECARIES WEIGHT.

21. This weight is used by apothecaries and druggists in mixing their medicines. They, however, buy and sell their drugs by avoirdupois weight. The pound and ounce are the same as the pound and ounce in troy weight. The difference between the two weights consists in the different divisions and subdivisions of the ounce.

TABLE.

Grains.	Scruples.	Drams.	Ounces.	Pound.
gr.	℈	ʒ	℥	℔
20	= 1			
60	3	= 1		
480	24	8	= 1	
5760	288	96	12	= 1

FOREIGN WEIGHTS.

22. The foreign weights differ somewhat from ours.

1 pound avoirdupois, English = 27.7274 cubic inches distilled water.

1 pound troy, English = 22.815689 cubic inches distilled water.

OLD FRENCH SYSTEM.

1 livre = 16 onces = 1.0780 lb. avoirdupois.
1 once = 8 gros = 1.0780 oz. "
1 gros = 72 grains = 58.9548 grains troy.
1 grain = 0.8188 "

QUEST.—21. By whom is apothecaries weight used? By what weight do druggists buy and sell their drugs? In what respects is the weight similar to troy? In what is the difference? Repeat the table of apothecaries weight. 22. What is the value of the English pound avoirdupois? Of the English pound troy? What is the old French system of weights?

NEW FRENCH SYSTEM.

23. The basis of this system of weights is the weight in vacuo of a cubic decimetre of distilled water. This weight is called a kilogramme, and is the unit of the French system. It is equal to 2.204737 pounds avoirdupois. (For the value of a decimetre, see table of linear measure, French, page 29.) The one-thousandth part of a kilogramme is called a gramme. and the one-thousandth part of a gramme is called a milligramme.

The divisions are made on the decimal principle, and are of the following denominations:

TABLE.

Milligramme.	Centigramme.	Decigramme.	Gramme.	Decagramme.	Hectogram'e.	Kilogr'me.	Quintal.	Millier. (ton sea-water (French))
10	= 1							
100	10	= 1						
1000	100	10	= 1					
10000	1000	100	10	= 1				
100000	10000	1000	100	10	= 1			
1000000	100000	10000	1000	100	10	= 1		
10000000	1000000	100000	10000	1000	100	10		
100000000	10000000	1000000	100000	10000	1000	100	=1	
1000000000	100000000	10000000	1000000	100000	10000	1000	10	=1

COMPARISON OF WEIGHTS.

English,	1 pound	= 1.000936 pounds avoirdupois.
French,	1 kilogramme	= 2.204737 " "
Spanish,	1 pound	= 1.0152 " "
Swedish,	1 pound	= 0.9376 " "
Austrian,	1 pound	= 1.2351 " "
Prussian,	1 pound	= 1.0333 " "

QUEST.—23. What is the basis of the new French system? Repeat the table. How do the weights of different countries compare with ours?

ENGLISH WOOL WEIGHT.

24. The following is the table of wool weight in England. As yet, many of the denominations have not been much used in this country; but as we are now exporting wool to England, they must soon be generally introduced.

TABLE.

Pounds.	Cloves.	Stones.	Tods.	Weys.	Sacks.	Last.
7	= 1					
14	2	= 1				
28	4	2	= 1			
182	26	13	$6\frac{1}{2}$	= 1		
364	52	26	13	2	= 1	
4368	624	312	156	24	12	= 1

CLOTH MEASURE.

25. Cloth measure is used by woollen and linen drapers. Hollands are measured in English ells, and tapestry by the French ell; woollens, linens, silks, tape, &c., by the yard.

TABLE.

Inches.	Nails.	Quarters.	Ells Flemish.	Yards.	Ells English.	Ell French.
in.	*na.*	*qr.*	*E. Fl.*	*yd.*	*E. E.*	*E. Fr.*
$2\frac{1}{4}$	= 1					
9	4	= 1				
27	12	3	= 1			
36	16	4	$1\frac{1}{3}$	= 1		
45	20	5	$1\frac{2}{3}$	$1\frac{1}{4}$	= 1	
54	24	6	2	$1\frac{1}{2}$	$1\frac{1}{5}$	= 1

QUEST.—24. Is English wool weight yet in use in this country? Repeat the table of wool weight. 25. By whom is cloth measure used? How are hollands measured? Tapestry? Repeat the table.

LONG MEASURE.

26. This measure is used to measure distances, lengths, breadths, heights, depths, &c. Gunter's chain is generally used by surveyors in measuring land. A standard measure has been adopted by the United States, copies of which are distributed in various parts of the country. This standard is a brass rod, one yard or 3 feet long.

TABLE.

Inches.	Gunter's Link.	Feet.	Yards.	Fathoms.	Rods.	Gunter's Chain.	Fur-longs.	Mile.
in.	*l.*	*ft.*	*yd.*		*rd.*	*c.*	*fur.*	*mi.*
$7\frac{23}{25}$	=1							
12	$1\frac{17}{33}$	=1						
36	$4\frac{6}{11}$	3	=1					
72	$9\frac{1}{11}$	6	2	=1				
198	25	$16\frac{1}{2}$	$5\frac{1}{2}$	$2\frac{3}{4}$	=1			
792	100	66	22	11	4	=1		
7920	1000	660	220	110	40	10	=1	
63360	8000	5280	1760	880	320	80	8	=1

FOREIGN MEASURES OF LENGTH.

27. The imperial standard yard of Great Britain is the one from which our yard is taken. It is referred to a natural standard, viz., to the distance between the axis of suspension and the centre of oscillation of a pendulum which shall vibrate seconds in vacuo, in London, at the level of the sea. This distance is declared to be 39.1393 *imperial inches;* that is, 1 imperial yard and 3.1393 inches.

QUEST.—26. For what is long measure used? For what is Gunter's chain used? Repeat the table of long measure. 27. What is the standard of the English imperial yard? What is its length?

OLD FRENCH SYSTEM.

1 point = 0.0074 U. S. inches.
1 line = 12 points = 0.08884 "
1 inch = 12 lines = 1.06604 "
1 foot = 12 inches = 12.7925 "
1 ell = 43 in. 10 lines = 46.728 " = 1.298 yd
1 toise = 6 feet = 76.755 " = 2.132 "
1 perch (Paris) = 18 feet.
1 perch (royal) = 22 feet.
1 league (common) 25 to a degree = 2280 toises = 4861 yards = 2.76 miles.
1 league (post) = 2000 toises = 4264 yards = 2.42 miles.
1 fathom (*Brasse*) = 5 feet French = 63.963 inches, or $5\frac{1}{3}$ feet English, nearly.
1 cable length = 100 toises, = 120 fathoms French = $106\frac{2}{3}$ fathoms English.

TABLE FOR REDUCING OLD FRENCH MEASURES TO UNITED STATES MEASURES.

(According to Mr. Hassler's comparison.)

French feet.	U. States inches.	French ft. or inches.	U. S. feet or inches.	French lines.	U. States inches.	French points.	U. States inches.
1	12.7925	1	1.0660	1	0.0888	1	0.0074
2	25.5850	2	2.1321	2	0.1777	2	0.0148
3	38.3775	3	3.1981	3	0.2665	3	0.0222
4	51.1700	4	4.2642	4	0.3554	4	0.0296
5	63.9625	5	5.3302	5	0.4442	5	0.0370
6	76.7550	6	6.3963	6	0.5330	6	0.0444
7	89.5475	7	7.4623	7	0.6219	7	0.0518
8	102.3400	8	8.5283	8	0.7107	8	0.0592
9	115.1325	9	9.5944	9	0.7995	9	0.0666
10	127.9250	10	10.6604	10	0.8884	10	0.0740
11	140.7175	11	11.7265	11	0.9772	11	0.0814

NEW FRENCH SYSTEM.

28. The basis of the new French system of measures is the

QUEST.—What is the old French long measure? 28. What is the basis of the new French system?

measure of the meridian of the earth, a quadrant of which is 10,000,000 *metres*, measured at the temperature of 32° Fahr. The multiples and divisions of it are decimals, viz.: 1 metre = 10 decimetres = 100 centimetres = 1000 millimetres = 39.3809171 United States inches, or 3.28174 feet.

Road measure. Myriametre = 10,000 metres. Kilometre = 1000 metres. Decametre = 10 metres. Metre = 0.51317 toise.

TABLE FOR REDUCING METRES TO INCHES.

(According to Mr. Hassler's comparisons; 1 metre = 39.3809171 inches.)

Metres.	Inches.	Metres.	Inches.	Metres.	Inches.	Metres.	Inches.
0.001	0.039381	0.026	1.023904	0.051	2.008427	0.076	2.992950
2	0.078762	27	1.063285	52	2.047808	77	3.032331
3	0.118143	28	1.102666	53	2.087189	78	3.071712
4	0.157524	29	1.142047	54	2.126570	79	3.111093
5	0.196905	0.030	1.181428	55	2.165950	0.080	3.150474
6	0.236286	31	1.220809	56	2.205331	81	3.189855
7	0.275666	32	1.260189	57	2.244712	82	3.229236
8	0.315047	33	1.299570	58	2.284093	83	3.268617
9	0.354428	34	1.338951	59	2.323474	84	3.307998
0.010	0.393809	35	1.378332	0.060	2.362855	85	3.347379
11	0.433190	36	1.417713	61	2.402236	86	3.386759
12	0.472571	37	1.457094	62	2.441617	87	3.426140
13	0.511952	38	1.496475	63	2.480998	88	3.465521
14	0.551333	39	1.535856	64	2.520379	89	3.504902
15	0.590714	0.040	1.575237	65	2.559760	0.090	3.544282
16	0.630095	41	1.614618	66	2.599141	91	3.583663
17	0.669476	42	1.653999	67	2.638522	92	3.623044
18	0.708855	43	1.693379	68	2.677903	93	3.662425
19	0.748237	44	1.732760	69	2.717283	94	3.701806
0.020	0.787618	45	1.772141	0.070	2.756664	95	3.741187
21	0.826999	46	1.811522	71	2.796045	96	3.780568
22	0.866380	47	1.850903	72	2.835426	97	3.819949
23	0.905761	48	1.890284	73	2.874807	98	3.859330
24	0.945142	49	1.929665	74	2.914188	99	3.898711
25	0.984593	0.050	1.969046	75	2.953569	0.100	3.938092

QUEST.—What are the multiples and divisions of it? Repeat the table of road measure

Austrian, 1 foot = 12.448 U. S. inches = 1.03737 foot.
Prussian, *Rhineland*, } 1 foot = 12.361 " " = 1.0300 "
Swedish, 1 foot = 11.690 " " = 0.974145 "
Spanish, { 1 foot = 11.034 " " = 0.9195 "
league (royal) = 25000 Span. ft. = $4\frac{1}{3}$ miles } nearly.
" (common) = 19800 " = $3\frac{1}{2}$ " }

SQUARE MEASURE.

29. Square measure is used for measuring all kinds of superficies, such as land, paving, flooring, plastering, and every thing which has length and breadth.

TABLE.

Square inches.	Square links.	Square feet.	Squa e yards.	Poles or perches.	Square chains.	Roods.	Acres.	Square mile.
Sq. in.	*Sq. l.*	*Sq. ft.*	*Sq. yd.*	*P.*	*Sq. c.*	*R.*	*A.*	*M.*
$62\frac{454}{625}$	= 1							
144	$2\frac{322}{1089}$	= 1						
1296	$20\frac{80}{121}$	9	= 1					
39204	625	$272\frac{1}{4}$	$30\frac{1}{4}$	= 1				
627264	10000	4356	484	16	= 1			
1568160	25000	10890	1210	40	$2\frac{1}{2}$	= 1		
6272640	100000	43560	4840	160	10	4	=1	
4014489600	64000000	27878400	3097600	102400	6400	1560	640	=1

FRENCH SUPERFICIAL MEASURE, OLD SYSTEM.

1 square inch = 1.1364 U. S. square inches.
1 arpent (Paris) = 100 square perches, (Paris,) or 900 square toises = 4088 square yards, or $\frac{5}{6}$ths of an acre, nearly.
1 arpent (woodland) = 100 square perches (royal) = 6108 square yards, or 1 acre, 1 rood, 1 perch.

QUEST.—29. For what is square measure used? Repeat the table. What is the old French system of square measure?

NEW SYSTEM.

1 are = 100 square metres = 119.665 square yards.
10 ares = 1 decare. 10 decares = 1 hecatare.

CUBIC OR SOLID MEASURE.

30. Forty cubic feet of round timber, or 50 solid feet of square timber, make 1 ton. A cord of wood is a pile 4 feet high, 4 feet wide, and 8 feet long, and consequently contains 128 solid feet. A *cord foot* is one foot in length of the pile which makes a cord. It contains 16 solid feet.

TABLE.

Cubic inches.	Cubic feet.	Cubic yards.	Cubic rods.	Cubic furlongs.	Cub. mile.
S. in.	*S. ft.*	*S. yd.*	*S. rd.*	*S. fur.*	*S. mi.*
1728	= 1				
46656	27	= 1			
7762392	$4492\frac{1}{8}$	$166\frac{3}{8}$	= 1		
496793088000	287496000	10648000	64000	= 1	
254358061056000	147197952000	5451776000	32768000	512	= 1

FRENCH SOLID MEASURE.

cubic foot = 2093.470 cubic inches of the U. States.
1 cubic decimetre = 61.074664 " " "
1 stere = 1 cubic metre = 61074.664 cubic inches = 35.375 cubic feet = 1.309 cubic yards.

LIQUID MEASURE OF THE UNITED STATES.

31. The standard gallon of the United States is the wine gallon, which measures 231 cubic inches, and contains, as

QUEST.—What the new system? 30. How much timber makes a ton? What is a cord of wood? How many solid feet does it contain? What is a cord foot? How many solid feet does it contain? Repeat the table of cubic or solid measure. Repeat the table of French solid measure. 31 What is the standard gallon of the United States?

determined by Mr. Hassler, 8.3388822 pounds avoirdupois of distilled water.

TABLE.

Cubic inches.	Gills.	Pints.	Quarts.	Gallon.
S. in.	*gi.*	*pt.*	*qt.*	*gal.*
$7\frac{7}{32}$	$=1$			
$28\frac{7}{8}$	4	$=1$		
$57\frac{3}{4}$	8	2	$=1$	
231	32	8	4	$=1$

DRY MEASURE.

32. The standard bushel of the United States is the *Win chester* bushel, which measures 2150.4 cubic inches, and con tains 77.627413 pounds avoirdupois of distilled water.

TABLE.

Cubic inches.	Pints.	Quarts.	Gallons.	Pecks.	Bushels.
S. in.	*pt.*	*qt.*	*gal.*	*pk.*	*bu.*
$33\frac{3}{5}$	$=1$				
$67\frac{1}{5}$	2	$=1$			
$268\frac{4}{5}$	8	4	$=1$		
$537\frac{3}{5}$	16	8	2	$=1$	
$2150\frac{2}{5}$	64	32	8	4	$=1$

FOREIGN MEASURES.

33. The British imperial gallon contains 10 pounds avoirdupois of distilled water weighed in air, and measures 277.274 cubic inches. The same measure is now used for liquids as for dry articles which are not measured by heaped measure.

QUEST.—Repeat the table of liquid measure. 32. What is the standard bushel of the United States? How many cubic inches does it contain? Repeat the table of dry measure. 33. What is the standard of the British imperial gallon?

For the latter, the bushel is heaped in the form of a cone, not less than 6 inches high, the base being 9½ inches.

French, 1 litre = 1 cubic decimetre = 61.074 U. S. cubic inches = 1.057 U. S. quarts, wine measure = 1.761 imperial pints of Great Britain.
1 boisseau = 13 litres = 793.364 cubic inches = 3.4349 gallons.
1 pinte = 0.931 litre = 56.817 cubic inches = 0.98397 quarts.

Spanish, 1 wine arroba = 4.2455 gallons.
1 fanega (corn measure) = 1.593 bushels.

ENGLISH ALE AND BEER MEASURE.

34. The following is the English beer measure. By it all malt liquors and water are measured.

TABLE.

Cubic inches.	Pints.	Quarts.	Imp. gallons.	Firkins.	Kilderkins.	Barrels.	Hogsheads.	Puncheons.	Butts.	Tun.
S. in.	*pt.*	*qt.*	*gal.*	*fir.*	*K.*	*bar.*	*hhd.*	*pun.*	*B.*	*tun.*
34.659¼	= 1									
69.318½	2	= 1								
277.274	8	4	= 1							
2495.466	72	36	9	= 1						
4990.932	144	72	18	2	=1					
9981.864	288	144	36	4	2	=1				
14972.796	432	216	54	6	3	1½	=1			
19963.728	576	288	72	8	4	2	1⅓	=1		
29945.592	864	432	108	12	6	3	2	1½	=1	
59891.184	1728	864	216	24	12	6	4	3	2	=1

QUEST.—What is heaped measure? What is the French measure? 34. What is measured by English beer measure? Repeat the table

ENGLISH WINE MEASURE.

35. The following is the English wine measure. All the denominations of it are not generally used in this country. By this measure all wines, brandy, rum, and distilled liquors are bought and sold.

TABLE.

Cubic inches.	Gills.	Pints.	Quarts.	Imp. gallons.	Ankers.	Runlets.	Barrels.	Tierces.	Hogsheads.	Puncheons.	Pipes, or butts.	Tun.
S. in.	*gi.*	*pt.*	*qt.*	*gal.*	*ank.*	*run.*	*bar.*	*tier.*	*hhd.*	*pun.*	*pi.*	*tun.*
$8.664\frac{13}{16}$	$=1$											
$34.659\frac{1}{4}$	4	$=1$										
$69.318\frac{1}{2}$	8	2	$=1$									
277.274	32	8	4	$=1$								
2772.740	320	80	40	10	$=1$							
4990.932	576	144	72	18	$1\frac{4}{5}$	$=1$						
8734.131	1008	252	126	$31\frac{1}{2}$	$3\frac{3}{20}$	$1\frac{3}{4}$	$=1$					
11645.508	1344	336	168	42	$4\frac{1}{5}$	$2\frac{1}{3}$	$1\frac{1}{3}$	$=1$				
17468.262	2016	504	252	63	$6\frac{3}{10}$	$3\frac{1}{2}$	2	$1\frac{1}{2}$	$=1$			
23291.016	2688	672	336	84	$8\frac{2}{5}$	$4\frac{2}{3}$	$2\frac{2}{3}$	2	$1\frac{1}{3}$	$=1$		
34936.524	4032	1008	504	126	$12\frac{3}{5}$	7	4	3	2	$1\frac{1}{2}$	$=1$	
69873.048	8064	2016	1008	252	$25\frac{1}{5}$	14	8	6	4	3	2	$=1$

ENGLISH CORN OR DRY MEASURE.

36. Dry measure is used for all dry commodities, such as wheat, barley, beans, coal, oysters, &c. The following is the English table, all the denominations of which are not in general use in this country. The standard bushel is a cylinder 18.789 inches in the interior diameter, and 8 inches in depth, and consequently contains 2218.192 cubic inches.

QUEST.—35. What liquids are measured by wine measure? Repeat the table. 36. What articles are measured by corn measure? What is the standard bushel? How many cubic inches does it contain?

TABLE.

Cubic inches.	Pints.	Quarts.	Pottles.	Imperial gallons.	Pecks.	Bushels.	Strikes.	Cooms.	Quarters.	Weys.	Last.
S. in.	*pt.*	*qt.*		*gal.*	*pk.*	*bu.*			*qr.*		
34.659¼	= 1										
69.318½	2	= 1									
138.637	4	2	= 1								
277.274	8	4	2	= 1							
554.548	16	8	4	2	= 1						
2218.192	64	32	16	8	4	=1					
4436.384	128	64	32	16	8	2	=1				
8872.768	256	128	64	32	16	4	2	=1			
17745.536	512	256	128	64	32	8	4	2	=1		
88727.680	2560	1280	640	320	160	40	20	10	5	=1	
177455.360	5120	2560	1280	640	320	80	40	20	10	2	=1

OLD AND NEW ENGLISH COAL MEASURE.

37. By act of Parliament passed in 1831, all coals sold within 25 miles of the Post Office in London, are to be sold by weight. One sack weighs 2 cwt. or 224 lbs.; consequently, 10 sacks make 1 ton. Twelve sacks make a London chaldron of 36 bushels, while it takes 79½ bushels to make a Newcastle chaldron, as shown by the table.

TABLE.

Pounds weight.	Pecks.	Bushels.	Sacks.	Vats, or strikes.	London chaldron.	Newc. chald.	Keels.	Scows.	1 ship load.
$18\frac{2}{3}$	= 1								
$74\frac{2}{3}$	4	= 1							
224	12	3	= 1						
672	36	9	3	= 1					
2688	144	36	12	4	= 1				
5936	318	$79\frac{1}{2}$	$26\frac{1}{2}$	$8\frac{5}{6}$	$2\frac{5}{24}$	=1			
47488	2544	636	212	$70\frac{2}{3}$	$17\frac{2}{3}$	8	=1		
56448	3024	756	252	84	21	$9\frac{27}{53}$	$1\frac{10}{23}$	=1	
949760	50880	12720	4240	$413\frac{1}{3}$	$353\frac{1}{3}$	160	20	$13\frac{31}{33}$	= 1

QUEST.—Repeat the table. 37. What was established by act of Parliament? Repeat the table of coal measure.

MEASURE OF TIME.

Thirds.	Seconds.	Minutes.	Hours.	Days.	Weeks.	Months.	Year.
thirds.	*sec.*	*m.*	*hr.*	*da.*	*wk.*	*mo.*	*yr.*
60	= 1						
3600	60	= 1					
216000	3600	60	= 1				
5184000	86400	1440	24	= 1			
36288000	604800	10080	168	7	=1		
145152000	2419200	40320	672	28	4	=1	
1893456000	31557600	525960	8766	$365\frac{1}{4}$	$52\frac{5}{28}$	$13\frac{5}{112}$	= 1

38. The whole days only are reckoned. The odd six hours, by accumulating for 4 years, make one day, so that every fourth year contains 366 days. This is called the Bissextile, or Leap year. The leap years may always be known by this, that the numbers which express them are exactly divisible by 4. Thus, 1840, 1844, 1848, &c., are all leap years.

Although the year is reckoned at 365*da.* 6*hr.*, it is in fact but 365*da.* 5*hr.* 48*m.* 48*sec.*, and the difference by acumulating for 100 years makes about 1 day, so that the centennial years, though divisible by 4, are not leap years.

The year is also divided into 12 calendar months, which contain an unequal number of days.

		Names.	*No of Days.*
1	month	January - - -	31
2	"	February - - -	28
3	"	March - - -	31
4	"	April - - - -	30
5	"	May - - - -	31
6	"	June - - - -	30
7	"	July - - - -	31
8	"	August - - -	31
9	"	September - - -	30
10	"	October - - -	31
11	"	November - - -	30
12	"	December - - -	31
		Total	365

QUEST.—38. Repeat the table of time. What is the length of a year? What is done with the quarter of a day? How do you determine the leap years? What years that are divisible by four are not leap years?

The additional day, when it occurs, is added to the month of February, so that this month has 29 days in the Leap year.

Thirty days hath September,
April, June, and November;
All the rest have thirty-one,
Excepting February, twenty-eight alone.
But Leap year coming once in four,
February then has one day more.

TABLE, SHOWING THE NUMBER OF DAYS FROM ANY DAY OF ONE MONTH TO THE SAME DAY OF ANY OTHER MONTH IN THE SAME YEAR.

From any day of	To the same day.												From any day of
	Jan.	Feb.	Mar.	April	May	June	July	Aug.	Sept.	Oct.	Nov.	Dec.	
January	365	31	59	90	120	151	181	212	243	273	304	334	Jan.
February	334	365	28	59	89	120	150	181	212	242	273	303	Feb.
March	306	337	365	31	61	92	122	153	184	214	245	275	March
April	275	306	334	365	30	61	91	122	153	183	214	244	April
May	245	276	304	335	365	31	61	92	123	153	184	214	May
June	214	245	273	304	334	365	30	61	92	122	153	183	June
July	184	215	243	274	304	335	365	31	62	92	123	153	July
August	153	184	212	243	273	304	334	365	31	62	92	122	August
Sept.	122	153	181	212	242	273	303	334	365	30	61	91	Sept.
October	92	123	151	182	212	243	273	304	335	365	31	61	Oct.
Nov.	61	92	120	151	181	212	242	273	304	334	365	30	Nov.
Dec.	31	62	90	121	151	182	212	243	274	304	335	365	Dec.

39. The months counted *from any day of*, are arranged in the left hand vertical column; those counted *to the same day*, are in the upper horizontal line: the days between these periods are found in the angle of intersection, in the same way as in a common table of multiplication. If the end of February be included between the two points of time, a day must be added in leap years. Suppose, for example, it were

---

QUEST.—38. What are the calendar months? How many days does each contain? What is done with the odd day in leap year? Repeat the verse which indicates the number of days in each month of the year. 39. What is the object of this table?

required to know the number of days from the fourth of March to the fifteenth of August. In the left hand vertical column find March, and then referring to the intersection of a horizontal line drawn from March, with the column under August, we find 153, which is the number of days from the fourth (*or any other*) day of March to the fourth (*or same*) day of August; but as we want the time to the fifteenth of August, 11 days (*the difference between* 4 *and* 15) must be added to 153, which shows that 164 is the number of days between the fourth of March and the fifteenth of August.

Again, required the number of days between the tenth of October and the third of June, in the following year. Opposite to October and under June, we find 243, which is the number of days from the tenth of October to the tenth of June; but as we sought the time to the third only, which is 7 days earlier, we must deduct 7 from 243, leaving 236, the number of days required; and so of others.

DIVISION OF THE CIRCLE—MEASURE OF TIME.

The geographical division of any line drawn round the circumference of the Earth.	Diurnal motion of the Earth reduced to time.
60 seconds, 1 minute - - - - -	= 4 seconds.
60 minutes, 1 degree - - - - -	= 4 minutes.
15 degrees, ½ sign of the zodiac - - -	= 1 hour.
30 degrees, 1 sign of the zodiac - - -	= 2 hours.
90 degrees, 1 quadrant - - - -	= 6 hours.
4 quadrants, or 360 degrees, 1 great circle -	= 24 hours.

40. Every circle is supposed to be divided into 360 equal parts called degrees, each degree into 60 equal parts called minutes, and each minute into 60 equal parts called seconds. For astronomical purposes, the circumference of the circle is also supposed to be divided into 12 equal parts, each of which

QUEST.—How do you find the number of days from the fourth of March to the fifteenth of August? What is the number of days from the tenth of October to the third of June? Also the same in a leap year? 40. How is any circle supposed to be divided? What is a sign, or sign of the zodiac? Repeat the table.

is called a sign. The characters which mark these divisions are as follows:

c	s	°	′	″
circumference,	sign,	degree,	minutes,	seconds.

TABLE OF PARTICULARS.

41. For various things

12 things	make	1 dozen.
12 dozen	- - - -	1 gross.
12 gross, or 144 dozen	-	1 great gross.
	ALSO,	
20 things	make	1 score.
112 pounds	- - - -	1 quintal of fish.
24 sheets of paper	- -	1 quire.
20 quires	- - - -	1 ream.
2 reams	- - - -	1 bundle.

BOOKS.

A sheet folded in two leaves is called a folio.		
" folded in four leaves	-	a quarto, or 4to.
" folded in eight leaves	-	an octavo, or 8vo.
" folded in twelve leaves	-	a duodecimo, or 12mo.
" folded in eighteen leaves	-	an 18mo.

DIMENSIONS OF DRAWING PAPER.

Demy,	1 ft. 7½ in.	by	1 ft. 3½ in.
Medium,	1 ft. 10 in.	by	1 ft. 6 in.
Royal,	2 ft. 0 in.	by	1 ft. 7 in.
Super Royal,	2 ft. 3 in.	by	1 ft. 7 in.
Imperial,	2 ft. 5 in.	by	1 ft. 9¼ in.
Elephant,	2 ft. 3¾ in.	by	1 ft. 10¼ in.
Columbia,	2 ft. 9¾ in.	by	1 ft. 11 in.
Atlas,	2 ft. 9 in.	by	2 ft. 2 in.
Double Elephant,	3 ft. 4 in.	by	2 ft. 2 in.
Antiquarian,	4 ft. 4 in.	by	2 ft. 7 in.

QUEST.—41. Repeat the table of particulars. Also for books. What are the dimensions of drawing paper?

REMARKS ON THE FORMATION OF NUMBERS.

42. We have seen (Art. **10**) that when figures are written by the side of each other, thus,

8562041304723

the language implies that the unit in each place is equal to ten units of the place next to the right; or that ten units of any one place make one unit of the place next to the left.

43. When figures are written thus,

£	*s.*	*d.*	*far.*
4	17	10	3

the language implies, that four units of the lowest denomination make one of the second; twelve of the second, one of the third; and twenty of the third, one of the fourth.

44. When figures are written thus,

T.	*cwt.*	*qr.*	*lb.*	*oz.*	*dr.*
27	17	2	27	11	10

the language implies, that 16 units of the lowest denomination make one of the second; 16 of the second, one of the third; 28 of the third make one of the fourth; four of the fourth, one of the fifth; and 20 of the fifth, one of the sixth.

All the other denominate numbers are formed on the same principle; and in all of them we pass from a lower to the next higher denomination by considering how many units of the one make one unit of the other.

45. In our written language, each of its elementary letters has a particular signification, which must be learned as a first step. We next learn to place these letters in the form of words, and then what may be done by using these words in connection with each other.

Quest.—42. When figures are written by the side of each other, what does the language imply? 43. When figures are written with the mark *£ s. d. far.* placed over them, what does that language imply? 44. When figures are written with *T. cwt. qr. lb. oz. dr.* placed over them, what relation exists between the orders of their units? How do we always pass from one denomination of denominate numbers to another? 45. How do we learn a common language?

So in figures: we first learn what each figure expresses by itself, and then what it is made to express in all the various ways in which it may be written. We thus learn the language of figures.

46. Let us give a few examples of the changes which are produced in the signification, by changing the places of letters and figures.

In common language, *was*, is a known word.
But the same letters also give *saw*, an instrument.
Also, 375 expresses, three hundred and seventy-five;
but 573 expresses, five hundred and seventy-three.

It may be well to observe that the same letter has the same name, and generally represents the same sound wherever it may fall in a word. So, likewise, the same figure always expresses the same *number* of units, wherever it may be placed. Thus, in the example above: in the first number, 5 expresses *five* units of the first order, and 3, *three* units of the third. In the second number, 5 expresses *five* units of the third order, and 3, *three* units of the first order. The *value* of the unit, however, always depends on the *place* of the figure.

OF REDUCTION.

47. Reduction is changing the denomination of a number from one unit to another, without altering the value of the number. Thus, if we have 2 tens, and wish to reduce them to the denomination of units of the first order, we multiply by 10, or add one 0; this gives 20 units of the first order, which are equal to 2 tens.

Quest.—How must we learn the language of figures? 46. Give some examples of the changes in signification which are produced by altering the places of letters and figures. Has the same letter always the same name and sound? Has the same figure always the same name? Does it always express the same number of units? Does the value of the unit expressed remain the same? On what does it depend? 47. What is reduction? How are tens reduced to units of the first order?

If, on the contrary, we wish to reduce 300 to units of the second order, we divide by 10, and the quotient is 30 units of the second order, which are equal to 300. Had we wished to reduce to units of the third order, we should have divided by 100, giving 3 for the quotient: hence, reduction of denominate numbers is divided into two parts;

1st. To reduce a number from a higher denomination to a lower; and

2d. To reduce a number from a lower denomination to a higher.

The first reduction is effected by beginning with the number in the highest denomination. *Multiply this number by the value of its unit expressed in units of the next lower denomination, and add to the product the number in that denomination. Proceed in the same manner through all the denominations to the lowest.*

The second reduction is effected thus: *Divide the given number by so many as make one of the denomination next higher; set aside the remainder, if any, and proceed in the same manner through all the denominations to the highest.*

Thus, in the first, if we wish to reduce

£	*s.*	*d.*
3	14	4

to pence, we first multiply the £3 by 20, which gives 60 shillings. We then add 14, making 74 shillings. We next multiply by 12, and the product is 888 pence. To this we add 4*d.*, and we have 892 pence, which are of the same value as £3 14*s.* 4*d.*

If, on the contrary, we wished to change 892 pence to pounds, shillings, and pence, we should first divide by 12: the quotient is 74 shillings, and 4*d.* over. We again divide by 20, and the quotient is £3, and 14*s.* over: hence, the result is £3 14*s.* 4*d.*, which is equal to 892 pence.

QUEST.—How will you reduce units of the first order to those of the second? How to those of the third? To those of the fourth? Into how many parts is reduction of denominate numbers divided? How do you effect the first reduction? How do you effect the second?

The reductions, in all the denominate numbers, are made in the same manner.

EXAMPLES.

1. In £5 5*s*., how many shillings, pence, and farthings?

£	*s*.	
5	5	
20		
105		shillings.
12		
1260		pence.
4		
5040		

Here the reduction is from a greater to a less unit.

In 5040 farthings, how many pence, shillings, and pounds?

4)5040 farthings.
12)1260 pence.
2|0)10|5 shillings.
£5 5*s*.

In this example, the reduction is from a less to a greater unit.

2. In 55 guineas, how many shillings, pence, and farthings?

3. Reduce £54 11*s*. 9¼*d*. to farthings.

4. Reduce £77 11*s*. 10½*d*. to halfpence.

5. Reduce £94 14*s*. 8*d*. to pence.

6. Reduce £47 14*s*. 4*d*. to twopences.

7. Reduce £34 11*s*. 9*d*. to threepences, and to pence.

8. In £108 11*s*. 6*d*., how many sixpences?

9. How many crowns, half-crowns, shillings, sixpences, pence, and farthings are there in £54?

10. Reduce £74 13*s*. 9*d*. into shillings, threepences, and farthings.

11. In 11520 farthings, how many pence, shillings, and pounds?

12. In 17880 pence, how many pounds?

13. Reduce 100000 farthings into guineas.

14. In 50400 halfpence, how many pounds?

15. In 12050 shillings, how many crowns and pounds?

16. Reduce 311040 pence into shillings, crowns, and pounds.

17. Reduce 17*lb.* 5*oz.* troy weight to grains.

18. Reduce 6720 grains to ounces.

19. In 14 ingots, or bars of silver, each weighing 27*oz.* 10*pwt.*, how many grains? How many in one?

20. How many grains of silver in 4*lb.* 6*oz.* 12*pwt.* and 7*gr.*?

21. How many pounds, ounces, pennyweights, and grains of gold in 704121 grains?

22. How many of each denomination in 351262 grains?

23. In 25℔, apothecaries weight, how many ounces, drams, scruples, and grains?

24. In 907920 grains, how many ounces and pounds?

25. In 15℔ 1℥ 1ʒ 1℈ 2*gr.*, how many grains?

26. In 174947 grains, how many pounds?

27. In 16℔ 10℥ 1ʒ 14℈ 7*gr.*, how many grains?

28. In 12 tons, avoirdupois weight, how many pounds?

29. In 31360*lb.* of iron, how many tons?

30. In 375*cwt.* 292*lb.* of copper, how many pounds?

31. Reduce 740900*oz.* into *cwts.* and *tons.*

32. In 9*T.* 19*cwt.* 3*qr.* 27*lb.* 14*oz.*, how many ounces?

33. In 14478406*oz.*, how many *tons*, *cwt.*, *qrs.*, *lb.*, *oz.*, and *drs.*?

34. In 314 yards of cloth, how many nails?

35. In 576 French ells, how many yards?

36. Reduce 97*yds.* 3*qrs.* to English ells.

37. In 57 pieces of cloth, each 35 ells Flemish, how many ells English and nails?

38. In 14 bales of cloth, each 17 pieces, each piece 56 ells Flemish, how many yards, ells English, and ells French?

39. In 471 miles, long measure, how many furlongs and poles?

40. In 123200 yards, how many miles?

41. In 50 miles, how many yards, feet, inches, and barleycorns?

42. Reducc 37*mi.* 7*fur.* 37*rd.* 6*yd.* 5*ft.* to feet.

43. How many barleycorns will reach round the earth, each degree being $69\frac{1}{2}$ miles? and how many quarters of

barley are contained in such a number of barleycorns, admitting 7922 barleycorns to fill a pint?

44. In 77*A.* 1*R.* 14*P.*, land measure, how many perches?

45. In 17280 perches, how many acres?

46. In 50*A.* 3*R.* 10*P.* 9*sq. yd.* 789*sq. ft.*, how many square feet?

47. In 175 square chains, how many square rods?

48. In 14976 perches, or square rods, how many acres?

49. In 83789263*P.*, how many square miles?

50. In 28 tons of round timber, how many solid inches?

51. In 155 cords of wood, how many solid feet?

52. In 17 cords of wood, how many solid inches?

53. In 56320 solid feet, how many cords?

54. Reduce 349938 cord feet to cords.

55. In 32*hhds.*, wine measure, how many quarts?

56. In 3276 gallons, how many tuns?

57. In 75*hhds.*, how many pints?

58. In 77*hhds.* of brandy, how many half-ankers?

59. In 10*tuns* 2*hhds.* 18*gals.* of wine, how many gills?

60. In 98 hogsheads of ale, how many pints?

61. In 38 butts of porter, how many pints?

62. In 516 barrels of beer, how many half-pints?

63. How many gallons of beer are contained in 50 barrels?

64. In 44 quarters of corn, how many pecks?

65. In 30720 quarts, how many lasts?

66. How many sacks in 103 London chaldrons and 12 bushels of coal?

67. How many seconds in a year of 365*da.* 6*hr.*?

68. How many seconds in 6 years of 365*da.* 23*hr.* 57*m.* 39*sec.* each?

69. In 7569520118 seconds, how many years of 365 da each?

70. In 5927040 minutes, how many weeks?

ADDITION.

48. *The sum of two or more numbers, is a number which contains as many units, and no more, as are found in all the numbers added; and*

ADDITION *is the process of finding the sum or sum total of two or more numbers.*

If 3 be added to 5 their sum will be 8, and the unit of the number 8 will be the same as the unit of the numbers 3 and 5. The numbers 3 and 5, which are thus added, must have the same unit; for, if 3 denoted tens, and 5 expressed units of the first order, their sum would neither be 8 tens nor 8 simple units. So if 3 expressed yards, and 5 feet, their sum would neither be 8 yards nor 8 feet.

49. Small numbers may be added mentally; but it is not convenient to add large numbers without first writing them down. How are they to be written?

If we place one above the other, units of the same kind will fall in the same vertical line, and the units of the same order will fall directly under each other in the sum.

OPERATION.

	3
	5
Sum	8

Again, let it be required to add together 324 and 635. In the first number there are 4 units, 2 tens, and 3 hundreds. In the second, 5 units, 3 tens, and 6 hundreds. Let the figures of each order of units be placed under those of the same order, and added: their sum will be 9 units, 5 tens, and 9 hundreds, or nine hundred and fifty-nine.

OPERATION.

	hundreds.	tens.	units.
	3	2	4
	6	3	5
Sum	9	5	9

QUEST.—48. What is the sum of two or more numbers? What is addition? What numbers can be blended into one sum? 49. How may small numbers be added? How are numbers written down for addition?

50. Add together the numbers 894 and 637.

	OPERATION.
Write the numbers thus - - - - - -	894
	637
And draw a line beneath them - - -	——
sum of the column of units - - - - -	11
sum of the column of tens - - - - -	12
sum of the column of hundreds - - -	14
Sum total	1531

In this example, the sum of the units is 11, which cannot be expressed by a single figure. But 11 units are equal to 1 ten and 1 unit; therefore, we set down 1 in the place of units, and 1 in the place of tens. The sum of the tens is 12. But 12 tens are equal to 1 hundred, and 2 tens; so that 1 is set down in the hundred's place, and 2 in the ten's place. The sum of the hundreds is 14. The 14 hundreds are equal to 1 thousand, and 4 hundreds; so that 4 is set down in the place of hundreds, and 1 in the place of thousands. The sum of these numbers, 1531, is the sum sought.

The example may be done in another way, thus: Having set down the numbers, as before, we say, 7 and 4 are 11: we set down 1 in the units place, and write the 1 ten under the 3 in the column of tens. We then say, 1 to 3 is four, and 9 are 13. We set down the 3 in the tens place, and write the 1 hundred under the 6 in the column of hundreds. We then add the 1, 6, and 8 together, for the hundreds, and find the entire sum 1531, as before.

OPERATION.
894
637
11
1531

When the sum in any one of the denominations exceeds 10, or an exact number of tens, the excess must be written down, and a number equal to the number of tens added to the next higher denomination. This is called *carrying to the next column or higher denomination.* The number to be car-

QUEST.—50. What is the sum of the units? What of the tens? What of the hundreds? What the entire sum?

ried may be written under that column or remembered and added in the mind.

51. What is the sum of the numbers 375, 6321, and 598?

In this example, the small figure placed under the 4, shows how many are to be carried from the first denomination to the second, and the small figure under the 9, how many are to be carried from the second to the third. In like manner, in the examples below, the small figure under each column shows how many are to be carried to the next higher denomination. Beginners had better set down the numbers to be carried as in the examples.

OPERATION.

375
6321
598
7294
11

	(2.)		(3.)		(4.)
	96972		9841672		81325
	3741		793139		6784
	9299		888923		2130
Sum	110012	Sum	11523734	Sum	90239
	2221		221111		1110

52. Let it be required to find the sum of £14 7*s.* 8*d.* 3*far.*, and £6 18*s.* 9*d.* 2*far.*

We write down the numbers, as before, so that units of the same value shall fall under each other. Beginning with the lowest denomination, we find the sum to be 5 farthings. But as 4 farthings make a penny, we set down the 1 farthing over, and carry 1 to the column of pence. The sum of the pence then becomes 18, which make 1 shilling and 6 over. Set down the 6, and carry 1 to the column of shillings, the sum of which becomes 26; that is, 1 pound and 6 shillings. Setting down the 6 shillings and

OPERATION.

£	*s.*	*d.*	*far.*
14	7	8	3
6	18	9	2
21	6	6	1

QUEST.—How may the units to be carried be disposed of? 51. How will you remember how many are to be carried from one column to another? 52. Explain the manner of adding pounds, shillings, and pence, and of passing from one denomination to another.

carrying 1 to the column of pounds, we find the entire sum to be £21 6*s.* 6*d.* 1*far.*

53. Hence, for the addition of all numbers,

Write down the numbers so that units of the same denomination shall fall in the same column, and draw a line beneath them.

Add up the units of the lowest denomination, and divide their sum by so many as make one of the denomination next higher. Set down the remainder and carry the quotient to the next higher denomination, and proceed in the same manner through all the denominations to the last.

PROOF OF ADDITION.

54. The proof of an arithmetical operation is a second operation, by means of which the first is shown to be correct.

Addition may be proved by adding all the columns downward. It may also be proved by dividing the numbers to be added into two parts, adding each of the parts separately, and then adding their sums. If the last sum is the same as that of all the numbers first found, the work may be considered right.

EXAMPLES.

	182796
	143274
	32160
	47047
Sum	405277

	182796	32160
	143274	47047
Partial sums	326070	79207

	326070	1st partial sum.
	79207	2d "
Proof	405277	

QUEST.—53. What is the general rule for addition? 54. What is the proof of an arithmetical operation? What is the first method of proving addition? What the second?

	(1.)	(2.)	(3.)
	34578	22345	23456
	3750	67890	78901
	87	8752	23456
	328	340	78901
	17	350	23456
	327	78	78901
Sums total			
Partial sums	4509	77410	283615
Proofs	39087	99755	307071

(4.)	(5.)	(6.)
672981043	1278976	8416785413
67126459	7654301	6915123460
39412767	876120	31810213
7891234	723456	7367985
109126	31309	654321
84172	4871	37853
72120	978	2685

7. Add together six tens, fourteen hundreds, seven thousands, nine ten thousands, forty-five millions, and six thousand seven hundred and fifty-one.

8. What is the sum of six hundreds, eight units of the fifth order, thirteen of the sixth, twenty of the second, forty of the third, and two billions, three millions, four trillions, two hundred and twenty-one thousand seven hundred and fifty-five?

9. What is the sum of eight hundred units of the first order, sixty of the second, one thousand of the third, ninety-nine of the fourth, one hundred of the fifth, six trillions, one billion, forty-nine thousand eleven hundred and sixty-one?

10. What is the sum of three hundred and forty units of the third order, seven thousand six hundred and fifty of the fourth, three millions of the second, and six trillions seven hundred and ninety-nine of the first?

11. Collect together into one sum, two hundred and seventy-eight millions four thousand six hundred and sixty-nine; seventy-six billions four hundred and fifty-eight millions four

hundred and seventy-five thousand five hundred and two; fifty billions three hundred millions; four hundred and seventy-two millions four thousand five hundred and fifty-five; nine millions seven hundred thousand three hundred and two; twelve millions three hundred thousand four hundred and sixty-one; two hundred millions four hundred thousand and four; eight hundred millions seven hundred and forty-nine thousand seven hundred and ninety-nine; two hundred and six millions four hundred and forty thousand and thirty-four.

12. Find the sum total of five billions six hundred and forty-nine millions three hundred and seven thousand and sixty; nine hundred and forty millions three hundred and seventy-four thousand six hundred and eighty-one; nine billions eight hundred and seventy-six millions five hundred and forty-three thousand two hundred and ten; one hundred and twenty-three millions four hundred and fifty-six thousand seven hundred eighty-nine; five billions three hundred millions seven hundred and seventy-seven thousand seven hundred and seven.

13. Add together seven hundred and four billions three hundred and sixty-millions five hundred and thirteen thousand and forty-two; sixty-four billions seven hundred and ninety-three millions six hundred and twenty-nine thousand five hundred and forty-eight; six hundred and ninety-nine billions six hundred and ninety-nine millions eight hundred and sixty-five thousand seven hundred and seventy-five.

14. Collect together and find the sum of fifty-eight billions nine hundred and eighty-two millions four hundred and eighty-seven thousand six hundred and fifty-four; seven hundred and forty billions three hundred and fifty millions five hundred and forty thousand seven hundred and sixty; four hundred and twenty-five billions seven hundred and three millions four hundred and two thousand six hundred and three; thirty-four billions twenty millions forty thousand and twenty; five hundred and sixty billions eight hundred millions seven hundred thousand and four hundred.

(15.)	(16.)	(17.)	(18.)
$87,046	$950,60	$109,049	$8704,067
21,846	107,27	691,027	7504,61

19. What is the sum of 6 eagles 15 dollars 75 cents 5 mills, + 4 eagles 100 dollars 30 cents 8 mills, + 607 dollars 8 cents 1 mill, + 407 eagles 604 dollars 89 cents 9 mills?

20. What is the sum of 47 eagles 207 dollars 51 cents 8 mills, + 4 eagles 49 dollars 1 cent 1 mill, + 1000 eagles 40009 dollars 16 cents 9 mills, + 691 eagles 9791 dollars 14 cents 2 mills?

(21.)			(22.)			(23.)			(24.)		
£	*s.*	*d.*	£	*s.*	*d.*	£	*s.*	*d.*	£	*s.*	*d.*
149	14	7¼	14	11	3¼	14	19	4¼	14	10	4½
37	11	9¾	19	18	10	17	11	10	77	18	3
64	14	7	77	11	3¼	39	18	11¼	14	13	9½
104	19	11½	49	14	7	19	14	9	67	12	4¾
64	13	10	16	18	4½	19	15	11¼	9	11	10
174	19	11¾	17	15	10	18	19	10	18	10	5
47	14	10¼	1	14	9¼	77	19	11¼	17	19	4
39	15	11½	6	18	10¾	14	11	10¼	19	10	4

(25.)			(26.)			(27.)			(28.)		
lb.	*oz.*	*pwt.*	*oz.*	*pwt.*	*gr.*	*lb.*	*oz.*	*pwt.*	*oz.*	*pwt.*	*gr.*
174	11	19	174	19	23	71	11	19	74	19	23
74	10	13	714	11	14	64	8	14	64	14	17
944	9	14	714	0	18	77	0	0	74	19	11
74	11	19	74	1	22	14	3	11	66	13	9
944	10	13	948	2	21	64	2	9	74	14	11
74	11	3	74	1	12	74	1	14	14	10	3
14	9	4	715	2	14	77	2	13	19	11	14
77	10	11	714	18	16	19	2	14	17	10	13

(29.)

℔	℥	ʒ
47	11	7
94	10	6
74	10	4
75	9	3
69	0	2
57	1	2
18	2	1
74	1	2

(30.)

℥	ʒ	℈
149	7	2
714	3	0
619	2	1
74	6	2
169	5	2
74	1	2
777	6	1
948	5	2

(31.)

ʒ	℈	*gr.*
749	2	19
607	1	18
714	2	17
400	0	0
74	1	13
715	2	14
64	1	18
174	2	19

(32.)

℔	℥	ʒ
84	11	7
74	10	6
37	5	4
19	4	3
74	1	2
79	2	6
19	2	4
74	9	5

(33.)

T.	*cwt.*	*qr.*
174	19	3
74	14	2
714	13	1
718	16	2
734	15	2
714	14	1
70	13	2

(34.)

cwt.	*qr.*	*lb.*
174	3	27
714	2	24
149	1	14
719	2	16
407	1	23
149	2	17
714	2	18

(35.)

qr.	*lb.*	*oz.*
44	27	15
74	26	14
19	14	13
74	19	14
66	27	13
74	19	10
14	18	11

(36.)

lb.	*oz.*	*dr.*
17	15	15
27	14	11
16	13	9
74	14	14
70	0	0
64	13	10
74	14	11

(37.)

yd.	*qr.*	*na.*
74	3	3
64	2	1
74	1	3
49	2	1
74	1	2
44	3	1
74	2	0
14	1	2

(38.)

E.E.	*qr.*	*na.*
77	4	3
14	3	2
74	2	1
49	1	2
74	2	1
74	3	2
44	1	2
74	2	3

(39.)

E. Fr.	*qr.*	*na.*
749	5	3
704	4	2
108	3	1
705	4	0
708	3	1
474	5	2
174	0	1
194	3	2

(40.)

E. Fl.	*qr.*	*na.*
714	2	3
615	1	2
714	1	3
724	2	2
149	1	2
718	2	3
419	1	1
710	1	2

(41.)			(42.)			(43.)			(44.)		
L.	*mi.*	*fur.*	*Fur.*	*rd.*	*yd.*	*Rd.*	*yd.*	*ft*	*Ft.*	*in.*	*bar*
17	2	7	147	39	$5\frac{1}{4}$	177	$5\frac{1}{4}$	2	174	11	2
14	1	6	614	37	$4\frac{3}{4}$	714	$4\frac{3}{4}$	1	49	10	1
74	1	7	714	19	$3\frac{1}{2}$	714	$1\frac{1}{2}$	2	74	11	2
69	2	4	674	17	$1\frac{1}{4}$	615	0	1	64	9	1
74	1	0	719	27	$2\frac{3}{4}$	714	$1\frac{3}{4}$	2	74	10	1
69	2	1	197	19	$1\frac{1}{2}$	719	$1\frac{1}{2}$	1	64	11	2
74	1	2	714	14	$3\frac{1}{4}$	437	$2\frac{3}{4}$	1	74	10	0
94	0	3	704	19	$4\frac{3}{4}$	614	$1\frac{1}{2}$	2	64	9	1

(45.)			(46.)			(47.)			(48.)		
A.	*R.*	*P.*	*A.*	*R.*	*P.*	*A.*	*R.*	*P.*	*A.*	*R.*	*P.*
77	3	39	714	3	39	14	3	39	174	3	39
64	2	37	619	1	18	74	1	19	714	1	27
74	1	24	714	2	27	64	2	14	618	2	12
64	2	19	619	1	34	74	1	18	719	1	14
74	1	18	719	2	37	47	2	24	734	2	11
64	2	17	719	1	24	18	1	14	715	1	24
14	1	13	615	2	14	74	2	19	639	2	14
74	2	11	74	1	18	34	1	14	714	3	24

(49.)			(50.)			(51.)			(52.)		
Tun	*hhd.*	*gal.*	*Pun.*	*gal.*	*qt.*	*Tierce*	*gal.*	*qt.*	*Gal.*	*qt.*	*pt.*
714	3	62	714	83	3	74	41	3	14	3	1
614	2	61	615	81	2	64	40	2	74	2	1
174	1	39	714	74	1	74	19	1	39	2	1
164	2	47	614	18	2	64	39	2	17	1	0
274	1	49	713	75	0	74	40	1	19	2	0
175	2	37	614	17	1	69	19	1	77	1	1
375	1	49	715	14	3	17	39	2	39	3	1
714	2	61	719	28	2	18	41	1	14	1	1

(53.)

Bar.	*fir.*	*gal.*
71	3	8
14	2	7
16	1	4
17	1	3
29	2	2
17	1	7
41	2	6
37	1	5

(54.)

Bar.	*fir.*	*gal.*
73	3	7
69	2	6
14	1	7
39	2	2
19	1	6
49	2	6
37	1	4
19	1	2

(55.)

Hhd.	*gal.*	*qt.*
714	47	3
614	44	1
374	43	2
157	41	1
719	42	1
374	41	2
174	12	1
19	13	2

(56.)

Hhd.	*gal.*	*qt.*
714	53	3
415	47	2
714	19	1
614	27	1
715	51	2
714	37	2
615	19	1
714	18	2

(57.)

L. ch.	*bu.*	*pk.*
14	31	3
74	31	2
64	30	1
74	27	2
64	19	2
74	31	1
64	11	1
95	10	2

(58.)

Weys	*qr.*	*bu.*
174	3	7
375	1	6
400	0	5
371	1	4
634	2	3
719	1	2
149	2	1
375	1	3

(59.)

Qr.	*bu.*	*pk.*
149	7	3
715	3	2
649	1	3
479	2	1
675	1	3
149	2	1
375	1	2
649	1	3

(60.)

Scows.	*L. ch.*	*bu*
74	20	35
49	19	33
64	17	35
74	14	10
39	13	9
47	16	3
19	17	4
37	18	34

(61.)

Yr.	*mo.*	*wk.*
737	12	3
347	11	2
618	10	1
374	9	2
175	3	1
714	12	3
615	10	1
714	3	1

(62.)

Mo.	*wk.*	*da.*
64	3	6
74	1	5
34	2	3
74	1	4
63	2	1
74	1	2
64	2	1
74	1	3

(63.)

Da.	*hr.*	*min.*
714	23	59
74	14	54
64	21	55
74	13	53
69	12	14
74	12	19
37	11	17
16	12	19

(64.)

Hr.	*min.*	*sec.*
647	59	59
137	54	54
375	56	56
714	17	19
615	54	54
714	17	13
613	34	56
624	27	39

APPLICATIONS.

1. In 1843, the number of acres of the public lands sold in the several states and territories was as follows:—In Ohio, 13338 acres, Indiana 50545, Illinois 409767, Missouri 436241, Alabama 178228, Mississippi 34500, Louisiana 102986, Michigan 12594, Arkansas 47622, Wisconsin 167746, Iowa 143375, Florida 8318. What was the whole number of acres sold in the United States?

2. The number of acres of the public lands sold in 1834 was 4658218; in 1835, 12564478; in 1836, 25167833. The number sold in 1840 was 2236889; in 1841, 1164796; in 1842, 1129217. How many acres were sold in the first three, and how many in the last three years?

3. In 1844, the school fund of Connecticut was invested as follows: in bonds and mortgages, $1695407,44; in bank stock, $221700; in cultivated lands, and buildings, $78367; in wild lands, $52493,75; in stock in Massachusetts, $210; in cash, $3245,58. What was the whole amount of the fund?

4. The salaries of the English cabinet ministers are as follows: of the First Lord of the Treasury, £5000; of the Lord High Chancellor, £14000; of the Lord President of the Council, £2000; of the Lord Privy Seal, £2000; of the Secretaries of State for the Home, Foreign, and Colonial Departments, £15000; of the Chancellor of the Exchequer, £5000; of the First Lord of the Admiralty, £4500; of the Paymaster-general, £2500; of the President of the Board of Control, £2000. Required the sum of the salaries of the cabinet.

5. What was the whole number of pieces coined in the United States' mint in 1835, there having been 371534 half-eagles, 131402 quarter-eagles, 5352006 half-dollars, 1952000 quarter-dollars, 1410000 dimes, 2760000 half-dimes, 3878000 cents, and 141000 half-cents? Required also the value of the whole number of coins executed in that year.

6. The value of the imports during Mr. Monroe's second

administration was, in 1821, $62585724; in 1822, $83241541; in 1823, $77579267; in 1824, $80549007. The value of the exports in 1821, was $64974382; in 1822, $72160281; in 1823, $74699030; in 1824, $75986657. What was the amount of imports and the amount of exports in that term?

7. What was the population of the British provinces in North America in 1834, the population of Lower Canada being stated at 549005, of Upper Canada 336461, of New Brunswick 152156, of Nova Scotia and Cape Breton 142548, of Prince Edward's Island 32292, of Newfoundland 75000?

8. What was the population of Brazil in 1819, there having been of whites 843000; of free people of mixed blood, 426000; of Indians, 259400; of free negroes, 159500; of negro slaves, 1728000; of slaves of mixed blood, 202000?

9. The imports into France, in 1826, were valued at 564728392 francs; in 1827, at 565804228 francs; in 1828, 607677321 francs; in 1829, 616353397 francs; in 1830, 638338433 francs; in 1831, 512825551 francs; in 1832, 652872341 francs; in 1833, 693275752 francs. What was the value of the imports for those years?

10. The number of emigrants in 1837, from Great Britain to British North America, was: from England, 5027; from Scotland, 2394; and from Ireland, 22463. The number to the United States the same year was, from England, 31769; from Scotland, 1130; from Ireland, 33871. Required the number of emigrants to each place, and the entire number.

11. The consumption of coffee in Great Britain is stated to be 10500 tons a year; in the Netherlands and Holland, 40500 tons; in Germany and the countries round the Baltic, 32000 tons; in France, Spain, Italy, Turkey in Europe, and the Levant, 35000 tons; in America, 20500 tons. What is the entire consumption of coffee in these countries?

12. The number of regular troops furnished by each of the states in the revolution, was as follows: New Hampshire, 12497; Massachusetts, 67907; Rhode Island, 5908; Connecticut, 31939; New York, 17781; New Jersey, 10726;

Pennsylvania, 25678; Delaware, 2386; Maryland, 13912; Virginia, 26678; North Carolina, 7263; South Carolina, 6417; Georgia, 2679. What was the number of regular troops engaged during the war?

13. The revenue of the post-office at Albany, for the fourth quarter of 1845, was $2697; at Baltimore, $10339; at Brooklyn, N. Y., $1279; at Bangor, Me., $1107; at Buffalo, $2339; at Cincinnati, $6103; at Detroit, $1007; at Hartford, $1239; at Louisville, $1946; at Mobile, $4199; at Nashville, $1194; at Newark, N. J., $1026; at Norfolk, Va., $1175; at Petersburg, Va., $1090; at Philadelphia, $21642; at Pittsburg, Pa., $3612; at Providence, $3046; at Rochester, N. Y., $2606; at Springfield, Mass., $1031; at Troy, N. Y., $1883. What was the total amount of revenue received from these post-offices?

14. The list of vessels in the British navy, on the 1st of January, 1846, was as follows: sailing vessels in 'commission' and in 'ordinary,' 361; sail vessels building, 42; steam frigates, 11; steam frigates building, 12; other steam vessels, 88; steam vessels building, 8; steam packets, 25; receiving and quarantine vessels, transports, &c., 134. What is the whole number of vessels, and what the number of each kind?

15. The deposites of gold for coinage at the mint in Philadelphia, in 1842, were: from mines in the United States, $273587; coins of the United States, old standard, $27124; foreign coins, $497575; foreign bullion, $158780; jewellery, $20845. The deposites of silver were: bullion from North Carolina, $6455; foreign bullion, $153527; Mexican dollars, $1085374; South American dollars, $26372; European coins, $272282; plate, $23410. What was the amount of gold deposited? What of silver? And what the entire sum?

16. Of the public lands, there were ceded by the states of Virginia, New York, Massachusetts, and Connecticut, 169609819 acres; by Georgia, 58898522 acres; by North and South Carolina, 26432000 acres; and 987852332 acres were purchased of France and Spain. Required the number of acres ceded and purchased.

17. The population of New York city in 1840 was 312710; of Philadelphia, 258037; of Baltimore, 134379; of New Orleans, 102193; of Boston, 93383; of Cincinnati, 46338; of Brooklyn, 36233; of Albany, 33721; of Charleston, 29261; of Louisville, 21210; of Richmond, 20153; of St. Louis, 16469. What was the whole number of inhabitants in these twelve cities?

18. The following table exhibits the population of the several states and territories, at the taking of each census to 1840. What was the population of the United States in each of those years?

States.	1790.	1800.	1810.	1820.	1830.	1840.
Maine - - -	96540	151719	228705	298335	399955	501793
New Hampshire	141899	183762	214360	244161	269328	284574
Vermont -	85416	154465	217713	235764	280652	291948
Massachusetts	378717	423245	472040	523287	610408	737699
Rhode Island -	69110	69122	77031	83059	97199	108830
Connecticut -	238141	251002	262042	275202	297665	309978
New York -	340120	586756	959949	1372812	1918608	2428921
New Jersey -	184139	211949	249555	277575	320823	373303
Pennsylvania -	434373	602365	810091	1049458	1348233	1724033
Delaware - -	59098	64273	72674	72749	76748	78085
Maryland - -	319728	341548	380546	407350	447040	470019
Virginia - -	748308	880200	974642	1065379	1211405	1239797
North Carolina	393751	478103	555500	638829	737987	753419
South Carolina	249073	345591	415115	502741	581185	594398
Georgia - -	82548	162101	252433	340987	516823	691392
Alabama - -	–	–	20845	127901	309527	590756
Mississippi -	–	8850	40352	75448	136621	375651
Louisiana - -	–	–	76556	153407	215739	352411
Arkansas - -	–	–	–	14273	30388	97574
Tennessee - -	30791	105602	261727	422813	681904	829210
Kentucky - -	73077	220955	406511	564317	687917	779828
Ohio - - -	–	45365	230760	581434	937903	1519467
Michigan - -	–	–	4762	8896	31639	212267
Indiana - -	–	4875	24520	147178	343031	685866
Illinois - - -	–	–	12282	55211	157455	476183
Missouri - -	–	–	20845	66586	140445	383702
Dist. Columbia	–	14093	24023	33039	39834	43712
Florida - -	–	–	–	–	34730	54477
Wisconsin - -	–	–	–	–	–	30945
Iowa - - -	–	–	–	–	–	43112

19. The slave population of the states and territories, according to each census, is shown in the following table. Required the number of slaves in the United States at each enumeration.

States.	1790.	1800.	1810.	1820.	1830.	1840.
Maine - - -	0	0	0	0	0	0
New Hampshire -	158	8	0	0	0	1
Vermont - -	17	0	0	0	0	0
Massachusetts -	0	0	0	0	0	0
Rhode Island -	952	381	103	48	17	5
Connecticut - -	2759	951	310	97	25	17
New York - -	21324	20343	15017	10088	75	4
New Jersey - -	11423	12422	10851	7657	2254	674
Pennsylvania -	3737	1706	795	211	403	64
Delaware - -	8887	6153	4177	4509	3292	2605
Maryland - -	103036	105635	111502	107398	102294	89737
Virginia - -	203427	345796	392518	425153	469757	448987
North Carolina -	100572	133296	168824	295017	235601	245817
South Carolina -	107094	146151	196365	258475	315401	327038
Georgia - - -	29264	59404	105218	149656	217531	280944
Alabama - -	–	–	–	41879	117549	253532
Mississippi - -	–	3489	17088	32814	65659	195211
Louisiana - -	–	–	34660	69064	109588	168452
Arkansas - -	–	–	–	1617	4576	19935
Tennessee - -	3417	13584	44535	80107	141603	183059
Kentucky - -	11830	40343	80561	126732	165213	182258
Ohio - - -	–	–	–	–	–	3
Michigan - -	–	–	24	–	32	0
Indiana - - -	–	135	237	190	0	3
Illinois - - -	–	–	168	117	*747	331
Missouri - -	–	–	3011	10222	25081	58240
Dist. Columbia -	–	3244	5395	6377	6119	4694
Florida - - -	–	–	–	–	15501	25717
Wisconsin - -	–	–	–	–	–	11
Iowa - - -	–	–	–	–	–	16

* Indented colored servants.

SUBTRACTION.

55. SUBTRACTION is the process of finding the difference between two numbers.

When the numbers are unequal, the larger of the two is called the *minuend*, and the less is called the *subtrahend*, and their difference, whether they are equal or unequal, is called the *remainder*.

When the numbers are small, their difference is apparent, and the subtraction may be made mentally.

EXAMPLES.

1. From 869 subtract 327: that is, from 8 hundreds 6 tens and 9 units, it is required to take 3 hundreds 2 tens and 7 units.

OPERATION.

869	Minuend.
327	Subtrahend.
542	Remainder.

We begin at the right hand figure of the lower line, and say, 7 from 9 leaves 2: set down the 2 under the 7. Proceeding to the next column, we say, 2 from 6 leaves 4; set down the 4, and then say, 3 from 8 leaves 5. Hence, the remainder or true difference between the numbers is 542.

2. From 843 subtract 562.

OPERATION.

```
 10
843
562
 1
---
281
```

Beginning with the lowest denomination, we say, 2 from 3 leaves 1. At the next step we meet a difficulty, for we cannot subtract 6 from 4. If, now, we add 10 tens to the 4, (which are written in small figures above,) and 10 tens to the 6 directly under it, it is plain that the difference will not be affected, since both the

QUEST.—55. What is subtraction? How many numbers are considered in subtraction? What are they called? When the numbers are small, how may the subtraction be made? Ex. 2. How do we get over the difficulty in subtracting the tens? If equal numbers be added to the minuend and subtrahend, will their difference be changed?

numbers are equally increased. But adding 10 tens to the 6 is the same thing as adding 1 to the 5 hundreds: hence, *we may consider* 10 *to be added to any figure of the minuend, provided we add* 1 *to the next figure of the subtrahend to the left.*

We can now go on with the subtraction; for we say, 6 from 14 leaves 8. Then, 1 carried to 5 makes 6: and 6 from 8 leaves 2. Hence the remainder is 281; and all similar examples are done in the same manner.

3.	*T.*	*cwt.*	*qr.*	*lb.*	*oz.*
		20		28	
From	6	14	2	20	12
take	4	17	1	21	10
	1		1		
Remainder	1	17	0	27	2

In this example we say, 10 ounces from 12 leaves 2. At the next denomination we meet a difficulty, for we cannot subtract 21 from 20. We add to the 20 so many units as make 1 unit of the next higher denomination—that is, 28, and suppose at the same time 1 unit to be added to that denomination in the subtrahend. We then say, 21 from 48 leaves 27: then 2 from 2 leaves 0. In the hundreds we again have to add, after which we say, 17 from 34 leaves 17; then we take 5 from 6, and have the true remainder.

56. Hence, to find the difference between two numbers:

Set down the less number under the greater, so that units of the same denomination shall fall in the same column, and beginning with the lowest denomination, subtract each from the one above it. When the units in any one denomination of the subtrahend exceed those of the same denomination in the minuend, suppose so many units added in the minuend as make one unit of the next higher denomination; after which add one to the next denomination of the subtrahend, and subtract as before.

QUEST.—If you add 10 to any figure of the minuend, what will you add to the subtrahend? Ex. 3. How is the subtraction made in this example? 56. What is the rule for subtraction?

PROOF.

57. Add the remainder to the subtrahend, and if the sum is equal to the minuend, the work may be regarded as right. Or, subtract the remainder from the minuend, and the remainder thus found should be equal to the subtrahend.

EXAMPLES.

	(1.)	(2.)	(3.)
	10 10	10 10 10 10	20 12 4
From	87407	27431	£14 16*s.* 7½*d.*
take	6079	19872	6 17 9¾
	1 1	1 1 1 1	1 1 1
Rem.			
Proof			

4. From 47348406051320047 take 13456507031079054.

5. From 19493899900056075 take 14954298990056076.

6. From 500714960079690650 take 742350986470501.

7. From 149348761340526465 take 48973024012394.

	(8.)	(9.)	(10.)	(11.)
From	$374,674	$270,604	$137,04	$9496,004
take	195,097	191,280	127,97	8496,049
Rem.				

12. What is the difference between $487,25 and $379,674?

13. What is the difference between $670,04 and one hundred and four dollars and 6 mills?

14. What is the difference between $1000 and $14,003?

(15.)			(16.)			(17.)			(18.)		
lb.	*oz.*	*pwt.*	*oz.*	*pwt.*	*gr.*	*lb.*	*oz.*	*pwt.*	*oz.*	*pwt.*	*gr.*
14	11	9	74	12	13	175	3	10	17	10	20
11	10	14	64	14	17	159	11	14	14	11	23

QUEST.—57. What is the first method of proving subtraction? What is the second?

(19.)			(20.)			(21.)			(22.)		
℔	℥	ʒ	℥	ʒ	℈	ʒ	℈	*gr.*	℔	℥	ʒ
144	10	5	27	4	1	27	1	14	74	10	5
64	11	7	14	7	2	14	0	19	65	11	6

(23.)			(24.)			(25.)			(26.)		
T.	*cwt.*	*qr.*	*Cwt.*	*qr.*	*lb.*	*Qr.*	*lb.*	*oz.*	*lb.*	*oz.*	*dr.*
14	12	2	17	1	25	143	22	12	174	11	10
1	14	3	14	2	27	74	19	14	39	12	13

(27.)			(28.)			(29.)			(30.)		
Yd.	*qr.*	*na.*	*E.E.*	*qr.*	*na.*	*E. Fr.*	*qr.*	*na.*	*E. Fl.*	*qr.*	*na.*
174	2	1	174	3	1	171	1	3	12	1	1
39	3	2	49	4	2	74	5	2	10	2	3

(31.)			(32.)			(33.)			(34.)		
L.	*mi.*	*fur.*	*Fur.*	*rd.*	*yd.*	*Rd.*	*yd.*	*ft.*	*Ft.*	*in.*	*bar.*
21	2	4	13	34	$3\frac{3}{4}$	14	$3\frac{3}{4}$	1	17	11	2
3	2	6	12	39	$5\frac{1}{4}$	9	$4\frac{1}{2}$	2	14	11	1

(35.)			(36.)			(37.)			(38.)		
A.	*R.*	*P.*	*A.*	*R.*	*P.*	*A.*	*R.*	*P.*	*A.*	*R.*	*P.*
12	1	32	112	1	31	12	1	25	19	1	20
1	3	14	74	2	37	10	3	39	14	2	21

(39.)			(40.)			(41.)			(42.)		
Tun	*hhd.*	*gal.*	*Pun.*	*gal.*	*qt.*	*Tier.*	*gal.*	*qt.*	*Gal.*	*qt.*	*pt.*
27	2	54	147	14	2	14	1	2	24	3	0
19	3	62	79	83	3	12	41	3	17	0	1

(43.)			(44.)			(45.)			(46.)		
Bar.	*fir.*	*gal.*	*Bar.*	*fir.*	*gal.*	*Hhd.*	*gal.*	*qt.*	*Hhd.*	*gal.*	*qt.*
14	3	5	147	1	3	271	1	2	143	1	2
12	3	7	39	3	8	49	47	3	79	52	3

(47.)			(48.)			(49.)			(50.)		
L.ch.	*bu.*	*pk.*	*Weys*	*qr.*	*bu.*	*Qr.*	*bu.*	*pk.*	*Scows.*	*L.ch.*	*bu.*
74	31	3	17	3	1	147	6	2	47	1	12
47	31	2	14	3	7	94	7	3	14	20	35

(51.)			(52.)			(53.)			(54.)		
Yr.	*mo.*	*wk.*	*Mo.*	*wk.*	*da.*	*Da.*	*hr.*	*min.*	*Hr.*	*min.*	*sec.*
17	11	2	147	2	3	167	21	50	147	50	51
14	12	3	19	2	4	19	23	54	94	59	57

PROMISCUOUS EXAMPLES.

55. A horse in his furniture is worth £52 10*s.*; out of it, £24 10*s.* 6*d.* How much does the price of the furniture exceed that of the horse?

56. What sum added to £11 14*s.* $9\frac{1}{4}$*d.* will make £133 11*s.* and $9\frac{1}{2}$*d.*?

57. A tradesman failing, was indebted to A £105 19*s.* 11*d.*, to B 150 guineas, to C £34 18*s.* 10*d.*, to D £500 19*s.*, to E £700 14*s.* 9*d.* When this happened, he had cash by him to the amount of £50, goods to the amount of £350 14*s.* 9*d.*, his household furniture was worth £24 11*s.*, his book-debts amounted to £94 14*s.* 8*d.* If these things were faithfully given up to his creditors, what did they lose by him?

58. The great bell at *Oxford*, the heaviest in *England*, weighs 7*T.* 11*cwt.* 3*qr.* 4*lb.*; *St. Paul's* bell at *London* weighs 5*T.* 2*cwt.* 1*qr.* 22*lb.*; and *Tom* of *Lincoln* weighs 4*T.* 16*cwt*

3qr. 18lb. How much are these bells, together, inferior in weight to the great bell at *Moscow*, the largest in the world, which weighs *198T. 2cwt. 1qr.*?

59. An apprentice, who is 14 years, 11 months, 13 weeks, 14 hours, 38 minutes old, is to serve his master till he is 21 years of age. How long has he to serve?

60. What is the difference of latitude and longitude between *Calcutta* in the *East Indies*, (lat. 22° 34′ N., long. 88° 34′ E.,) and *Lima*, in *South America*, (lat. 12° 1′ S., long. 76° 44′ W.)?

61. NEWTON (Sir Isaac) was born at *Woolsthorp*, a hamlet in the parish of *Colsworth*, in *Lincolnshire*, on Sunday, the 25th December, 1642; and died at *Kensington*, in *Middlesex*, on Monday, the 20th March, 1727. EULER (Leonard) was born at *Basil*, in *Switzerland*, on Tuesday, the 15th April, 1707; and died at *Petersburg*, in *Russia*, on Sunday, the 7th September, 1783. LAGRANGE (Joseph Louis) was born at *Turin*, in *Italy*, on Friday, the 30th January, 1736; and died at *Paris*, on Saturday, the 10th April, 1813. LAPLACE (Pierre Simon, marquis of) was born at *Beaumont-en-Auge*, in *France*, on Thursday, the 23d March, 1749; and died at *Paris*, on Tuesday, the 27th March, 1827. How old was each of these eminent philosophers and mathematicians at the time of his decease? and how many years was it from the time each died to January 1st, 1846.

62. In 1840 the amount of tobacco sent from the United States to England, was 26255 hogsheads, and to Holland, 29534 hogsheads. How much more was sent to Holland than to England?

63. The population of the northern district of New York in 1840 was 1683068, and the population of the southern district was 745853. How many more inhabitants were there in the northern than in the southern district, and what was the population of the state?

64. The population of England in 1841 was 14995508, the population of Scotland 2628957, and of Wales 911321. How

much did the population of England and Wales combined exceed that of Scotland, and what was the entire population of great Britain?

65. The value of the gold coined at the mint in Philadelphia in 1842 was $960017,50; the value of that coined at Charlotte, N. C., was $159005; at Dahlonega, Ga., $309648; and at New Orleans, $405500. How much more was coined at Philadelphia than at the three other places?

66. The whole amount received for the public lands to 1843, was $170940942,62. There have been paid for the Indian title, the Florida and Louisiana purchase, including interest, $68524991,32; and for surveying and selling, including salaries of officers, $9966610,14. Required the net amount derived from the sale of the public lands.

67. The revenue of Great Britain for the year 1843 was £50071943, and for the previous year, £44329865. Required the increase.

68. The value of the merchandise imported into the United States during the year ending June 30th, 1844, was $108435035; of which $24766881 was admitted free of duty, $31352863 paid specific duties, and the remainder paid duties ad valorem. What amount paid ad valorem duties?

69. The value of the products of the sea exported from the United States in 1844, was $3350501; the value of the products of the forest, exported the same year, was $5808712. How much more was exported of the products of the forest than of the sea?

70. The imports from England to the United States in 1844, amounted to $41476081, from Scotland $527239, and from Ireland $88084. The value of the exports to England, the same year, was $46940156, to Scotland $1953473, and to Ireland $42591. How much did our exports to Great Britain and Ireland exceed the imports?

71. What was the balance in the treasury of the state of Tennessee, in October, 1844, the income for the year ending

that month having been $271823,08; a surplus had been left the preceding year of $38875,21; and the expenditure was $261416,26?

72. The cost of the internal improvements of the state of Ohio, was $15283783,64, of which the Ohio canal cost $4695203,69; the Miami canal, $1237552,16; the Miami Extension, $2856635,96; and the Wabash and Erie canal, $3028340,05. What was the cost of the other works of the state?

73. St. Augustine was founded Sept. 8th, 1565. Jamestown was founded May 13th, 1607. The Battle of Princeton was fought Jan. 3d. 1777. Cornwallis surrendered, Oct. 19th, 1781. Washington was first inaugurated April 30th, 1789: he died, Dec. 14th, 1799. The French Berlin decree was issued Nov. 21st, 1806, and the British orders in council, Nov. 11th, 1807. The United States declared war against Great Britain June 18th, 1812. The Guerriere was captured by the Constitution Aug. 19th, 1812. The frigate United States captured the Macedonian, Oct. 25th, 1812. York in Upper Canada was captured by the Americans, and General Pike killed, April 27, 1813. Fort George was captured May 27th, 1813. The British were repulsed from Sackett's Harbor by the Americans commanded by General Brown, May 28th, 1813. The Battle of Lake Erie was fought Sept. 10th, 1813. The Battle of Chippewa was gained by a detachment of the American army under General Scott, July 5th, 1814. The Battle of Niagara, or Lundy's Lane, was fought July 25th, 1814. General Brown conducted the sortie from Fort Erie, Sept. 17th, 1814. The battle of New Orleans was fought Jan. 8th, 1815. Adams and Jefferson died July 4th, 1826. The compromise bill was introtroduced into the senate Feb. 12th, 1833. General Lafayette died May 20th, 1833. The Cherokees began to remove May 26th, 1838. What time has elapsed from the date of each of these events to March 17th, 1846?

MULTIPLICATION.

58. If the number 1 be multiplied by 2, that is, taken *two times*, the result will be 2; and 2 is said to be two times greater than 1.

If 1 be multiplied by 3, that is, taken three times, the result will be 3; and 3 is said to be *three times* greater than 1.

If 2 be multiplied by 2, that is, taken 2 times, the result will be 4; and 4 is said to be *two times* greater than 2.

If 3 be multiplied by 4, the result will be 12; and 12 is said to be *four times* greater than 3.

In the first case, 1 was taken 2 times; in the second it was taken 3 times; in the third 2 was taken 2 times; and in the fourth 3 was taken 4 times.

Multiplication *is a short process of taking one number as many times as there are units in another.* Hence, it is a short method of performing addition.

The number to be taken is called the *multiplicand.*

The number denoting how many times the multiplicand is to be taken, is called the *multiplier.*

The number arising from taking the multiplicand as many times as there are units in the multiplier, is called the *product.*

The multiplicand and multiplier, together, are called *factors,* or *producers* of the *product.*

There are three numbers in every multiplication. First, the multiplicand; second, the multiplier; and third, the product.

Quest.—58. If 1 be multiplied by 2, what is the result? How many times greater is this result than 1? If 3 be multiplied by 4, what is the result? How many times greater is the result than 3? What is multiplication? What is the number to be taken called? What is the number showing how many times the multiplicand is to be taken, called? What is the result called? What are the multiplier and multiplicand taken together called? How many numbers are there in every multiplication? What are they called?

59. Now, since the product is the result which arises from taking the multiplicand as many times as there are units in the multiplier, it follows that,

1st. If the multiplier is unity, the product will be equal to the multiplicand.

2d. If the multiplier contains several units, the product will be as many times greater than the multiplicand, as the multiplier is greater than unity.

3d. If the multiplier be less than unity, that is, if it be a proper fraction, then the product will be as many times *less* than the multiplicand as the multiplier is less than unity.

60. Let it be required to multiply any two numbers together, say 6 by 4.

If we make, in a horizontal line, as many stars as there are units in the multiplicand, and make as many such lines as there are units in the multiplier, it is evident that all the stars will represent the number of units which result from taking the multiplicand as many times as there are units in the multiplier.

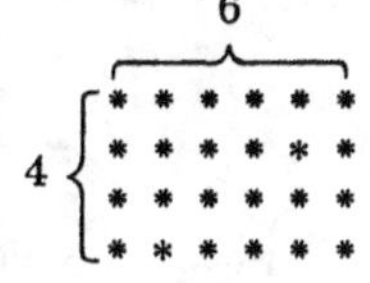

Let us now change the multiplier into the multiplicand, and let the multiplicand become the multiplier. Then make, in a vertical line, as many stars as there are units in the new multiplicand, and as many vertical lines as there are units in the new multiplier, and it will be again evident that all the stars will represent the number of units in the product. Hence,

Either of the factors may be used as the multiplier without altering the product. For example,

$3 \times 7 = 7 \times 3 = 21$: also, $6 \times 3 = 3 \times 6 = 18$.
$9 \times 5 = 5 \times 9 = 45$: also, $8 \times 6 = 6 \times 8 = 48$.
and, $8 \times 7 = 7 \times 8 = 56$: also, $5 \times 7 = 7 \times 5 = 35$.

QUEST.—59. If the multiplier is unity, how will the product compare with the multiplicand? How will it compare if the multiplier is greater than unity? How when it is less? 60. If the multiplicand be made the multiplier, will the product be altered?

61. A composite number is one that may be produced by the multiplication of two or more numbers, which are called the *components* or *factors*. Thus, $2 \times 3 = 6$. Here 6 is the composite number, and 2 and 3 are the factors, or components. The number $16 = 8 \times 2$: here 16 is a composite number, and 8 and 2 are the factors; and since $4 \times 4 = 16$, we may also regard 4 and 4 as factors or components of 16.

Let it be required to multiply 8 by the composite number 6, of which the factors are 2 and 3.

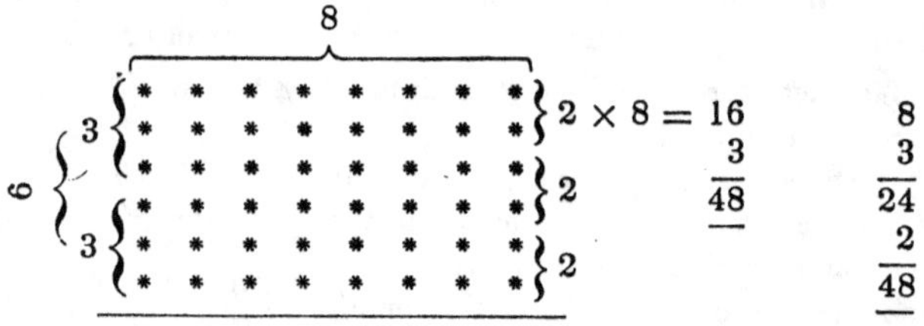

If we write 6 horizontal lines with 8 stars in each, it is evident that the product of $8 \times 6 = 48$, the number of units in all the lines.

But let us first connect the lines in sets of 2 each, as on the right; there will then be in each set $8 \times 2 = 16$, or 16 units in each set. But there are 3 sets; hence, $16 \times 3 = 48$, the number of units in all the sets.

If we divide the lines into sets of 3 each, as on the left, the number of units in each set will be equal to $8 \times 3 = 24$, and there being 2 sets, $24 \times 2 = 48$, the whole number of units. As the same may be shown for any composite number, we may conclude that,

When the multiplier is a composite number, we may multiply by each of the factors in succession, and the last product will be the entire product sought.

QUEST.—61. What is a composite number? What are the separate parts called? What are the components or factors in the number 12? In 16? In 20? How do you proceed when the multiplier is a composite number?

62. Let it be required to multiply 236 by 4; that is, to take 6 units, 3 tens, and 2 hundreds, each 4 times.

First set down the 236, then place the 4 under the unit's place 6, and draw a line beneath it. Then multiply the 6 units by 4: the product is 24 units; set them down. Next multiply the 3 tens by 4: the product is 12 tens; set down the 2 under the tens of the 24, leaving the 1 to the left, which is the place of the hundreds. Next multiply the 2 by 4: the product is 8, which being hundreds, is set down under the 1. The sum of these numbers, 944, is the entire product.

OPERATION.

```
 236
   4
 ---
  24 units.
 12  tens.
 8   hundreds.
 ---
 944
```

The product can also be found, thus: say 4 times 6 are 24; set down the 4, and then say, 4 times 3 are 12 and 2 to carry are 14; set down the 4, and then say, 4 times 2 are 8 and 1 to carry are 9. Set down the 9, and the product is 944 as before.

OPERATION.

```
236
  4
---
944
```

63. Let it be required to multiply 627 by 84.

Multiply by the 4 units, as in the last example. Then multiply by the 8 tens. The first product 56, is 56 tens; the 6, therefore, must be set down under the 0, which is the place of tens, and the 5 carried to the product of the 2 by 8. Then multiply the 6 by 8, carry the 2 from the last product, and set down the result 50. The sum of the numbers, 52668, is the required product.

OPERATION.

```
  627
   84
-----
 2508
5016
-----
52668
```

64. Let it be required to multiply £3 8*s.* 6*d.* 3*far.* by 6, in which each of the denominate numbers is to be taken 6 times.

QUEST.—62. Explain the manner of multiplying 236 by 4. 63. Explain the manner of multiplying 627 by 84.

We first say, 6 times 3 are 18; that is, 18 farthings, which by dividing by 4 are found equal to 4*d*., and 2 farthings over. Set down the 2 farthings, and then say, 6 times 6 are 36, and 4 to carry make 40; that is, 40 pence, which after dividing by 12, are found equal to 3 shillings and 4 pence. Set down the 4*d*., and then say, 6 times 8 are 48 and 3 are 51; that is, 51 shillings, which are equal to £2 and 11 shillings over. Set down the 11 shillings, and say, 6 times 3 are 18, and 2 to carry make 20, which write under the pounds.

£	*s.*	*d.*	*far.*
3	8	6	3
			6
20	11	4	2

65. Hence, to multiply one number by another,

Multiply every order of units in the multiplicand, in succession, beginning with the lowest, by each figure in the multiplier, and divide each product so formed by so many as make one unit of the next higher denomination: write down each remainder under units of its own order, and carry the quotient to the next product.

PROOF OF MULTIPLICATION.

66. Write the multiplier in the place of the multiplicand, and find the product as before; if the two products agree, the work may be supposed right: Or,

Divide the product by one of the factors, and the quotient will be the other factor.

EXAMPLES.

(1.)	(2.)	(3.)	(4.)
847046	9807602	570409	216987
8	7	8	6

QUEST.—64. Explain the manner of multiplying £3 8*s.* 6*d.* 3*far.* by 6 65. What is the general rule for multiplication? 66. What is the first proof of multiplication? What is the second?

(5.)		(5.)
Multiply	471493475	471493475
by	4395	4395
	2357467375	1885973900
	4243441275	1414480425
	1414480425	4243441275
	1885973900	2357467375
	2072213822625	2072213822625

NOTE 1. Although we generally begin the multiplication by the figure of the lowest denomination, yet we may multiply in any order, if we only preserve the places of the different orders of units. In the example to the right, we began with the order of thousands.

NOTE 2. Although either factor may be used as the multiplier, (Art. 60,) still it is best to use that one which contains the fewest places of figures, as is shown in the last example. For, if we change the process and use the multiplicand as the multiplier, there will be nine multiplications instead of four.

6. Multiply 430714934 by 743. *Ans.* ——
7. Multiply 37157437 by 14972. *Ans.* ——
8. Multiply 47157149 by 37049. *Ans.* ——
9. Multiply 57104937 by 40709. *Ans.* ——
10. Multiply 79861207 by 890416. *Ans.* ——
11. Multiply 9084076 by 9908807. *Ans.* ——
12. Multiply 2748 by 200. *Ans.* 549600.

When there are naughts on the right hand of the significant figures of the multiplier or multiplicand, we may at first neglect them in the multiplication; but then the first significant figure of the product will be of a higher order than the first, and all the ciphers must be added in order to reduce the product to units of the first order.

13. Multiply 67046 by 10: also by 100.
14. Multiply 57049 by 100: also by 1000.
15. Multiply 4980496 by 1000: also by 10000
16. Multiply 90720400 by 100: also by 10000
17. Multiply 74040900 by 1: also by 10
18. Multiply 674936 by 100: also by 100000.

19. Multiply 478400 by 270400. *Ans.* ——
20. Multiply 367000 by 37409000. *Ans.* ——
21. Multiply 7849000 by 84694000. *Ans.* ——
22. Multiply 89999000 by 97770400. *Ans.* ——
23. Multiply 9187416300 by 274987650000. ——
24. Multiply 86543291213456 by 12637482965. ——
25. Multiply 76729835645873 by 217834569. ——
26. Multiply 92413627858476 by 90587963412. ——
27. Multiply 87956743982714 by 819254837609. ——
28. Multiply 23869572491872 by 4007865347912. ——
29. Multiply 68 by the composite number 72. ——

In this example we multiply in succession by the factors 9 and 8.

30. Multiply 3657 by the factors of 64.
31. Multiply 37046 by the factors of 121.
32. Multiply 2187406 by the factors of 144.

67. In multiplying Federal money care must be taken to point off as many places for cents and mills as there are in the multiplicand

1. Multiply 14 dollars 16 cents and 8 mills, by 5, 6, and 7.

\$14,168	\$14,168	\$14,168
5	6	7

(2.)	(3.)	(4.)
\$870,46	\$894,120	\$2141,096
9	14 = 7 × 2	36 = 6 × 6.

5. What will 95 pounds of tea cost, at \$1,04 per pound?
6. What will 105 yards of cloth cost, at \$3,25 per yard?
7. What will four firkins of butter cost, each containing 97 pounds, at $25\frac{1}{2}$ cents per pound?

QUEST.—67 What precaution is necessary in multiplying Federal money?

8. What will five casks of wine cost, each containing 59 gallons, at \$2,756 per gallon?

9. A bale of goods contains 106 pieces, costing \$55 and $37\frac{1}{2}$ cents each: what is the cost of the entire bale?

10. What is the value of 695 hats, at \$3,654 each?

11. What will be the cost of 97046 oranges, at $2\frac{1}{2}$ cents each?

12. What will be the cost of 6742 sheep, at \$$2\frac{1}{4}$ each?

13. What will be the cost of 59 barrels of apples, at \$$2\frac{3}{4}$ per barrel?

14. What will be the cost of 6741 barrels of corn, at \$3,254 per barrel?

BILLS OF PARCELS.

15. New York, May 1st, 1846.

Mr. James Spendthrift

Bought of Benj. Saveall.

18 pounds of tea at 85 cents per pound - - - -
35 pounds of coffee at $15\frac{1}{2}$ cents per pound - - -
27 yards of linen at 66 cents per yard - - - -

Rec'd payment, *Benj. Saveall.*

---

16. Albany, June 2d, 1846.

Mr. Jacob Johns *Bought of Gideon Gould.*

48 pounds of sugar at $9\frac{1}{2}$ cents per pound - - - -
6 hogsheads of molasses, 63 gals. each, at 27 cents a gallon - - - -
8 casks of rice, 285 pounds each, at 5 cts. per pound
9 chests of tea, 86 pounds each, at 96 cts. per pound

Total cost

Rec'd payment, For Gideon Gould,

Charles Clark.

17. Hartford, November 21st, 1846.

Gideon Jones *Bought of Jacob Thrifty.*

78 chests of tea, at $55,65 per chest - - - - -
251 bags of coffee, 100 pounds each, at 12½ cts. per pound - - - - - - -
317 boxes of raisins, at $2,75 per box - - - -
1049 barrels of shad, at $7,50 per barrel - - - -
76 barrels of oil, 32 gallons each, at $1,08 per gal.

Amount

Received the above in full, *Jacob Thrifty.*

	(18.)			(19.)				(20.)		
	£	*s.*	*d.*	*T.*	*qr.*	*lb.*	*oz.*	*yds.*	*ft.*	*in.*
Multiply	20	6	8½	3	3	27	15	16	2	9
by			4				8			9

21. What will 4 yards of cloth cost at 7*s.* 6½*d.* per yard?
22. 5 bushels at 5*s.* 10*d.*
23. 6 yards at 6*s.* 9*d.*
24. 7 ells at 5*s.* 11½*d.*
25. 8*oz.* at 7*s.* 10*d.*
26. 8*lb.* at 7*s.* 5¼*d.*
27. 10 gallons at 16*s.* 4½*d.*
28. 11*cwt.* at £1 9*s.* 10½*d.*
29. 12 sheep at £1 17*s.* 9*d.*

30. In 9 pieces of kersey, each 14*yds.* 3*qrs.* 2*na.*, how many yards?

31. What is the weight of 12 tankards, each weighing 11*oz.* 10*pwt.* 19*gr.*?

32. In 11 pieces of cloth, each 17*yds.* 3*qrs.* 3*na.*, how many yards?

68. In multiplying denominate numbers, if the multiplier is a composite number, and greater than 12, it is best to multiply by the factors in succession.

QUEST.—68. If the multiplier is a composite number, how should you multiply in denominate numbers?

33. What will 15 gallons of wine cost at 5*s*. 3½*d*. per gallon?

34. 18*hhds*. at £3 14*s*. 5*d*.

35. 24*yds*. at 7*s*. 5½*d*.

36. 35*cwt*. at £1 17*s*. 8½*d*.

37. 36*T*. at. £5 15*s*. 11¼*d*.

38. 84 chaldrons at £1 16*s*. 9¾*d*.

39. 108 bushels at 7*s*. 9½*d*.

40. 132 ells at 18*s*. 9¼*d*.

41. 144 butts at £5 13*s*. 9½*d*.

42. In 32 wedges of gold, each 21*lb*. 7*oz*. 14*gr*., how many pounds?

43. In 21 fields, each 3*A*. 2*R*. 19*P*., how many acres?

69. When the multiplier is greater than 12 and is not a composite number,

Take the nearest composite number to the given multiplier, and multiply by its factors in succession. Then multiply by the difference, and add the product when the composite number is less than the multiplier, and subtract it when greater.

44. What is the cost of 23 yards of cloth, at 14*s*. 9*d*. per yard?

OPERATION.

	£	*s*.	*d*.		
		(14	9)	$\times(7\times3)+2$	
			7		
	5	3	3	price of 7*yds*.	
			3		
	15	9	9	price of 21.	
Add	1	9	6	price of 2.	
Ans.	£16	19	3	price of 23.	

	£	*s*.	*d*.		
Or this,		(14	9)	$\times(6\times4)-1$	
			6		
	4	8	6	price of 6.	
			4		
	17	14	0	price of 24.	
Subtract		14	9	price of	1.
Ans.	£16	19	3		23

45. What is the cost of 31 yards at 12*s*. 7¼*d*.?

46. 39 dozen of handkerchiefs at 16*s*. 9¼*d*.

47. 139 pairs of stockings at 4*s*. 9¼*d*.

48. 86*lb*. of silk at 19*s*. 4*d*.

49. 111 sacks of flour at £1 4*s*. 9*d*.

50. 156*cwt*. at £4 9*s*. 6*d*.

QUEST.—69. How do you multiply when the multiplier is greater than 12 and not a composite number?

51. In 57 years, each 13 months, 1 day, 6 hours, how many months?

52. What is the weight of 29*hhds.* of sugar, each weighing 7*cwt.* 2*qr.* 18*lb.*?

53. In 67 parcels of tea, each 25*lb.* 7*oz.* 13*dr.*, how many cwt., &c.?

54. What will 394 yards cost at 17*s.* 5½*d.* per yard?

OPERATION.

	£	*s.*	*d.*	
		17	5½	
			10	
9 ×	8	14	7	price of 10*yds.*
			10	
	87	5	10	price of 100.
			3	
	261	17	6	price of 300.
	78	11	3	price of 90.
	3	9	10	price of 4.
	£343	18	7	price of 394.

55. 357 calves at £7 10*s.* 7*d.*

56. 549 yards at 12*s.* 9½*d.*

57. 754*lb.* of tea at 6*s.* 10*d.*

58. 198*lb.* of indigo at 6*s.* 3¼*d.*

59. 754 weys at £20 5*s.* 10*d.*

60. 178 ells at 5*s.* 9¼*d.*

61. 198*bbls.* at £1 14*s.* 9*d.*

62. 744 chaldrons at £1 18*s.* 8*d.*

70. When the multiplier has a fraction annexed to it, multiply first by the whole number, and then add such a part of the multiplicand as the fraction is of unity.

63. What will 56½ chaldrons cost at £1 14*s.* 9*d.* per chaldron?

	£	*s.*	*d.*	
	1	14	9	
			7	
	12	3	3	price of 7.
			8	
	97	6	0	price of 56.
		17	4½	price of ½.
Ans.	£98	3	4½	price of 56½.

64. What will be the cost of 4⁵⁄₉ yards at 7*s.* 6*d.* per yard?

£	*s.*	*d.*	
	7	6	
		4	
1	10	0	price of 4.
	4	2	price of ⁵⁄₉.
1	14	2	price of 4⁵⁄₉.

	s.	*d.*
	7	6
		5
9)	37	6
	4	2

QUEST.—70. How do you multiply when the multiplier has a fraction annexed?

65. 1788$\frac{1}{5}$ gallons at 6*s*. 4*d*. *Ans.* ——

66. 3714$\frac{1}{8}$*cwt*. at £4 11*s*. 9*d*. *Ans.* ——

67. 7149$\frac{3}{7}$ chaldrons at £1 14*s*. 9*d*. *Ans.* ——

68. 547$\frac{5}{8}$ lasts at £5 5*s*. *Ans.* ——

69. 1749$\frac{1}{2}$ firkins at 14*s*. 9$\frac{1}{4}$*d*. *Ans.* ——

70. 754$\frac{3}{4}$*cwt*. at 17*s*. 5$\frac{1}{4}$*d*. *Ans.* ——

BILLS OF PARCELS.

71. New Orleans, Jan. 2d, 1846.

James Lamb, Esq.

Bought of John Simpson.

			£ s. d.
7$\frac{1}{2}$	*lbs.* of	green tea at 10*s*. 4*d*. per *lb*. - - - - -	
14$\frac{1}{2}$	do.	finest bloom at 14*s*. 8*d*. per *lb*. - - - -	
10$\frac{3}{4}$	do.	fine green at 16*s*. 5*d*. per *lb*. - - -	
21	do.	hyson at 10*s*. 10$\frac{1}{2}$*d*. per *lb*. - - - -	
19	do.	good hyson at 13*s*. 9$\frac{1}{2}$*d*. per *lb*. - - -	
8$\frac{1}{4}$	do.	bohea at 6*s*. 9*d*. per *lb*. - - - - - -	
			£

---

72. Louisville, March 19th, 1846.

George Veres, Esq.

Bought of Charles West.

	£ s. d.
A loin of lamb, weight 7$\frac{1}{4}$ *lb*., at 10$\frac{3}{4}$*d*. per *lb*. - -	
A fillet of veal, weight 16$\frac{3}{4}$ *lb*., at 6$\frac{1}{2}$*d*. per *lb*. - -	
A buttock of beef, weight 37$\frac{1}{2}$ *lb*., at 4$\frac{1}{2}$*d*. per *lb*. -	
A pig, weight 12$\frac{3}{4}$ *lb*., at 7$\frac{1}{4}$*d*. per *lb*. - - - - -	
A leg of pork, weight 16$\frac{1}{4}$ *lb*., at 5$\frac{1}{2}$*d*. per *lb*. - -	
A leg of mutton, weight 13$\frac{3}{4}$ *lb*., at 4$\frac{3}{4}$*d*. per *lb*. -	
	£

DIVISION.

71. DIVISION is the process of finding how many times one number called the *dividend* is greater or less than another number called the *divisor;* and the number which expresses how many times the dividend is greater or less than the divisor, is called the *quotient.* Hence, the quotient is as many times greater or less than unity, as the dividend is greater or less than the divisor.

72. When the entire quotient can be expressed by a whole number, the dividend is said to contain the divisor an exact number of times; but when it cannot be so expressed, the part of the dividend which remains undivided is called the *remainder.*

73. Since the quotient shows how many times the dividend exceeds the divisor, it follows, that if the divisor be taken as many times as there are units in the quotient, the product will be equal to the dividend. And hence, if the divisor and quotient be multiplied together, and the remainder, if any, added to the product, the result will be equal to the dividend.

EXAMPLES.

1. Divide 86 by 2.

OPERATION.

Divisor. Dividend.
2)86
43 quotient.

Place the divisor on the left of the dividend, draw a curved line between them, and a straight line under the dividend.

Now, there are 8 tens and 6 units to be divided by 2. We say, 2 in 8, 4 times, which being 4 tens we write the 4 under

QUEST.—71. What is division? What is the quotient? How many times is it greater or less than unity? 72. When can the entire quotient be expressed by a whole number? When it cannot, what do you call the part of the dividend which is over? 73. If the divisor and quotient be multiplied together, what will the product be equal to?

the tens. We then say, 2 in 6, 3 times, which are three units, and must be written under the 6. The quotient, therefore, is 4 tens and 3 units, or 43. Remark that each order of units in the dividend, on being divided, gives the same order of units in the quotient.

2. Divide 729 by 3.

In this example there are 7 hundreds, 2 tens, and 9 units, all to be divided by 3. Now, we say, 3 in 7, 2 times; that is, 2 hundreds, and 1 hundred over. Set down the 2 hundreds under the 7. Now of the 7 hundreds there is 1 hundred or 10 tens not yet divided. We put the 10 tens with the 2 tens, making it 12 tens, and then say, 3 in 12, 4 times; that is, 4 tens times; therefore write the 4 in the quotient, in the ten's place; then say, 3 in 9, 3 times. The quotient, therefore, is 243.

OPERATION.

```
3)729
  243
```

3. Divide 729 by 9.

In this example we say, 9 in 7 we cannot, but 9 in 72, 8 times, which are 8 tens: then, 9 in 9, 1 time.

The quotient is therefore 81.

OPERATION.

```
9)729
   81
```

4. Divide 8040 by 8.

In this example we say, 8 in 8, 1 time, and set 1 in the quotient. We then say, 8 in 0, 0 times, and set the 0 in the quotient: then say, 8 in 4, 0 times, and set the 0 in the quotient: then, say, 8 in 40, 5 times; that is, 5 units times, and therefore we set the 5 in the unit's place of the quotient. Therefore the true quotient is 1005.

OPERATION.

```
8)8040
  1005
```

5. Let it be required to divide 36458 by 5.

In this example, we find the quotient to be 7291 and a remainder 3. This 3 ought in fact to be divided by the divisor 5, but the division cannot be effected, since 3 does not contain 5. The division must then be indicated by placing 5 under the 3, thus, $\frac{3}{5}$.

OPERATION.

```
5)36458
   7291–3 remain
```

The entire quotient, therefore, is $7291\frac{3}{5}$, which is read, seven thousand two hundred and ninety-one, and *three divided by five*. Therefore,

Where there is a remainder after the division, it may be written after the quotient, and the divisor placed under it.

74. When the divisor is 12 or less than 12, the operation may be performed as in the last examples, and this method of dividing is called SHORT DIVISION.

6. Divide 2756 by 26.

OPERATION.

```
Divisor. Dividend. Quotient.
26)2756(106
   26
   ---
    156
    156
    ---
```

We first say, 26 in 27 hundreds, once, and set down 1 in the quotient, in the hundred's place. Multiplying by 1, subtracting, and bringing down the 5, we say, 26 in 15 tens, 0 tens times, and place the 0 in the quotient. Bringing down the 6, we find that the divisor is contained in 156, 6 times. Hence, the entire quotient is 106.

7. Divide 11772 by 327.

OPERATION.

```
327)11772(36
    981
    ----
    1962
    1962
    ----
    0000
```

Having set down the divisor on the left of the dividend, it is seen that 327 is not contained in the first three figures on the left, which are 117 hundreds. But by observing that 3 is contained in 11, 3 times and something over, we conclude that the divisor is contained at least 3 times in the first *four* figures of the dividend, which are 1177 tens. Set down the 3, which are tens, in the quotient, and multiply the divisor by it: we thus get 981 tens, which being less than 1177, the quotient figure is not too great: we subtract the 981 tens from the first four figures of the dividend, and find a remainder 196 tens, which being less than the divisor, the quotient fig-

---

QUEST.—74. When the divisor is 12 or less than 12, what is the division called?

ure is not too small. Reduce this remainder to units and add in the 2, and we have 1962.

As 3 is contained in 19, 6 times, we conclude that the divisor is contained in 1962 as many as 6 times. Setting down 6 in the quotient and multiplying the divisor by it, we find the product to be 1962. Therefore the entire quotient is 36, or the divisor is contained 36 times in the dividend.

8. Divide £133 9*s*. 8*d*. by 4.

OPERATION.
4)£133 9*s*. 8*d*.
£33 7*s*. 5*d*.

Here we again take the least number of units of the highest order which will contain the divisor, viz., 13 tens of the denomination of tens of pounds. Dividing by 4, we find the quotient to be 3 tens of the same denomination, and 1 ten over. We reduce these tens to units and add in the 3, and thus obtain 13 pounds, which being divided by 4 gives 3 pounds and 1 over. Reducing this £1 to shillings and adding in the 9, gives 29, which being divided by 4 gives 7 shillings and 1 over. Reducing this to pence and adding in 8*d*., and again dividing by 4, we have £33 7*s*. 5*d* for the entire quotient.

9. Divide £6 8*s*. 8*d*. by 8.

8)£6 8*s*. 8*d*.
16*s*. 1*d*.

Here we have to pass to shillings before making the first division.

75. Combining the principles illustrated in the foregoing examples we have, for the division of numbers, the following:

Beginning with the highest order of units of the dividend, pass on to the lower orders until the fewest number of figures be found that will contain the divisor: divide these figures by it for the first figure of the quotient: the unit of this figure will be the same as that of the lowest order used in the dividend.

Multiply the divisor by the quotient figure so found, and subtract the product from the dividend, observing to place units of the same order under each other. Reduce the remainder to units of the next lower order, and add in the units of that order

QUEST.—75. What is the rule for the division of numbers?

found in the dividend: this will furnish a new dividend. Proceed in a similar manner until units of every order shall have been divided.

76. There are always three numbers in every operation of division, and sometimes four. First, the dividend; second, the divisor; third, the quotient; and fourth, the remainder, when the numbers are not exactly divisible.

77. There are five operations in division. First, to write down the numbers; second, find how many times; third, multiply; fourth, subtract; and fifth, reduce to the next lower order.

EXAMPLES.

1. Divide 1203033 by 3679.

OPERATION.

```
3679)1203033(327
     11037
      9933
      7358
     25753
     25753
```

By the first operation, 300 times the divisor is taken from the dividend; or, what is the same thing, the divisor is taken from the dividend 300 times. By the second, it is taken 2 tens or twenty times; and by the third, it is taken 7 units times; therefore, it is taken in all 327 times: hence,

78. Division is a short method of performing subtraction; and the quotient found according to the rules always shows how many times the divisor may be subtracted from the dividend.

Prove the above work by multiplying the divisor and quotient together.

2. Divide 714394756 by 1754. *Ans.* ——
3. Divide 47159407184 by 3574. *Ans.* ——
4. Divide 5719487194715 by 45705. *Ans.* ——
5. Divide 4715714937149387 by 17493. *Ans.* ——
6. Divide 671493471549375 by 47143. *Ans.* ——
7. Divide 571943007145 by 37149. *Ans.* ——

Quest.—76. How many numbers are considered in division? What are they? 77. How many operations are there in division? Name them. 78. How may division be defined; and what does the quotient show?

8. Divide 1714347149347 by 57143. *Ans.* ——
9. Divide 49371547149375 by 374567. *Ans.* ——
10. Divide 171493715947143 by 571007. *Ans.* ——
11. Divide 6754371495671594 by 678957.
12. Divide 7149371478 by 121.
13. Divide 71900715708 by 57149.
14. Divide 15241578750190521 by 123456789.
15. Divide 121932631112635269 by 987654321.
16. Divide 14714937148475 by 123456.
17. Divide 8890896691492249389482962974 by 987675.

PROOF OF MULTIPLICATION.

79. When two numbers are multiplied together, the multiplicand and multiplier are both factors of the product; and if the product be divided by one of the factors, the quotient will be the other factor. Hence, *if the product of two numbers be divided by the multiplicand, the quotient will be the multiplier; or, if it be divided by the multiplier, the quotient will be the multiplicand.*

1. The multiplicand is 61835720, the product 8162315040: what is the multiplier?

2. The multiplier is 270000, the product 1315170000000: what is the multiplicand? *Ans.* ——

3. The product is 68959488, the multiplier 96: what is the multiplicand?

4. The multiplier is 1440, the product 10264849920: what is the multiplicand? *Ans.* ——

5. The product is 6242102428164, the multiplicand 6795634: what is the multiplier? *Ans.* ——

80. When the divisor is a composite number.

Divide the dividend by one of the factors of the divisor, and then divide the quotient thus arising by the other factor: the last quotient will be the one sought.

QUEST.—79. How may multiplication be proved by division? 80. How do you divide when the divisor is a composite number?

EXAMPLES.

1. Let it be required to divide 1407 dollars equally among 21 men. Here the factors of the divisor are 7 and 3.

Let the 1407 dollars be first divided equally among 7 men. Each share will be 201 dollars. Let *each one* of the 7 men divide his share into 3 equal parts, each one of the three equal parts will be 67 dollars, and the whole number of parts will be 21; here the true quotient is found by dividing continually by the factors.

OPERATION

```
7)1407
3)201  1st quotient.
   67  quotient sought
```

2. Divide 18576 by 48 = 4 × 12. *Ans.* ——

3. Divide 9576 by 72 = 9 × 8. *Ans.* ——

4. Divide 19296 by 96 = 12 × 8. *Ans.* ——

81. It sometimes happens that there are remainders after division—they are to be treated as follows:

The first remainder, if there be one, forms a part of the true remainder. The product of the second remainder, if there be one, by the first divisor, forms a second part. Either of these parts, when the other does not exist, forms the true remainder, and their sum is the true remainder when they both exist together, and similarly when there are more than two remainders.

1. What is the quotient of 751 grapes, divided by 16?

```
          ( 4)751
4 × 4 = 16{ 4)187 . . . 3
          (    46 . . . 3 × 4 = 12
                                 3
                                15 the true remainder.
```

Ans. $46\frac{15}{16}$.

In 751 grapes there are 187 sets, (say bunches,) with 4 grapes or units in each bunch, and 3 units over. In the 187 bunches there are 46 piles, 4 bunches in a pile, and 3 bunches over. But there are 4 grapes in each bunch; therefore, the

QUEST.—81. How do you dispose of the remainders, if there are any, after division?

number of grapes in the 3 bunches is equal to $4 \times 3 = 12$, to which add 3, the grapes of the first remainder, and we have the entire remainder 15.

2. Divide 4967 by 32.

4)4967
$4 \times 8 = 32$ 8)1241 . . . 3, 1st remainder.
155 . . . $1 \times 4 + 3 = 7$ the true remainder.

Ans. $155\frac{7}{32}$.

3. Divide 956789 by $7 \times 8 = 56$.

4. Divide 4870029 by $8 \times 9 = 72$.

5. Divide 674201 by $10 \times 11 = 110$.

6. Divide 445767 by $12 \times 12 = 144$.

7. Divide 375197351937 by $349272 = 12 \times 11 \times 9 \times 7 \times 7 \times 6$.

12)375197351937
11)31266445994 . . . 9 - - - - - - = 9
9)2842404181 . . . 3 -- 12×3 - - - = 36
7)315822686 . . . 7 -- $12 \times 11 \times 7$ - - = 924
7)45117526 . . . 4 -- $12 \times 11 \times 9 \times 4$ - = 4752
6)6445360 . . . 6 -- $12 \times 11 \times 9 \times 7 \times 6$ = 49896
Quotient = 1074226 . . . 4 -- $12 \times 11 \times 9 \times 7 \times 7 \times 4$ = 232848
Remainder = 288465

8. Divide 7349473857 by 27. *Ans.* ——

9. Divide 749347549 by 144. *Ans.* ——

10. Divide 649305743 by 55. *Ans.* ——

11. Divide 4730715405 by 121. *Ans.* ——

12. Divide 3704099714 by 108. *Ans.* ——

13. Divide 4710437154 by 132. *Ans.* ——

14. Divide 1071540075 by 99. *Ans.* ——

15. Divide 468248 by $3 \times 4 \times 2 \times 5 \times 6$.

16. Divide 98765432101234567890 by $12 \times 11 \times 10 \times 9 \times 8 \times 7 \times 6 \times 5 \times 4 \times 3 \times 2$.

82. When the divisor has one or more 0's at the right, it may be regarded as a composite number, of which one factor is 1 with as many 0's on the right as there are 0's at the right of the divisor, and the remaining figures express the other factor. *Strike off the 0's and the same number of figures from the right of the dividend—this is dividing by one of the factors; then proceed to divide by the other.*

1. Divide 14715967899 by 145000.

145000)14715967899(101489$\frac{62899}{145000}$ Quotient.

```
145000)14715967899(101489 62899/145000 Quotient.
       145
        215
        145
         709
         580
         1296
         1160
          1367
          1305
           62899 Rem.
```

Or thus,

```
145000)14715967899(101489 62899/145000
           215
            709
            1296
             1369
              62899 Rem.
```

NOTE. In the second operation of this example, the products of the divisor by each quotient figure are subtracted mentally, and the remainders only written down. Let the pupil perform many examples in division in this way.

2. Divide 571436490075 by 36500.
3. Divide 194718490700 by 73000.
4. Divide 795498347594 by 47150.
5. Divide 1495070807149 by 31500.
6. Divide 6714934714934 by 754000.
7. Divide 1071491471430715 by 754000.
8. Divide 14714937493714957 by 157900.
9. Divide 7149374947194715 by 1749000.
10. Divide 714947349 by 90.
11. Divide 1714937148 by 14400.
12. Divide 696161034987219318OO by 975005700.

QUEST.—82. What is the process when the divisor has 0's annexed?

13. Divide 656458931996524171800 by 700489070.

14. Divide 7149437149547 by 3714900.

EXAMPLES IN DENOMINATE NUMBERS.

1. A gentleman's income is £1260 15*s.* 5*d.* a year: what is that per day, 365 days being contained in one year?

	£	*s.*	*d.*		£	*s.*	*d.*	
365)	1260	15	5	(	3	9	1	= *Ans.*
	1095						10	
	165				34	10	10	× 6
	20						10	
365)	3315	(9*s.*			345	8	4	
	3285						3	
	30				1036	5	0	
	12				207	5	0	
365)	365	(1*d.*			17	5	5	
	365				1260	15	5	Proof.
	0							

2. Divide £47 19*s.* 4*d.* by 3. *Ans.* ——
3. Divide £37 14*s.* 10*d.* by 24. *Ans.* ——
4. Divide £49 19*s.* 11¼*d.* by 66. *Ans.* ——
5. Divide £34 14*s.* 9¼*d.* by 149. *Ans.* ——
6. Divide £1774 19*s.* 10¼*d.* by 179. *Ans.* ——
7. Divide 47*yd.* 3*qr.* 2*na.* by 5. *Ans.* ——
8. Divide 37*A.* 3*R.* 14*P.* by 9. *Ans.* ——
9. Divide 714*lb.* 10*oz.* 12*gr.* by 89. *Ans.* ——
10. Divide 374*cwt.* 3*qr.* 10*lb.* by 48. *Ans.* ——
11. Divide 374*E.E.* 2*qr.* 3*na.* by 142. *Ans.* ——
12. If 60 sheep be sold for £112 10*s.*, what is the value of 1 sheep?
13. If 112*lb.* of cheese cost £2 18*s.* 8*d.*, what is that per pound?
14. If 17*cwt.* of lead cost £15 5*s.* 7¾*d.*, what costs 1*cwt.*?
15. Bought 7 yards of cloth for 16*s.* 4*d.*; what is that per yard?

16. If 63 oxen cost £2553 1*s.* 6*d.*, what costs 1 ox?

17. If 66*lb.* of butter cost £5 15*s.* 6*d.*, what costs 1*lb.*?

18. If 528*lb.* of tobacco cost £23 13*s.*, what costs 1*lb.*?

19. If a tun, or 252 gallons, of wine cost £60, what costs 1 gallon?

20. A prize of 1000 guineas is to be divided among 150 sailors; what is each man's share, after deducting $\frac{1}{6}$ part for the officers?

21. If 125 ingots of silver, each of an equal weight, weigh 1347*oz.* 11*pwt.* 14*gr.*, what is the weight of 1 ingot?

22. If 475*cwt.* 1*qr.* 14*lb.*, be the weight of 27*hhds.* of tobacco, what is the weight of 1*hhd.*?

23. Bought 6 pieces of tapestry, containing 237*E. Fl.* 2*qr.* 2*na.*; what is the length of 1 piece?

APPLICATIONS.

1. In 1842, nine mills in Lowell manufactured 434000 pounds of cotton per week. How much was manufactured by each mill, supposing the amount was exactly the same?

2. The number of inhabitants in the city of New York in 1840 was 312710, and the expenses of the city government $1645779,30. If this was raised by an equal tax upon every inhabitant, how much would each have to pay?

3. The number of hogsheads of tobacco exported from the United States in the 20 years preceding 1841, was 1792000, and their estimated value was $131346514. What was the average value by the hogshead?

4. The amount of coffee imported in 1840 was 94996095 pounds, and its value was estimated at $8546222. What was its worth per pound?

5. The number of scholars attending the public schools of the state of Maine, in 1839, was 201024, and the amount expended for the support of the schools was $258113,43. What was the cost to the state for the tuition of each scholar?

6. The militia force of the United States, according to the Army Register for 1845, was 1426868, and the number of commissioned officers belonging to it was 69450. How many soldiers did that allow to each officer?

7. The whole coinage of the United States for the 51 years preceding 1845, amounted to $110177761,38. Suppose an equal amount had been coined each year, what would it have been?

8. In 1843 there were sold 1605264 acres of the public lands. The sum received for them was $2016044,30, and the sum paid into the national treasury, after deducting expenses, was $1997351,57. What was the average cost per acre to the purchasers, and what was the average price per acre received by the government?

9. The net amount of duties on imports for eighteen years preceding 1843, was $452539300,81. How much was collected in each year, supposing the sums to have been equal?

10. There was inspected in Onondaga county, N. Y., in 1844, 4003554 bushels of salt. The duties collected on these amounted to $240305. What was the duty on each bushel?

11. There were thirty-five banks in New Hampshire in 1844, whose whole resources were $5836014,07. If this sum was equally divided, how much would belong to each?

12. The population of Europe in 1837 was estimated at 233884800, and the number of square miles at 3708871. How many inhabitants would this give to each square mile?

13. In 1843, there were 3173 public schools in Massachusetts, which were attended during the winter by 119989 scholars. How many would this allow to each school?

14. The number of male scholars attending the public schools of Pennsylvania was reported, in 1843, to be 161164 and the number of female scholars 127598. The number of male teachers employed by the state was 5264, and the number of female teachers 2330. How many scholars would this give to each teacher?

15. The value of the exports from the United States in 1841, was $104691534. If an equal amount had been exported each day of the year excepting Sundays, what would it have been?

OF THE PROPERTIES OF THE 9's.

83. Besides the methods already explained of proving the operations in figures, there is yet another called the method by *casting out the 9's*. That method we will now explain.

84. An excess of units over exact 9's, is the remainder after the number has been divided by 9: hence, any number less than 9 must be treated as an excess over exact 9's.

Let us write down the numbers to be added, as at the right. Now, if we divide each number by 9, and place the quotients to the right, and the remainders in the column still to the right, we shall have, in the middle column, the exact number of 9's contained in each number, and in the column at the right, the excesses over exact 9's. By adding these columns, we find 15 in the column of remainders, which is equal to one 9 and 6 over: hence, there are 764 exact 9's and 6 over. But it is evident that the sum of all the numbers, viz., 6882, must contain exactly the same number of 9's and the same excess over exact 9's, as are found in the numbers taken separately, since a whole is equal to the sum of all its parts any way taken: therefore, *in the sum of any numbers whatever, the number of exact 9's and the excess over 9's are equal, respectively, to the aggregate of exact 9's and the excess of 9's in the numbers taken separately.*

OPERATION.

	Quotients after dividing by 9.	Remainders, or excesses over 9.
3870	430	0
2708	300	8
304	33	7
9)6882	764	6
764–6		

QUEST.—83. What other methods of proof are there for arithmetical operations, besides those already explained? 84. What is an excess of 9's? How do the exact number of 9's and the excess of 9's in any sum compare with the exact 9's and the excess of 9's in the several numbers?

85. We will now explain a short process of finding the excess over an exact number of 9's in any number whatever; and to do this, we must look a little into the formation of numbers.

In any number, written with a single significant figure, as 4, 50, 600, 8000, &c., the excess over exact 9's will always be equal to the number of units expressed by the significant figure; for, in any such number we shall always have $4 = 4$

Also, - - - - - - $50 = (9 \quad +1)\times 5$

- - - - - - - $600 = (99 \quad +1)\times 6$

- - - - - - - $8000 = (999+1)\times 8$

&c. &c. &c.

Each of the numbers 9, 99, 999, &c., expresses an exact number of 9's; and hence, when multiplied by 5, 6, 8, &c., the several products will each contain an exact number of 9's; therefore, the excess over exact 9's, in each number, will be expressed by 4, 5, 6, 8, &c.

If, then, we write any number whatever, as

6253,

we may read it 6 thousand 2 hundred 50 and 3. Now, the excess of 9's in the 6 thousand is 6; in 2 hundred it is 2; in 50 it is 5; and in 3 it is 3: hence, in them all, it is 16, which makes one 9 and 7 over: therefore, 7 is the excess over exact 9's in the number 6253. Hence, *the excess over exact 9's in any number whatever, may be found by adding together the significant figures, and rejecting the exact 9's from the sum.*

NOTE.—It is best to reject or drop the 9 as soon as it occurs: thus we say, 3 and 5 are 8 and 2 are 10; then dropping the 9, we say, 1 to 6 is 7, which is the excess; and the same for all similar operations.

1. What is the excess of 9's in 48701? In 67498?
2. What is the excess of 9's in 9472021? In 2704962?
3. What is the excess of 9's in 87049612? In 4987051?

QUEST.—85. What will be the excess over exact 9's in any number expressed by a single significant figure? How may the excess over exact 9's be found in any number whatever?

PROOF OF ADDITION BY CASTING OUT THE 9's.

86.—1. In the first of these numbers we find the excess of 9's to be 5; in the second 2; in the third 8; in the fourth 2; and in the fifth 8: hence, in them all it is 25, which leaves 7 for the excess over exact 9's. We also find 7 to be the excess over exact 9's in the sum 195460: hence the work is supposed to be right. Notwithstanding this proof, it is possible, after all, that the work may be erroneous. For example, if either figure in the sum is too large by one or more units, and any other figure is too small by the same number of units, the excess over exact 9's will not be affected. But as it would seldom happen that one error would be exactly balanced by another, the work when proved may be relied on as correct. Similar sources of error exist in the proof of all arithmetical operations.

OPERATION.

	Excess of 9's.
94874 . . .	5
46073 . . .	2
50498 . . .	8
3674 . . .	2
341 . . .	8
195460–7	7

2. Add together, 8754608, 4908721, 6027983, 89704543, 2142367, and 28949760, and prove the result by rejecting the 9's.

3. Add together 40799903, 874162, 32704931, 6704192, 2146748, 94004169, and prove the result by casting out the 9's.

PROOF OF SUBTRACTION BY CASTING OUT THE 9's.

87.—1. Since the sums of the remainder and subtrahend must be equal to the minuend, it follows that the excess of 9's in these two numbers must be equal to the excess of 9's in the minuend: hence, *to the excess of 9's in the remainder add the excess of 9's in the subtrahend, and the excess of 9's in the sum will be equal to the excess of 9's in the minuend.*

OPERATION.

874136 . . .	2
45302 . . .	5
828834 . . .	6

QUEST.—86. Explain the proof of addition by casting out the 9's. In what is the proof defective? 87. Explain the proof of subtraction by casting out the 9's.

2. From 874096 take 370494, and prove the work by rejecting the 9's.

3. From 47096702 take 1104967, and prove the work by rejecting the 9's.

PROOF OF MULTIPLICATION BY CASTING OUT THE 9's.

88. We will first remark, that if any number containing an exact number of 9's be multiplied by another whole number, the product will also contain an exact number of 9's.

Let it be required to multiply any two numbers together, as 641 and 232.

We first find the excess over exact 9's in both factors, and then separate each factor into two parts, one of which shall contain exact 9's, and the other the excess, and unite the two together by the sign plus. It is now required to take $639 + 2 = 641$, as many times as there are units in $225 + 7 = 232$.

OPERATION.

$$\begin{array}{r} 641 = 639 + \ \ 2 \\ 232 = 225 + \ \ 7 \\ \hline 4473 + 14 \\ 450 \\ 3195 \\ 1278 \\ 1278 \\ \hline 148698 + 14 \end{array}$$

Beginning with the 7, we have 14 for the product of 2 by 7, and 4473 for the product of 639 by 7; and this last contains an exact number of 9's. We then take 2, 225 times, which gives 450, which also contains an exact number of 9's. We next multiply 639 by the figures of 225, and each of the several products contains an exact number of 9's, since 639 contains an exact number. Hence, the entire sum 148698 contains an exact number of 9's, to which if we add the one 9 from the 14, we shall find the excess of 9's in the product to be 5; and as the same may be shown for any numbers, we conclude that, *the excess of 9's in any product must arise from the product of the excess of 9's in the factors.*

QUEST.—88. Explain the proof of multiplication by casting out the 9's. What does the excess of 9's in any product arise from?

But since the product of two numbers found in the ordinary way must contain the same number of 9's, and the same excess of 9's as a product found above, it follows that, *if the excesses of 9's in any number of factors be multiplied together, the excess of 9's in such product will be equal to the excess of 9's in the product of the factors.*

EXAMPLES.

	(1.)		(2.)	
Multiply	87603	. . . 6	818327	. . . 2
by	9865	1	9874	1
Prod.	864203595	. . . 6	8080160798	. . . 2

3. By multiplication we have

Ex. 4.		Ex. 8.		Ex. 4.		Ex. of product, 2.
7285	×	143	×	976	=	1016752880.

4. We also have

Ex. 5.		Ex. 4.		Ex. 0.		Ex. 0.
869	×	49	×	36	=	1532916.

When the excess of 9's in any factor is 0, the excess of 9's in the product is always 0.

PROOF OF DIVISION BY CASTING OUT THE 9's.

89. Since the divisor multiplied by the quotient must produce the dividend, it follows that if the excess of 9's in the divisor be multiplied by the excess of the 9's in the quotient, the excess of 9's in the product will be equal to the excess of 9's in the dividend.

1. The dividend is 8162315040, the divisor 61835720, and the quotient 132: is the work right?

2. The dividend is 10264849920, the divisor 1440, and the quotient 7128368: is the work right?

3. The dividend is 74855092410, the quotient 78795, and the divisor 949998: is the work right?

Let the pupils apply the property of the 9's to other examples.

QUEST.—If the excess of 9's in any number of factors be multiplied together, what will the excess of 9's in the product be equal to? 89. How do you prove division by casting out the 9's?

REMARKS.

90.—1. Numeration, Addition, Subtraction, Multiplication, and Division are called the five ground rules, because all the other operations of arithmetic are performed by means of them. Multiplication, however, is but a short method of performing addition, and division but an abridged method of subtraction.

2. A prime number is one which cannot be exactly divided by any number except itself and unity. Thus, 1, 3, 5, 7, 11, 13, 19, 23, &c., are prime numbers.

3. The product of two or more prime numbers will be exactly divisible only by one or the other of the factors.

4. If an even number be added to itself any number of times, the sum will be even; hence, if one of the factors of a product be an even number, the product will be even.

5. An odd number is not divisible by an even number; nor is a less number exactly divisible by a greater.

6. The quotient arising from the division of the sum of two or more numbers, by any divisor, is equal to the sum of the quotients which arise from the division of the parts separately.

7. Any number is divisible by 2, if the last significant figure is even; and is divisible by 4, if the last two figures are divisible by 4.

8. Any number whose last figure is 5 or 0, is exactly divisible by 5; and any number whose last figure is 0, is exactly divisible by 10.

QUEST.—90.—1. What are the five ground rules of arithmetic? What other rule in fact embraces the rule of multiplication? How may division be performed? 2. What is a prime number? 3. By what numbers only will the product of prime factors be divisible? 4. If an even number be multiplied by a whole number, will the product be odd or even? 5. Is an odd number divisible by an even number? 6. What is the quotient arising from the division of the sum of two or more numbers by any divisor equal to? 7. When is a number exactly divisible by 2? When by 4? 8. If the last figure of a number be 5 or 0, by what numbers may it be divided?

DIVISIONS OF ARITHMETIC.

91. The science of arithmetic, which treats of numbers, may be divided into four parts:

1st. That which treats of the properties of entire units, called the Arithmetic of Whole Numbers;

2d. That which treats of the parts of unity, called the Arithmetic of Fractions;

3d. That which treats of the relations of the unit 1 to the numbers which come from it, whether they be integers or fractions, and the relations of these numbers to each other; and

4th. The application of the science of numbers to practical and useful purposes.

A portion of the First part has already been treated under the heads of Numeration, Addition, Subtraction, Multiplication and Division.

The Second part comes next in order, and naturally divides itself into two branches; viz.,

Vulgar or Common Fractions, in which the unit is divided into any number of equal parts, and Decimal Fractions, in which the unit is divided according to the scale of tens.

The Third part relates to the comparison of numbers, with respect either to their difference or quotient. The Rule of Three, and Arithmetical and Geometrical Progression, make up this branch of Arithmetic.

The Fourth part embraces the applications of rules deduced from the science of numbers, to the ordinary transactions and business of life.

---

106. Of what does the science of arithmetic treat? Into how many parts may it be divided? Of what does the first part treat? Of what does the second part treat? What is it called? Of what does the third part treat? What does the fourth part embrace? Which part has been treated? Under how many heads? Into how many heads is the second part divided? What are they called? What distinguishes them? To what does the third part relate? What does the fourth part embrace?

OF VULGAR FRACTIONS.

92. The unit 1 represents an entire thing, as 1 apple, 1 chair, 1 pound of tea.

If we suppose 1 thing, as 1 apple, or 1 pound of tea, to be divided into two equal parts, each part is called *one half* of the thing.

If the unit be divided into 3 equal parts, each part is called *one third.*

If the unit be divided into 4 equal parts, each part is called *one fourth.*

If the unit be divided into 12 equal parts, each part is called *one twelfth;* and when it is divided into any number of equal parts, we have a similar expression for each of the parts. These equal parts of a unit are called Fractions.

How are these fractions to be expressed by figures? They are expressed by writing one figure under another. Thus,

$\frac{1}{2}$ is read	one half.		$\frac{1}{7}$ is read	one seventh.
$\frac{1}{3}$ " "	one third.		$\frac{1}{8}$ " "	one eighth.
$\frac{1}{4}$ " "	one fourth.		$\frac{1}{10}$ " "	one tenth.
$\frac{1}{5}$ " "	one fifth.		$\frac{1}{15}$ " "	one fifteenth.
$\frac{1}{6}$ " "	one sixth.		$\frac{1}{50}$ " "	one fiftieth.

It should, however, be observed, that $\frac{1}{2}$ is an *entire* half; $\frac{1}{3}$, an *entire* third, and the same for all the other fractions. Now, these fractions being *entire things*, may be regarded as units, and each is called a *fractional unit.*

93. It is thus seen that every fraction is expressed by two numbers. The number which is written above the line is called the *numerator*, and the one below it, the *denominator*, because it gives a denomination or name to the fraction.

For example, in the fraction $\frac{1}{2}$, 1 is the numerator, and 2

Quest.—92. What does the unit 1 represent? If we divide it into two equal parts, what is each called? If it be divided into three equal parts, what is each part? Into 4, 5, 6, &c., parts? What are such expressions called? How may the fractions be regarded? What are they called? 93. Of how many numbers is each fraction made up? What is the one above the line called? The one below the line?

the denominator. In the fraction $\frac{1}{3}$, 1 is the numerator, and 3 the denominator.

The denominator in every fraction shows into how many equal parts the unit, or single thing, is divided. For example, in the fraction $\frac{1}{2}$, the unit is divided into 2 equal parts; in the fraction $\frac{1}{3}$, it is divided into three equal parts; in the fraction $\frac{1}{4}$, it is divided into four equal parts, &c. In each of the above fractions *one* of the equal parts is expressed.

But suppose it were required to express 2 of the equal parts, as 2 halves, 2 thirds, 2 fourths, &c.

We should then write,

$\frac{2}{2}$	they are read	two halves.
$\frac{2}{3}$	" " "	two thirds.
$\frac{2}{4}$	" " "	two fourths.
$\frac{2}{5}$	" " "	two fifths, &c.

If it were required to express three of the equal parts, we should place 3 in the numerator; and generally, *the numerator shows how many of the equal parts are expressed in the fraction.*

For example, three eighths are written,

$\frac{3}{8}$	and read	three eighths.
$\frac{4}{9}$	" "	four ninths.
$\frac{6}{13}$	" "	six thirteenths.
$\frac{9}{20}$	" "	nine twentieths.

94. When the numerator and denominator are equal, the numerator will express all the equal parts into which the unit has been divided: and, *the value of the fraction is then equal to* 1. But if we suppose a second unit, of the same kind, to be divided into the same number of equal parts, those parts

---

QUEST.—What does the denominator show? What does the numerator show? In the fraction three-eighths, which is the numerator? Which the denominator? Into how many parts is the unit divided? How many parts are expressed? In the fraction nine-twentieths, into how many parts is the unit divided? How many parts are expressed? 94. When the numerator and denominator are equal, what is the value of the fraction?

may also be expressed in the same fraction with the parts of the first unit. Thus,

$\frac{3}{2}$	is read	three halves.
$\frac{7}{4}$	" "	seven fourths.
$\frac{16}{5}$	" "	sixteen fifths.
$\frac{18}{6}$	" "	eighteen sixths.
$\frac{25}{7}$	" "	twenty-five sevenths.

If the numerator of a fraction be divided by its denominator, the integer part of the quotient will express the number of entire units which have been used in forming the fraction, and the remainder will show, how many fractional units are over.

The unit, or whole thing, which is divided, in forming a fraction, is called the *unit* of the fraction; and one of the equal parts is called the *unit of the expression.* Thus, in the fraction $\frac{3}{4}$, 1 is the unit of the fraction, and $\frac{1}{4}$ the unit of the expression.

In every fraction, we must distinguish carefully, between the unit of the fraction and the unit of the expression. The first, is the whole thing from which the fraction is derived; the second, one of the equal parts of the fractional expression.

From what has been said, we conclude:

1st. *That a fraction is the expression of one or more equal parts of unity.*

2d. *That the denominator of a fraction shows into how many equal parts the unit or single thing has been divided, and the numerator expresses how many such parts are taken in the fraction.*

3d. *That the value of every fraction is equal to the quotient arising from dividing the numerator by the denominator.*

4th. *That, when the numerator is less than the denominator, the value of the fraction is less than* 1.

QUEST.—What is the value of the fraction three-halves? Of seven-fourths? Of sixteen-fifths? Of eighteen-sixths? Of twenty-five-sevenths? What is the first conclusion? What the 2d? What the 3d? What the 4th? What the 5th? What the 6th? What the 7th? What is the unit of the fraction three-fourths? What is the unit of the expression?

5th. *That, when the numerator is equal to the denominator, the value of the fraction is equal to* 1.

6th. *That, when the numerator is greater than the denominator, the value of the fraction will be greater than* 1.

7th. *That, the unit of every fraction is the whole thing from which it was derived; and the unit of the expression, one of the equal parts taken.*

95. There are six kinds of Vulgar Fractions: Proper, Improper, Simple, Compound, Mixed, and Complex.

A Proper Fraction is one in which the numerator is less than the denominator. The value of every proper fraction is less than 1, (Art. **94**).

The following are proper fractions:

$$\frac{1}{2}, \frac{1}{3}, \frac{1}{4}, \frac{3}{4}, \frac{3}{7}, \frac{5}{8}, \frac{9}{10}, \frac{8}{9}, \frac{5}{6}.$$

An Improper Fraction is one in which the numerator is equal to, or exceeds the denominator. Such fractions are called improper fractions because they are equal to, or exceed unity. When the numerator is equal to the denominator the value of the fraction is 1; in every other case the value of an improper fraction is greater than 1.

The following are improper fractions:

$$\frac{3}{2}, \frac{5}{3}, \frac{6}{5}, \frac{8}{7}, \frac{9}{8}, \frac{12}{6}, \frac{14}{7}, \frac{19}{7}.$$

A Simple Fraction is a single expression. A simple fraction may be either proper or improper.

The following are simple fractions:

$$\frac{1}{4}, \frac{3}{2}, \frac{5}{6}, \frac{8}{7}, \frac{9}{2}, \frac{8}{3}, \frac{6}{3}, \frac{7}{5}.$$

A Compound Fraction is a fraction of a fraction, or several fractions connected together with the word *of* between them.

Quest.—Write the fraction nineteen-fortieths:—also, 60 fourteenths—18 fiftieths—16 twentieths—17 thirtieths—41 one thousandths—85 millionths—106 fifths. 95. How many kinds of vulgar fractions are there? What are they? What is a proper fraction? Is its value greater or less than 1? What is an improper fraction? Why is it called improper? When is its value equal to 1? What is a simple fraction? What is a compound fraction? Give an example of a proper fraction. Of an improper fraction. Of a simple fraction.

The following are compound fractions:

$\frac{1}{2}$ of $\frac{1}{4}$, $\frac{1}{3}$ of $\frac{1}{2}$ of $\frac{1}{3}$, $\frac{1}{6}$ of 3, $\frac{1}{7}$ of $\frac{1}{8}$ of 4.

A MIXED NUMBER is made up of a whole number and a fraction. The whole numbers are sometimes called *integers.*

The following are mixed numbers:

$3\frac{1}{2}$, $4\frac{1}{3}$, $6\frac{2}{8}$, $5\frac{3}{5}$, $6\frac{5}{8}$, $3\frac{1}{7}$,

A COMPLEX FRACTION is one having a fraction or a mixed number in the numerator or denominator, or in both.

The following are complex fractions:

$$\frac{\frac{3}{4}}{14}, \quad \frac{2}{47\frac{1}{2}}, \quad \frac{\frac{3}{5}}{\frac{4}{7}}, \quad \frac{42\frac{5}{6}}{87\frac{4}{5}}.$$

96. The numerator and denominator of a fraction, taken together, are called the *terms* of the fraction. Hence, every fraction has two terms.

97. A whole number may be expressed fractionally by writing 1 below it for a denominator. Thus,

3	may be written	$\frac{3}{1}$	and is	read,	3 ones.
5	"	"	$\frac{5}{1}$	"	"	5 ones.
6	"	"	$\frac{6}{1}$	"	"	6 ones.
8	"	"	$\frac{8}{1}$	"	"	8 ones.

But 3 ones are equal to 3, 5 ones to 5, 6 ones to 6, and 8 ones to 8. Hence, the value of a number is not changed by placing 1 under it for a denominator.

QUEST.—What is a mixed number? Give an example of a compound fraction. Of a mixed fraction. Is four-ninths a proper or improper fraction? What kind of a fraction is six-thirds? What is its value? What kind of a fraction is nine-eighths? What is its value? What kind of a fraction is one-half of a third? What kind of a fraction is two and one-sixth? Four and a seventh? Eight and a tenth? What is a complex fraction? 96. What are the terms of a fraction? What are the terms of the fraction three-fourths? Of five-eighths? Of six-sevenths? 97. How may a whole number be expressed fractionally? Does this alter its value? Give an example.

98. If an apple be divided into 6 equal parts,

$\frac{1}{6}$ will express one of the parts,
$\frac{2}{6}$ " " two of the parts,
$\frac{3}{6}$ " " three of the parts,
&c., &c., &c.,

and generally, the denominator shows into how many equal parts the unit is divided, and the numerator how many of the parts are taken.

Hence, also, we may conclude that,

$\frac{1}{6} \times 2$; that is, $\frac{1}{6}$ taken 2 times $= \frac{2}{6}$,
$\frac{1}{6} \times 3$; that is, $\frac{1}{6}$ taken 3 times $= \frac{3}{6}$,
$\frac{1}{6} \times 4$; that is, $\frac{1}{6}$ taken 4 times $= \frac{4}{6}$,
&c., &c., &c.,

and consequently we have,

PROPOSITION I. *If the numerator of a fraction be multiplied by any number, the denominator remaining unchanged, the value of the fraction will be increased as many times as there are units in the multiplier.* Hence, *to multiply a fraction by a whole number, we simply multiply the numerator by the number.*

EXAMPLES.

1. Multiply $\frac{7}{8}$ by 5.
2. Multiply $\frac{3}{11}$ by 7.
3. Multiply $\frac{18}{13}$ by 9.
4. Multiply $\frac{11}{17}$ by 12.
5. Multiply $\frac{7}{16}$ by 11.
6. Multiply $\frac{9}{17}$ by 12.
7. Multiply $\frac{9}{11}$ by 14.
8. Multiply $\frac{13}{27}$ by 15.

99. If three apples be each divided into 6 equal parts, there will be 18 parts in all, and these parts will be expressed by the fraction $\frac{18}{6}$. If it were required to express but one-third of the parts, we should take, in the numerator, but one-

QUEST.—98. If an apple be divided into six equal parts, how do you express one of those parts? Two of them? Three of them? Four of them? Five of them? Repeat the proposition. How do you multiply a fraction by a whole number? 99. If 3 apples be each divided into 6 equal parts, how many parts in all? If 4 apples be so divided, how many parts in all? If 5 apples be so divided, how many parts? How many parts in 6 apples? In 7? In 8? In 9? In 10?

third of the eighteen parts; that is, the fraction $\frac{6}{6}$ would express one-third of $\frac{18}{6}$. If it were required to express one-sixth of the 18 parts, we should take one-sixth of 18, and $\frac{3}{6}$ would be the required fraction.

In each case the fraction $\frac{18}{6}$ has been diminished as many imes as there were units in the divisor. Hence,

PROPOSITION II. *If the numerator of a fraction be divided by any number, the denominator remaining unchanged, the value of the fraction will be diminished as many times as there are units in the divisor.* Hence, *a fraction may be divided by a whole number by dividing its numerator.*

EXAMPLES.

1. Divide $\frac{24}{17}$ by 6.
2. Divide $\frac{112}{256}$ by 8.
3. Divide $\frac{264}{741}$ by 12.
4. Divide $\frac{147}{456}$ by 7.
5. Divide $\frac{75}{126}$ by 5.
6. Divide $\frac{360}{147}$ by 12.
7. Divide $\frac{256}{127}$ by 32.
8. Divide $\frac{324}{492}$ by 36.

100. Let us again suppose the apple to be divided into 6 equal parts. If, now, each part be divided into 2 equal parts, there will be 12 parts of the apple, and consequently each part will be but half as large as before.

Three parts in the first case will be expressed by $\frac{3}{6}$, and in the second by $\frac{3}{12}$. But since the parts in the second are only half the parts in the first fraction, it follows that,

$$\frac{3}{12} = \text{one half of } \frac{3}{6}.$$

If we suppose the apple to be divided into 18 equal parts,

QUEST.—What expresses all the parts of the three apples? What expresses one-half of them? One-third of them? One-sixth of them? One-ninth of them? One-eighteenth of them? What expresses all the parts of four apples? One-half of them? One-third of them? One-fourth of them? One-sixth of them? One-eighth of them? One-twelfth of them? One-twenty-fourth of them? Put similar questions for 5 apples, 6 apples, &c. Repeat the proposition. How may a fraction be divided? 100. If a unit be divided into 6 equal parts and then into 12 equal parts, how does one of the last parts compare with one of the first? If the second division be into 18 parts, how do they compare? If into 24?

three of the parts will be expressed by $\frac{3}{18}$, and since the parts are but one-third as large as in the first case, we have

$$\tfrac{3}{18} = \text{one third of } \tfrac{3}{6}:$$

and since the same may be said of all fractions, we have

PROPOSITION III. *If the denominator of a fraction be multiplied by any number, the numerator remaining unchanged, the value of the fraction will be diminished as many times as there are units in the multiplier.* Hence, *a fraction may be divided by any number, by multiplying the denominator by that number.*

EXAMPLES.

1. Divide $\frac{67}{47}$ by 6.
2. Divide $\frac{34}{87}$ by 9.
3. Divide $\frac{127}{349}$ by 12.
4. Divide $\frac{327}{451}$ by 11.
5. Divide $\frac{327}{406}$ by 14.
6. Divide $\frac{1249}{321}$ by 15.
7. Divide $\frac{327}{365}$ by 5.
8. Divide $\frac{94}{487}$ by 8.

101. If we suppose the apple to be divided into 3 parts instead of 6, each part will be twice as large as before, and three of the parts will be expressed by $\frac{3}{3}$ instead of $\frac{3}{6}$. But this is the same as dividing the denominator 6 by 2; and since the same is true of all fractions, we have

PROPOSITION IV. *If the denominator of a fraction be divided by any number, the numerator remaining unchanged, the value of the fraction will be increased as many times as there are units in the divisor.* Hence, *a fraction may be multiplied by a whole number, by dividing the denominator by that number.*

QUEST.—What part of 24 is 6? If the second division be into 30 parts, how do they compare? If into 36 parts? Repeat the proposition. How may a fraction be divided by a whole number? 101. If we divide 1 apple into three parts and another into 6, how much greater will the parts of the first be than those of the second? Are the parts larger as you decrease the denominator? If you divide the denominator by 2, how do you affect the parts? If you divide it by 3? By 4? By 5? By 6? By 7? By 8? Repeat the proposition. How may a fraction be multiplied by a whole number?

EXAMPLES.

1. Multiply $\frac{3}{4}$ by 2, by 4.
2. Multiply $\frac{16}{32}$ by 4, 8, 16.
3. Multiply $\frac{34}{96}$ by 4, 6, 12.
4. Multiply $\frac{147}{112}$ by 16, 56.
5. Multiply $\frac{127}{49}$ by 7.
6. Multiply $\frac{151}{200}$ by 5, 10, 20.
7. Multiply $\frac{246}{336}$ by 8, by 16.
8. Multiply $\frac{449}{84}$ by 7, by 21.

102. It appears from Prop. I. that if the numerator of a fraction be multiplied by any number, the value of the fraction will be *increased* as many times as there are units in the multiplier. It also appears from Prop. III., that if the denominator of a fraction be multiplied by any number, the value of the fraction will be *diminished* as many times as there are units in the multiplier.

Therefore, when the numerator and denominator of a fraction are both multiplied by the same number, the increase from multiplying the numerator will be just equal to the decrease from multiplying the denominator; hence we have,

PROPOSITION V. *If both terms of a fraction be multiplied by the same number, the value of the fraction will remain unchanged.*

EXAMPLES.

1. Multiply the numerator and denominator of $\frac{5}{7}$ by 7.

We have, $$\frac{5}{7}=\frac{5\times 7}{7\times 7}=\frac{35}{49}.$$

2. Multiply the numerator and denominator of $\frac{17}{19}$ by 3, by 4, by 6, by 7, by 9, by 15, by 17.

3. Multiply both terms of the fraction $\frac{13}{17}$ by 9, by 12, by 16, by 7, by 5, by 11.

QUEST.—102. If the numerator of a fraction be multiplied by a number, how many times is the fraction increased? If the denominator be multiplied by the same number, how many times is the fraction diminished? If then the numerator and denominator be both multiplied at the same time, is the value changed? Why not? Repeat the proposition.

103. It appears from Prop. II. that if the numerator of a fraction be divided by any number, the value of the fraction will be *diminished* as many times as there are units in the divisor. It also appears from Prop. IV., that if the denominator of a fraction be divided by any number, the value of the fraction will be *increased* as many times as there are units in the divisor. Therefore, when both terms of a fraction are divided by the same number, the *decrease* from dividing the numerator will be just equal to the *increase* from dividing the denominator: hence we have,

PROPOSITION VI. *If both terms of a fraction be divided by the same number, the value of the fraction will remain unchanged.*

EXAMPLES.

1. Divide both terms of the fraction $\frac{8}{16}$ by 4: this gives $\frac{4)8}{4)16} = \frac{2}{4}$. *Ans.* $\frac{2}{4}$.

2. Divide each term by 8: this gives $\frac{8)8}{8)16} = \frac{1}{2}$.

3. Divide each term of the fraction $\frac{32}{128}$ by 2, by 4, by 8, by 16, by 32.

4. Divide each term of the fraction $\frac{60}{180}$ by 2, by 3, by 4, by 5, by 6, by 10, by 12, by 15, by 20, by 30, by 60.

GREATEST COMMON DIVISOR.

104. Any number greater than unity that will divide two or more numbers without a remainder, is called their common divisor: and the greatest number that will so divide them, is called their GREATEST COMMON DIVISOR.

QUEST.—103. If the numerator of a fraction be divided by a number, how many times will the value of the fraction be diminished? If the denominator be divided by the same number, how many times will the value of the fraction be increased? If they are both divided by the same number, will the value of the fraction be changed? Why not? Repeat the proposition. 104. What is a common divisor? What is the greatest common divisor of two or more numbers?

Before explaining the manner of finding this divisor, it is necessary to explain some principles on which the method depends.

One number is said to be a multiple of another when it contains that other an exact number of times. Thus, 24 is a multiple of 6, because 24 contains 6 an exact number of times. For a like reason 60 is a multiple of 12, since it contains 12 an exact number of times.

FIRST PRINCIPLE. Every number which exactly divides another number will also divide without a remainder any multiple of that number. For example, 24 is divisible by 8 giving a quotient 3. Now, if 24 be multiplied by 4, 5, 6, or any other number, the product so arising will also be divisible by 8.

SECOND PRINCIPLE. If a number be separated into two parts, any divisor which will divide each of the parts separately, without a remainder, will exactly divide the given number. For, the sum of the two partial quotients must be equal to the entire quotient; and if they are both whole numbers, the entire quotient must be a whole number; for the sum of two whole numbers cannot be equal to a fraction.

For example, if 36 be separated into the parts 16 and 20, the number 4, which will divide both numbers 16 and 20, will also divide 36; and the sum of the quotients 4 and 5 will be equal to the entire quotient 9.

THIRD PRINCIPLE. If a number be decomposed into two parts, then any divisor which will divide the given number and one of the parts, will also divide the other.

For, the entire quotient is equal to the sum of the two partial quotients; and if the entire quotient and one of the partial quotients be whole numbers, the other must also be a whole number; for no proper fraction added to a whole number can give a whole number.

QUEST.—When is one number said to be a multiple of another? What is the first principle? What is the second? What is the third?

1. Let it be required to find the greatest common divisor of the numbers 216 and 408.

OPERATION.

```
216)408(1
    216
    ---
    192)216(1
        192
        ---
         24)192(8
            192
            ---
```

It is evident that the greatest common divisor cannot be greater than the least number 216. Now, as 216 will divide itself, let us see if it will divide 408; for if it will, it is the greatest common divisor sought.

Making this division, we find a quotient 1 and a remainder 192; hence, 216 is not the greatest common divisor. Now we say, *that the greatest common divisor of the two given numbers is the common divisor of the less number* 216 *and the remainder* 192 *after the division.* For, by the second principle, any number which will exactly divide 216 and 192, will also exactly divide the number 408.

Let us now seek the common divisor between 216 and 192. Dividing the greater by the less, we have a remainder of 24; and from what has been said above, the greatest common divisor of 192 and 216 is the same as the greatest common divisor of 192 and 24, which we find to be 24. Hence, 24 is the greatest common divisor of the given numbers 216 and 408; and to find it

Divide the greater number by the less, and then divide the divisor by the remainder, and continue to divide the last divisor by the last remainder until nothing remains. The last divisor will be the greatest common divisor sought.

EXAMPLES.

1. Find the greatest common divisor of 408 and 740.
2. Find the greatest common divisor of 315 and 810.
3. Find the greatest common divisor of 4410 and 5670.
4. Find the greatest common divisor of 3471 and 1869.
5. Find the greatest common divisor of 1584 and 2772.

QUEST.—Give the rule for finding the greatest common divisor. How do you find the greatest common divisor of more than two numbers?

NOTE.—If it be required to find the greatest common divisor of more than two numbers, find first the greatest common divisor of two of them, then of that common divisor and one of the remaining numbers, and so on for all the numbers: the last common divisor will be the greatest common divisor of all the numbers.

6. What is the greatest common divisor of 492, 744, and 1044? *Ans.* ——

7. What is the greatest common divisor of 944, 1488, and 2088?

8. What is the greatest common divisor of 216, 408, and 740?

9. What is the greatest common divisor of 945, 1560, and 22683?

10. What is the greatest common divisor of 204, 1190, 1445, and 2006?

SECOND METHOD.

105. It has already been remarked (Art. 90), that a prime number is one which is only divisible by itself or unity, and that a composite number is the product of two or more factors (Art. 61). Now, every composite number may be decomposed into two or more prime factors. For example, if we have the composite number 36, we may write

$$36 = 18 \times 2 = 9 \times 2 \times 2 = 3 \times 3 \times 2 \times 2;$$

in which we see there are four prime factors, viz., two 3's and two 2's.

Again, if we have the composite number 150, we may write

$$150 = 15 \times 10 = 3 \times 5 \times 10 = 3 \times 5 \times 5 \times 2;$$

in which there are also four prime factors, viz., one 3, two 5's, and one 2. Hence, to decompose a number into its prime factors,

---

QUEST.—105. What is a prime number? What is a composite number? Into what may it be decomposed? What are the prime factors of 36?

Divide it continually by any prime number which will divide it without a remainder, and the last quotient, together with the several divisors, will be the prime factors sought.

EXAMPLES.

1. What are the prime factors of 180?

We first divide by the prime number 2, which gives 90; then by 3, then by 5, then by 3, and find the six prime factors 2, 3, 5, 3, and 2.

OPERATION.

```
2)180
 3)90
  5)30
   3)6
     2
```

$2 \times 3 \times 5 \times 3 \times 2 = 180$

2. What are the prime factors of 645? *Ans.* ——

3. What are the prime factors of 360? *Ans.* ——

106. It is plain that the greatest common divisor of two or more numbers, will always be the greatest common factor, and that such factor must arise from the product of equal prime numbers in each. Hence, to find the greatest common divisor of two or more numbers,

Decompose them into their prime factors, and the product of those factors which are common will be the greatest common divisor sought.

EXAMPLES.

1. What is the greatest common divisor of 1365 and 1995?

```
3)1365          3)1995
 5)455           5)665
  7)91            7)133
    13               19
```

Hence, 3, 5, 7, and 13 are prime factors.

Hence, 3, 5, 7, and 19 are the factors.

Hence, $3 \times 5 \times 7 = 105 =$ the greatest common divisor.

QUEST.—How do you decompose a number into its prime factors? 106. What is the greatest common divisor of two or more numbers? What does such factor arise from? How then do you find the greatest common divisor?

2. What is the greatest common divisor of 12321 and 54345?

3. What is the greatest common divisor of 3775 and 1000?

4. What is the greatest common divisor of 6327 and 23997?

5. What is the greatest common divisor of 24720 and 4155?

LEAST COMMON MULTIPLE.

125. A number is said to be *a common multiple* of two or more numbers, when it can be divided by each of them, separately, without a remainder.

The *least* common multiple of two or more numbers, is the least number which they will separately divide without a remainder. Thus, 6 is the least common multiple of 3 and 2, it being the least number which they will separately divide without a remainder.

A *factor* of a number, is any number greater than 1 that will divide it without a remainder; and a *prime* factor is any prime number that will so divide it.

Now, it is plain, that a dividend will contain its divisor an exact number of times, when it contains as factors, *every* factor of that divisor: and hence, the question of finding the *least* common multiple of several numbers is reduced to finding a number which shall contain all the prime factors of each number and *none others*. If the numbers have no common prime factor their product will be their least common multiple.

EXAMPLE 1.—Let it be required to find the least common multiple of 6, 8 and 12.

---

125. When is one number said to be a common multiple of two or more numbers? What is the least common multiple of two or more numbers? Of what numbers is 6 the least common multiple? What is the difference between a common multiple and the least common multiple? What is a factor of a number? What is a prime factor? What is a prime number? When will a dividend exactly contain its divisor? To what is the question of finding the least common multiple reduced?

We see, from inspection, that the prime factors of 6, are 2 and 3;—of 8; 2, 2 and 2;—and of 12; 2, 2 and 3.

2×3	$2\times2\times2$	$2\times2\times3$
6	8	12

Now, every factor, of each number, must appear in the least common multiple, and none others: hence, we must have all the factors of 8, and such other prime factors of 6 and 12 as are not found among the prime factors of 8, that is, the factor 3. Hence

$$2\times2\times2\times3=24, \text{ the least common multiple.}$$

To separate the prime factors, or to find the least common multiple of two or more numbers,

FIRST METHOD.

I. *Place the numbers on the same line, and divide by any prime number that will divide two or more of them without a remainder, and set down in a line below, the quotients and the undivided numbers.*

II. *Divide as before, until there is no number greater than 1 that will exactly divide any two of the numbers: then multiply together the numbers of the lower line and the divisors, and the product will be the least common multiple. If, in comparing the numbers together, we find no common divisor, their product is the least common multiple.*

2. Find the least common multiple of 3, 8, and 9.

We arrange the numbers in a line, and see that 3 will divide two of them. We then write down the quotients 1 and 3, and also the 8 which cannot be divided. Then, as there is no common divisor between any two of the numbers 1, 8,

OPERATION.

	3	$2\times2\times2$	3×3
3)	3 . .	8 . . .	9
	1 . .	8 . . .	3

$1\times8\times3\times3=72.$

QUEST.—Give the rule for finding the least common multiple. If the numbers have no common divisor, what is the least common multiple?

and 3, it follows that their product, multiplied by the divisor 3, will give the least common multiple sought.

3. Find the least common multiple of 6, 7, 8, and 10.

4. Find the least common multiple of 21 and 49.

5. Find the least common multiple of 2, 7, 5, 6, and 8.

6. Find the least common multiple of 4, 14, 28, and 98.

7. Find the least common multiple of 13 and 6.

8. Find the least common multiple of 12, 4, and 7.

9. Find the least common multiple of 6, 9, 4, 14, and 16.

10. Find the least common multiple of 13, 12, and 4.

11. What is the least common multiple of 11, 17, 19, 21, and 7?

SECOND METHOD.

108. To find the least common multiple by this method.

Decompose each number into its prime factors; after which, select from each number so decomposed the factors which are common to them all, if there be such: then select those which are common to two or more of the numbers, and so on until all the factors common to any two of them shall have been selected. Then multiply these several factors together, and also the factors which are not common, and the product will be the least common multiple.

EXAMPLES.

1. What is the least common multiple of 99 and 468?

The prime factors of 99 are 3, 3, and 11; and of 468, 3, 3, 2, 2, and 13: hence, the common factors are 3 and 3, which are to be multiplied by 11, 2, 2, and 13.

OPERATION.

$$99=3\times3\times11$$
$$468=3\times3\times2\times2\times13.$$

$$3\times3\times11\times2\times2\times13=5148.$$

QUEST.—108. How do you find the least common multiple by the second method?

2. What is the least common multiple of 12, 14, and 36?

Having decomposed the numbers into their prime factors, we see that 2 is common to them all. We then set it aside as a multiplier, and cross it in each number. We next set 3 and 2 aside, and cross them in a contrary direction. We then have 7 and 3 remaining, which we use as factors. It is plain that this method introduces into the common multiple every prime factor of each number.

OPERATION.

$12 = 3 \times 2 \times 2$

$14 = 2 \times 7$

$36 = 2 \times 2 \times 3 \times 3.$

$2 \times 3 \times 2 \times 3 \times 7 = 252.$

3. What is the least common multiple of 4, 9, 10, 15, 18, 20, 21?

4. What is the least common multiple of 8, 9, 10, 12, 25, 32, 75, 80?

5. What is the least common multiple of 1, 2, 3, 4, 5, 6, 7, 8, 9?

6. What is the least common multiple of 9, 16, 42, 63, 21, 14, 72?

7. What is the least common multiple of 7, 15, 21, 28, 35, 100, 125?

8. What is the least common multiple of 15, 16, 18, 20, 24, 25, 27, 30?

REDUCTION OF VULGAR FRACTIONS.

109. Reduction of Vulgar Fractions is the method of changing their forms without altering their value.

A fraction is said to be in its lowest terms, when there is no number greater than 1 that will divide the numerator and denominator without a remainder. The terms of the fraction have then no common factor.

QUEST.—109. What is reduction? When is a fraction said to be in its lowest terms? Is one-half in its lowest terms? Is the fraction two-fourths? Is three-fourths?

CASE I.

110. To reduce an improper fraction to its equivalent whole or mixed number.

Divide the numerator by the denominator; the quotient will be the whole number; and the remainder, if there be one, placed over the given denominator will form the fractional part.

It was shown in Art. **94**, that the value of every fraction is equal to the quotient arising from dividing the numerator by the denominator: hence the value of the fraction is not changed by the reduction.

EXAMPLES.

1. Reduce $\frac{84}{4}$ and $\frac{67}{9}$ to their equivalent whole or mixed numbers.

OPERATION.	OPERATION.
4)84	9)67
Ans. 21	*Ans.* $7\frac{4}{9}$

2. Reduce $\frac{99}{8}$ to a whole or mixed number. *Ans.* ——
3. In $\frac{19}{7}$ of yards of cloth, how many yards? *Ans.* ——
4. In $\frac{51}{9}$ of bushels, how many bushels? *Ans.* ——
5. If I give $\frac{1}{3}$ of an apple to each one of 15 children, how many apples do I give?
6. Reduce $\frac{327}{125}$, $\frac{3672}{153}$, $\frac{50287}{6941}$, $\frac{987625}{72301}$, to their whole or mixed numbers.
7. If I distribute 878 quarter-apples among a number of boys, how many whole apples do I use?
8. Reduce $\frac{4790}{25}$ to a whole or mixed number.
9. Reduce $\frac{1512}{108}$ to a whole or mixed number.
10. Reduce $\frac{375941}{999}$ to a whole or mixed number.
11. Reduce $\frac{3745174}{349}$ to a whole or mixed number.

QUEST.—110. How do you reduce a fraction to its equivalent whole or mixed number? Does this reduction alter its value? Why not? What are four-halves equal to? Eight-fourths? Sixteen-eighths? Twenty-fifths? Thirty-six-sixths? Four-thirds? What are nine-fourths equal to? Nine-fifths? Seventeen-sixths? Eighteen-sevenths?

CASE II.

111. To reduce a mixed number to its equivalent improper fraction.

Multiply the whole number by the denominator of the fraction; to the product add the numerator, and place the sum over the given denominator.

EXAMPLES.

1. Reduce $4\frac{4}{5}$ to its equivalent improper fraction.

Here, $4 \times 5 = 20$: then $20 + 4 = 24$; hence,

$\frac{24}{5}$ is the equivalent fraction.

This rule is the reverse of Case I. In the example $4\frac{4}{5}$ we have the integer number 4 and the fraction $\frac{4}{5}$. Now 1 whole thing being equal to 5 fifths, 4 whole things are equal to 20 fifths; to which, add the 4 fifths, and we obtain the 24 fifths.

2. Reduce $25\frac{3}{8}$ to its equivalent improper fraction.

$$25\frac{3}{8} = \frac{25 \times 8 + 3}{8} = \frac{203}{8} \textit{ Ans.}$$

3. Reduce $47\frac{5}{6}$ to its equivalent improper fraction.

4. Reduce $676\frac{37}{51}$, $874\frac{33}{9}$, $690\frac{47}{100}$, $367\frac{9}{104}$, to their equivalent improper fractions.

5. Reduce $847\frac{36}{175}$, $874\frac{876}{104}$, $67426\frac{368}{879}$, to their equivalent improper fractions.

6. How many 200ths in $675\frac{187}{200}$? *Ans.* ——

7. How many 151ths in $187\frac{41}{151}$? *Ans.* ——

8. Reduce $149\frac{5}{9}$ to an improper fraction. *Ans.* ——

9. Reduce $375\frac{94}{99}$ to an improper fraction. *Ans.* ——

QUEST.—111. How do you reduce a mixed number to its equivalent improper fraction? How many fourths are there in one? In two? In three? How many sixths in four and one-sixth? In eight and two-sixths? In seven and three-sixths? In nine and five-sixths? In ten and five-sixths? How many eighths in two and one-eighth? In three and three-eighths? In four and four-eighths? In five and six eighths? In seven and seven-eighths? In eight and seven-eighths?

10. Reduce $17494\frac{543}{99999}$ to an improper fraction.

11. Reduce $4834\frac{57}{95}$ to an improper fraction.

12. Reduce $1789\frac{5}{9}$ to an improper fraction.

13. Place 4 sevens in such a manner that they may be equal to 78.

CASE III.

112. To reduce a fraction to its lowest terms.

I. *Divide the numerator and denominator by any number that will divide them both without a remainder, and then divide the quotients arising in the same way, until there is no number greater than* 1 *that will divide them without a remainder.*

II. *Or, find the greatest common divisor of the numerator and denominator, and divide them by it. The value of the fraction will not be altered by the reduction.*

EXAMPLES.

1. Reduce $\frac{70}{175}$ to its lowest terms.

1ST METHOD.

$\frac{5)\,70}{5)175} = \frac{7)14}{7)35} = \frac{2}{5}$, which are the lowest terms of the fraction, since no number greater than 1 will divide the numerator and denominator without a remainder.

2D METHOD, BY THE COMMON DIVISOR.

```
                           70)175(2
                              140
                              ---
Greatest common div.        35)70(2
                              70
                              --
```

$\frac{35)\,70}{35)\,175} = \frac{2}{5}$. *Ans.*

2. Reduce $\frac{104}{312}$ to its lowest terms. *Ans.* ——

3. Reduce $\frac{1049}{8392}$ to its lowest terms. *Ans.* ——

QUEST.—112. When is a fraction in its lowest terms? (see Art. 109.) How do you reduce a fraction to its lowest terms by the first method? By the second? What are the lowest terms of two-fourths? Of six-eighths? Of nine-twelfths? Of sixteen-thirty-sixths? Of ten-twentieths? Of fifteen-twenty-fourths? Of sixteen-eighteenths? Of nine-eighteenths?

4. Reduce $\frac{275}{440}$ to its lowest terms. *Ans.* ——
5. Reduce $\frac{351}{795}$ to its lowest terms. *Ans.* ——
6. Reduce $\frac{172}{1118}$ to its lowest terms. *Ans.* ——
7. Reduce $\frac{63}{81}$ to its lowest terms by the 2d method.
8. Reduce $\frac{315}{405}$ to its lowest terms by the 2d method.
9. Reduce $\frac{1157}{623}$ to its lowest terms by the 2d method.
10. Reduce $\frac{792}{1386}$ to its lowest terms by the 2d method.
11. Reduce $\frac{374}{1030}$ to its lowest terms. *Ans.* ——
12. Reduce $\frac{410}{510}$ to its lowest terms. *Ans.* ——
13. Reduce $\frac{345}{1745}$ to its lowest terms. *Ans.* ——
14. Reduce $\frac{8343}{9747}$ to its lowest terms. *Ans.* ——
15. Reduce $\frac{549}{7143}$ to its lowest terms. *Ans.* ——

CASE IV.

113. To reduce a whole number to an equivalent fraction having a given denominator.

Multiply the whole number by the given denominator, and set the product over the said denominator.

EXAMPLES.

1. Reduce 6 to a fraction whose denominator shall be 4.

Here $6 \times 4 = 24$; therefore $\frac{24}{4}$ is the required fraction. It is plain that the fraction will in all cases be equal to the whole number, since it may be reduced to the whole number by Case I.

2. Reduce 15 to a fraction whose denominator shall be 9.

3. Reduce 139 to a fraction whose denominator shall be 175.

4. Reduce 1837 to a fraction whose denominator shall be 181.

Quest.—113. How do you reduce a whole number to an equivalent fraction having a given denominator? How many thirds in 1? In 2? In 3? In 4? If the denominator be 5, what fraction will you form of 5? Of 4? Of 9? Of 7? Of 8? With the denominator 6, what fraction will you form of 3? Of 4? Of 5? Of 6? Of 7? Of 9?

5. If the denominator be 837, what fractions will be formed from 327? From 889? From 575?

6. Reduce 167 to a fraction whose denominator shall be 89.

7. Reduce 3074 to a fraction whose denominator shall be 17.

CASE V.

114. To reduce a compound fraction to its equivalent simple one.

I. *Reduce all mixed numbers to their equivalent improper fractions.*

II. *Then multiply all the numerators together for a numerator and all the denominators together for a denominator: their product will form the fraction sought.*

EXAMPLES.

1. Let us take the fraction $\frac{3}{4}$ of $\frac{5}{7}$.

First, $\frac{3}{4} = 3 \times \frac{1}{4}$: hence the fractions may be written $\frac{3}{4}$ of $\frac{5}{7} = 3 \times \frac{1}{4}$ of $\frac{5}{7}$; that is, three times one-fourth of $\frac{5}{7}$. But $\frac{1}{4}$ of $\frac{5}{7} = \frac{5}{28}$: hence we have,

$$\frac{3}{4} \text{ of } \frac{5}{7} = 3 \times \frac{5}{28} = \frac{15}{28}:$$

a result which is obtained by multiplying together the numerators and denominators of the given fractions.

When the compound fraction consists of more than two simple ones, two of them can be reduced to a simple fraction as above, and then this fraction may be reduced with the next, and so on. Hence, the reason of the rule is manifest.

2. Reduce $2\frac{1}{4}$ of $6\frac{1}{2}$ of 7 to a simple fraction. *Ans.* ——

Quest.—114. What is a compound fraction? How do you reduce a compound fraction to a simple one? Does this alter the value of the fraction? What is one-half of one-half? One-half of one-third? One-third of one-fourth? One-sixth of one-seventh? Three-halves of one-eighth? Six-thirds of two-ones?

3. Reduce 5 of $\frac{1}{2}$ of $\frac{1}{7}$ of 6 to a simple fraction.

METHOD BY CANCELLING.

115. The work may often be abridged by striking out or *cancelling* common factors in the numerator and denominator.

EXAMPLES.

1. Reduce $\frac{5}{3}$ of $\frac{3}{6}$ of $\frac{6}{7}$ to a simple fraction.

Here, $$\frac{5}{\cancel{3}} \times \frac{\cancel{3}}{\cancel{6}} \times \frac{\cancel{6}}{7} = \frac{5}{7},$$

by cancelling or striking out the 3's and 6's in the numerator and denominator.

By cancelling or striking out the 3's we only divide the numerator and denominator of the fraction by 3; and in cancelling the 6's we divide by 6. Hence, *the value of the fraction is not affected by striking out like figures, which should always be done when they multiply the numerator and denominator.*

2. Reduce $\frac{6}{8}$ of $\frac{8}{9}$ of $\frac{9}{15}$ to its simplest terms.

Here, $$\frac{\overset{2}{\cancel{6}}}{\cancel{8}} \times \frac{\cancel{8}}{\cancel{9}} \times \frac{\cancel{9}}{\underset{5}{\cancel{15}}} = \frac{2}{5} \textit{ Ans.}$$

Besides cancelling the like factors 8 and 8 and 9 and 9, we also cancel the factor 3 common to 6 and 15, and write the quotients 2 and 5 above and below the numbers.

3. Reduce $\frac{3}{4}$ of $\frac{5}{6}$ of $\frac{5}{9}$ of $\frac{27}{101}$ of $\frac{5}{13}$ to its simplest terms.
4. Reduce $\frac{41}{110}$ of $\frac{3}{19}$ of $\frac{41}{108}$ of $\frac{3}{7}$ to its simplest terms.
5. Reduce $3\frac{5}{8}$ of $\frac{5}{9}$ of $\frac{27}{301}$ of 49 to its simplest terms.

CASE VI.

116. To reduce complex fractions to simple ones.

Reduce the numerator and denominator, when necessary, to simple fractions: then the numerator multiplied by the denominator with its terms inverted, will give the equivalent simple fraction.

QUEST.—115. How may the work often be abridged? 116. What is a complex fraction? How do you reduce a complex fraction to a simple one?

EXAMPLES.

1. Reduce the complex fraction $\frac{\frac{4}{7}}{\frac{2}{9}}$ to a simple fraction.

Now, if we multiply the numerator and denominator of this fraction by any number whatever, the value of the fraction will not be altered (Art. **102**). Let us then multiply them by the denominator with it terms inverted. This will give,

$$\frac{\frac{4}{7}\times\frac{9}{2}}{\frac{2}{9}\times\frac{9}{2}}=\frac{\frac{36}{14}}{1}=\frac{36}{14}.$$

It is plain that when the denominator is multiplied by the fraction which arises from inverting its terms, the product will be equal to unity. Hence, the required simple fraction will always be equal to the numerator of the given fraction multiplied by the denominator with its terms inverted.

All the cases in the reduction of fractions of this class are embraced in the following eight forms.

First. $\frac{\frac{1}{9}}{\frac{7}{8}}=\frac{1}{9}\times\frac{8}{7}=\frac{8}{63}.$

Second. $\frac{4}{\frac{7}{8}}=4\times\frac{8}{7}=\frac{32}{7}=4\frac{4}{7}.$

Third. $\frac{\frac{7}{10}}{5}=\frac{7}{10}\times\frac{1}{5}=\frac{7}{50}.$

Fourth. $\frac{4\frac{7}{8}}{9}=\frac{4\times8+7}{8}\times\frac{1}{9}=\frac{39}{72}=\frac{13}{24}.$

Fifth. $\frac{7}{8\frac{5}{9}}=\frac{7}{\frac{(72+5)}{9}}=\frac{7}{\frac{77}{9}}=7\times\frac{9}{77}=\frac{63}{77}=\frac{9}{11}$

Sixth. $\frac{\frac{7}{8}}{4\frac{1}{2}}=\frac{\frac{7}{8}}{\frac{9}{2}}=\frac{7}{8}\times\frac{2}{9}=\frac{14}{72}=\frac{7}{36}.$

Seventh. $\frac{5\frac{3}{4}}{\frac{8}{9}}=\frac{\frac{23}{4}}{\frac{8}{9}}=\frac{23}{4}\times\frac{9}{8}=\frac{207}{32}=6\frac{15}{32}.$

Eighth. $\frac{9\frac{7}{9}}{3\frac{5}{7}}=\frac{\frac{88}{9}}{\frac{26}{7}}=\frac{88}{9}\times\frac{7}{26}=\frac{616}{234}=\frac{308}{117}=2\frac{74}{117}.$

2. Reduce $\frac{47\frac{5}{8}}{95}$ to a simple fraction. *Ans.* ——

3. Reduce $\frac{34\frac{5}{7}}{84}$ to a simple fraction. *Ans.* ——

4. Reduce $\frac{44}{147\frac{5}{9}}$ to a simple fraction. *Ans.* ——

5. Reduce $\frac{247}{\frac{5}{7}}$ to a simple fraction. *Ans.* ——

6. Reduce $\frac{\frac{147}{504}}{1789}$ to a simple fraction. *Ans.* ——

7. Reduce $\frac{394\frac{74}{99}}{894\frac{547}{719}}$ to a simple fraction. *Ans.* ——

CASE VII.

117. To reduce fractions of different denominators to equivalent fractions having a common denominator.

I. *Reduce complex and compound fractions to simple ones, and all whole or mixed numbers to improper fractions.*

II. *Then multiply the numerator and denominator of each fraction by the product of the denominators of all the others.*

EXAMPLES.

1. Reduce $\frac{1}{2}$, $\frac{7}{3}$, and $\frac{4}{5}$ to a common denominator.

$1 \times 3 \times 5 = 15$ the new numerator of the 1st.
$7 \times 2 \times 5 = 70$ " " " 2d.
$4 \times 3 \times 2 = 24$ " " " 3d.
and $2 \times 3 \times 5 = 30$, the common denominator.

Therefore, $\frac{15}{30}$, $\frac{70}{30}$, and $\frac{24}{30}$ are the equivalent fractions.

It is plain that this reduction does not alter the values of the several fractions, since the numerator and denominator of each are multiplied by the same number (see Prop. V).

QUEST.—117. What is the first step in reducing fractions to a common denominator? What is the second? Does the reduction alter the values of the several fractions? Why not?

2. When the numbers are small the work may be performed mentally.

Thus, $\frac{1}{2}$, $\frac{1}{4}$, $\frac{2}{5} = \frac{20}{40}$, $\frac{10}{40}$, $\frac{16}{40}$.

Here we find the first numerator by multiplying 1 by 4 and 5; the second, by multiplying 1 by 2 and 5; the third, by multiplying 2 by 4 and 2; and the common denominator by multiplying 2, 4, and 5 together.

3. Reduce $2\frac{1}{3}$ and $\frac{1}{2}$ of $\frac{1}{7}$ to a common denominator.

$2\frac{1}{3} = \frac{7}{3}$; and $\frac{1}{2}$ of $\frac{1}{7} = \frac{1}{14}$:

consequently, $\frac{7}{3}$ and $\frac{1}{14} = \frac{98}{42}$ and $\frac{3}{42}$ are the answers.

4. Reduce $5\frac{1}{2}$, $\frac{6}{7}$ of $\frac{1}{3}$, and 4 to a common denominator.

5. Reduce $\frac{7}{8}$, $\frac{135}{75}$, and 37 to a common denominator.

6. Reduce 4, $\frac{31}{25}$, $\frac{62}{2}$ to a common denominator.

7. Reduce $7\frac{1}{2}$, $\frac{31}{18}$, $6\frac{1}{4}$ to a common denominator.

8. Reduce $4\frac{1}{9}$, $8\frac{1}{7}$, and $2\frac{1}{2}$ of 5 to a common denominator.

9. Reduce $\frac{1}{5}$, $\frac{4}{9}$, $\frac{5}{8}$, and $\frac{11}{16}$ to a common denominator.

10. Reduce $\frac{3}{7}$ of $\frac{2}{3}$ of $\frac{5}{8}$ and $\frac{3}{4}$ of $\frac{5}{7}$ of $\frac{3}{5}$ to a common denominator.

11. Reduce $5\frac{1}{8}$, $3\frac{5}{7}$, $4\frac{1}{9}$, and $6\frac{5}{8}$ to fractions having a common denominator.

12. Reduce $\frac{3}{5}$, $\frac{2}{3}$, $\frac{1}{7}$, and $\frac{1}{2}$ to a common denominator.

13. Reduce $\frac{4}{7}$, $\frac{5}{9}$, $\frac{3}{8}$, $\frac{7}{8}$, and 19 to a common denominator.

14. Reduce $\frac{1}{15}$, $\frac{\frac{4}{7}}{6}$, $\frac{3\frac{1}{5}}{24}$, $\frac{9}{3\frac{3}{8}}$, $\frac{\frac{4}{9}}{\frac{5}{9}}$, $\frac{\frac{6}{7}}{8}$, and $1\frac{1}{9}$ to simple fractions having a common denominator.

118. It is often convenient to reduce fractions to a common denominator by multiplying the numerator and denominator in each by such a number as shall make the denominators the same in all.

QUEST.—When the numbers are small, how may the work be performed? 118. By what second method may fractions be reduced to a common denominator?

EXAMPLES.

1. Let it be required to reduce $\frac{3}{8}$ and $\frac{5}{12}$ to a common denominator.

We see at once that if we multiply the numerator and de nominator of the first fraction by 3, and the numerator and denominator of the second by 2, they will have a common denominator.

The two fractions will be reduced to $\frac{9}{24}$ and $\frac{10}{24}$.

2. Reduce $\frac{4}{5}$ and $\frac{5}{3}$ to a common denominator.

If we multiply both terms of the first fraetion by 3, and both terms of the second by 5, we have

$$\frac{4}{5}=\frac{12}{15}, \text{ and } \frac{5}{3}=\frac{25}{15}.$$

3. Reduce $\frac{1}{6}$, $\frac{1}{12}$, and $\frac{3}{4}$ to a common denominator.
4. Reduce $\frac{3}{6}$, $\frac{8}{28}$, $\frac{4}{14}$ to a common denominator.
5. Reduce $\frac{5}{8}$, $3\frac{5}{6}$, and $\frac{3}{4}$ to a common denominator.
6. Reduce $6\frac{5}{12}$, $8\frac{9}{6}$, and $5\frac{7}{24}$ to a common denominator.
7. Reduce $7\frac{5}{6}$, $\frac{4}{9}$, $\frac{3}{12}$, and $\frac{2}{18}$ to a common denominator.

119. To reduce fractions to their *least common denominator.*

I. *Find the least common multiple of the denominators as in* Art. 107, *and it will be the least denominator sought.*

II. *Multiply the numerator of each fraction by the quotient which arises from dividing the common multiple by the denominator, and the products will be the numerators of the required fractions; under which write the least common multiple.*

EXAMPLES.

1. Reduce $\frac{3}{7}$, $\frac{5}{8}$, and $\frac{2}{6}$ to their least common denominator.

OPERATION.

$$\begin{array}{r|l} 2 & 7\ .\ .\ 8\ .\ .\ 6 \\ \hline & 7\ .\ .\ 4\ .\ .\ 3 \end{array}$$

and $3 \times 4 \times 7 \times 2 = 168$ the leas common denominator.

QUEST.—119. How do you reduce fractions to their least common denominator? Does this reduction affect the values of the fractions?

$$\frac{168}{7} \times 3 = 24 \times 3 = 72 \text{ 1st numerator.}$$

$$\frac{168}{8} \times 5 = 21 \times 5 = 105 \text{ 2d numerator.}$$

$$\frac{168}{6} \times 2 = 28 \times 2 = 56 \text{ 3d numerator.}$$

Ans. $\frac{72}{168}$, $\frac{105}{168}$, and $\frac{56}{168}$

2. Reduce $\frac{4}{5}$, $\frac{8}{9}$, and $\frac{3}{15}$ to their least common denominator.

3. Reduce $14\frac{5}{4}$, $6\frac{3}{8}$, and $5\frac{1}{2}$ to their least common denominator.

4. Reduce $\frac{3}{15}$, $\frac{4}{24}$, and $\frac{8}{9}$ to their least common denominator.

5. Reduce $\frac{67}{120}$, $\frac{6}{40}$, $\frac{5}{2}$ to their least common denominator.

6. Reduce $\frac{41}{50}$, $3\frac{61}{20}$, $4\frac{1}{2}$, and 8 to a common denominator.

7. Reduce $3\frac{1}{8}$, $4\frac{4}{12}$, $8\frac{6}{18}$, $14\frac{7}{16}$ to their least common denominator.

8. Reduce $\frac{1}{2}$, $\frac{2}{3}$, $\frac{3}{4}$, and $\frac{5}{6}$ to fractions having the least common denominator.

9. Reduce $\frac{2}{5}$, $\frac{4}{6}$, $\frac{5}{9}$, and $\frac{7}{10}$ to fractions having the least common denominator.

10. Reduce $\frac{1}{3}$, $\frac{3}{4}$, $\frac{5}{6}$, $\frac{7}{8}$, $\frac{11}{16}$, and $\frac{17}{24}$ to equivalent fractions having the least common denominator.

REDUCTION OF DENOMINATE FRACTIONS.

120. We have seen (Art. 14), that a denominate number is one in which the kind of unit is denominated or expressed. For the same reason, a denominate fraction is one which expresses the *kind of unit* that has been divided. Such unit is called the unit of the fraction. Thus, $\frac{2}{3}$ of a £ is a denominate fraction. It expresses that one £ is the unit which has been divided.

Quest.—120 What is a denominate number? What is a denominate fraction? What is the unit called? In two-thirds of a pound, what is the unit of the fraction?

The fraction $\frac{3}{8}$ of a shilling is also a denominate fraction, in which the unit is one shilling. The two fractions, $\frac{2}{3}$ of a £ and $\frac{3}{8}$ of a shilling, are of different denominations, the unit of the first being one pound, and that of the second, one shilling.

Fractions, therefore, are of the same denomination when they express parts of the same unit, and of different denominations when they express parts of different units.

REDUCTION *of denominate fractions consists in changing their denominations without altering their values.*

CASE I.

121. To reduce a denominate fraction from a lower to a higher denomination.

I. *Consider how many units of the given denomination make one unit of the next higher, and place* 1 *over that number forming a second fraction.*

II. *Then consider how many units of the second denomination make one unit of the denomination next higher, and place* 1 *over that number forming a third fraction, and so on to the denomination to which you would reduce. Then reduce the compound fraction to a simple one* (Art. **114**).

EXAMPLES.

1. Reduce $\frac{1}{3}$ of a penny to the fraction of a £.

OPERATION.

$\frac{1}{3}$ of $\frac{1}{12}$ of $\frac{1}{20} = £\frac{1}{720}$.

The given fraction is $\frac{1}{3}$ of a penny. But one penny is equal to $\frac{1}{12}$ of a shilling: hence $\frac{1}{3}$ of a penny is equal to $\frac{1}{3}$ of $\frac{1}{12}$ of a shilling. But one shilling is

QUEST.—In three-eighths of a shilling, what is the unit? In one-half of a foot, what is the unit? When are fractions of the same denomination? When of different denominations? Are one-third of a £ and one-fourth of a £ of the same or different denominations? One-fourth of a £ and one-sixth of a shilling? One-fifth of a shilling and one-half of a penny? What is reduction? How many shillings in a £? How many in £2? In 3? In 4? How many pence in 1*s.*? In 2? In 3? In 2*s.* 8*d.*? In 3*s.* 6*d.*? In 5*s.* 8*d.*? How many feet in 3 yards 2*ft.*? How many inches? 121. How do you reduce a denominate fraction from a lower to a higher denomination? What is the first step? What the second? What the third?

equal to $\frac{1}{20}$ of a pound: hence $\frac{1}{3}$ of a penny is equal to $\frac{1}{3}$ of $\frac{1}{12}$ of $\frac{1}{20}$ of a £ = £$\frac{1}{720}$. The reason of the rule is therefore evident.

2. Reduce $\frac{3}{8}$ of a barleycorn to the denomination of yards.

Since 3 barleycorns make an inch, we first place 1 over 3: then as 12 inches make a foot, we place 1 over 12, and as 3 feet make a yard, we next place 1 over 3.

OPERATION.

$\frac{3}{8}$ of $\frac{1}{3}$ of $\frac{1}{12}$ of $\frac{1}{3} = \frac{3}{864}$ *yards.*

3. Reduce $\frac{3}{4}$*oz.* avoirdupois to the denomination of tons.
4. Reduce $\frac{5}{9}$ of a pint to the fraction of a hogshead.
5. Reduce $\frac{4}{15}$ of a shilling to the fraction of a £.
6. Reduce $\frac{1}{3}$ of a farthing to the fraction of a £.
7. Reduce $\frac{3}{8}$ of a gallon to the fraction of a hogshead.
8. Reduce $\frac{5}{8}$ of a shilling to the fraction of a £.
9. Reduce $\frac{187}{127}$ of a minute to the fraction of a day.
10. Reduce $\frac{6}{8}$ of a pound to the fraction of a *cwt.*
11. Reduce $\frac{1}{2}$ of an ounce to the fraction of a ton.
12. Reduce $\frac{2880}{7}$ of a farthing to the fraction of a pound.
13. Reduce $\frac{5}{7}$ of a penny to the fraction of a pound
14. What part of a *lb.* troy is $\frac{5}{6}$ of a *pwt.*?
15. What part of a *cwt.* is $\frac{4}{5}$ of a *lb.* avoirdupois?
16. What part of a *hhd.* of wine is $\frac{11}{13}$ of a gallon?

CASE II.

122. To reduce a denominate fraction from a higher to a lower denomination.

I. *Consider how many units of the next lower denomination make* 1 *unit of the given denomination, and place* 1 *under that number forming a second fraction.*

II. *Then consider how many units of the denomination still lower make one unit of the second denomination, and place* 1

QUEST.—122. How do you reduce a denominate fraction from a higher to a lower denomination? What is the first step? What the second? What the third?

under that number forming a third fraction, and so on to the denomination to which you would reduce.

III. *Connect all the fractions together, forming a compound fraction. Then reduce the compound fraction to a simple one* (Art. 114.)

EXAMPLES.

1. Reduce $\frac{1}{7}$ of a £ to the fraction of a penny.

In this example $\frac{1}{7}$ of a pound is equal to $\frac{1}{7}$ of 20 shillings. But 1 shilling is equal to 12 pence; hence, $\frac{1}{7}$ of a £ $= \frac{1}{7}$ of $\frac{20}{1}$ of $\frac{12}{1} = \frac{240}{7}d.$ Hence the reason of the rule is manifest.

OPERATION.

$\frac{1}{7}$ of $\frac{20}{1}$ of $\frac{12}{1} = \frac{240}{7}d.$

2. Reduce $\frac{4}{7}$*cwt.* to the fraction of a pound.
3. Reduce $\frac{2}{15}$ of a £ to the fraction of a penny.
4. Reduce $\frac{1}{3}$ of a day to the fraction of a minute.
5. Reduce $\frac{3}{9}$ of an acre to the fraction of a pole.
6. Reduce $\frac{6}{7}$ of a £ to the fraction of a farthing.
7. Reduce $\frac{3}{504}$ of a hogshead to the fraction of a gallon.
8. Reduce $\frac{4}{10}$ of a bushel to the fraction of a pint.
9. Reduce $\frac{3}{7}$ of a day to the fraction of a second.
10. Reduce $\frac{5}{8}$ of a tun to the fraction of a gill.
11. Reduce $\frac{3}{7}$ of a pound to the fraction of a farthing.
12. Reduce $\frac{7}{960}$ of a pound to the fraction of a penny.
13. Reduce $\frac{1}{300}$ of a *lb.* troy to the fraction of a *pwt.*
14. Reduce $\frac{1}{196}$ of a *cwt.* to the fraction of a *lb.*
15. Reduce $\frac{7}{71}$ of a week to the fraction of a second.
16. Reduce $\frac{5}{9}$ of a ton to the fraction of an ounce.
17. Reduce $\frac{14}{17}$ of a yard to the fraction of a nail.
18. Reduce $\frac{11}{23}$ of a league to the fraction of a foot.
19. Reduce $\frac{17}{22}$ of a ℔ to the fraction of a scruple.

QUEST.—123. How much is one-half of a £? One-third of a shilling? One-half of a penny? How much is one-half of a *lb.* avoirdupois? One-fourth of a ton? One-fourth of a *cwt.*? One-half of a quarter? One-fourth of a quarter? One-seventh of a quarter? One-fourteenth of a quarter? One-twenty-eighth of a quarter?

CASE III.

123. To find the value of a fraction in integers of a less denomination.

I. *Reduce the numerator to the next lower denomination, and then divide the result by the denominator.*

II. *If there be a remainder, reduce it to the denomination still less, and divide again by the denominator. Proceed in the same way to the lowest denomination. The several quotients, being connected together, will form the equivalent denominate number.*

EXAMPLES.

1. What is the value of $\frac{2}{3}$ of a £?

We first bring the pounds to shillings. This gives the fraction $\frac{40}{3}$ of shillings, which is equal to 13 shillings and 1 over. Reducing this to pence gives the fraction $\frac{12}{3}$ of pence, which is equal to 4 pence.

OPERATION.

```
   2
  20
3)40
 13s. . . 1 Rem.
         12
       3)12
         4d.
Ans. 13s. 4d.
```

2. What is the value of $\frac{4}{5}$*lb.* troy? *Ans.* ——
3. What is the value of $\frac{5}{16}$ of a *cwt.*? *Ans.* ——
4. What is the value of $\frac{5}{8}$ of an acre? *Ans.* ——
5. What is the value of $\frac{1}{6}$ of a £? *Ans.* ——
6. What is the value of $\frac{5}{6}$ of a hogshead? *Ans.* ——
7. What is the value of $\frac{196}{504}$ of a hogshead? *Ans.* ——
8. What is the value of $\frac{2}{9}$ of a guinea? *Ans.* ——
9. What is the value of $\frac{3}{5}$ of a *lb.* troy? *Ans.* ——
10. What is the value of $\frac{7}{8}$ of a tun of wine? *Ans.* ——
11. What is the value of $\frac{1}{2}$ of $\frac{6}{5}$ of a *lb.* troy? *Ans.* ——

QUEST.—How do you find the value of a fraction in terms of integers of a less denomination?

12. What is the value of $\frac{3}{5}$ of a league? *Ans.* ——
13. What is the value of $\frac{5}{8}$ of $\frac{3}{5}$ of an acre? *Ans.* ——
14. What is the value of $\frac{5}{9}$ of 15 yards of cloth?
15. What is the value of $\frac{7}{9}$ of a tun of wine?
16. What is the value of $\frac{3}{11}$ of a butt of beer?
17. What is the value of $\frac{7}{26}$ of a year?
18. What is the value of $\frac{5}{9}$ of a chaldron of coal?
19. What is the value of $\frac{2}{5}$ of 13*s.* 4*d.*?
20. What is the value of $\frac{3}{7}$ of 15*cwt.* 3*qr.* 14*lb.*?
21. What is the value of $\frac{3}{8}$ of a cubic yard?
22. What quantity of ale is contained in $\frac{5}{9}$ of 15228 cubic inches, English measure?

CASE IV.

124. To reduce a denominate number to a denominate fraction of a given denomination.

Reduce the number to the lowest denomination mentioned in it: then if the reduction is to be made to a denomination still less, reduce as in Case II.; but if to a higher denomination, reduce as in Case I.

EXAMPLES.

1. Reduce 4*s.* 7*d.* to the fraction of a £.

OPERATION.

4*s.* 7*d.* = 55*d.*

Then, 55 of $\frac{1}{12}$ of $\frac{1}{20}$ = £$\frac{55}{240}$.

We first reduce the given number to the lowest denomination named in it, viz., pence. Then, as the reduction is to be made to pounds, a higher denomination, we reduce by Case I.

2. What part of a bushel is 2*pk.* 3*qt.*?

OPERATION.

2*pk.* 3*qt.* = 19*qt.*

19 of $\frac{1}{8}$ of $\frac{1}{4}$ = $\frac{19}{32}$*bu*

We first reduce to quarts, this being the lowest denomination. We then reduce to bushels by Case I.

QUEST.—124. How do you reduce a denominate number to a fraction of a given denomination?

3 Reduce 2 feet 2 inches to the fraction of a yard.
Ans. ——

4. Reduce 3 gallons 2 quarts to the fraction of a hogshead.
Ans. ——

5. Reduce 1*qr.* 7*lb.* to the fraction of a hundred.
Ans. *cwt.*

6. What part of a hogshead is 3*qt.* 1*pt.*? *Ans.* ——

7. What part of a mile is 6*ft.* 7*in.*? *Ans.* ——

8. What part of a mile is 1 inch? *Ans.* ——

9. What part of a month of 30 days, is 1 hour 1 minute 1 second? *Ans.* ——

10. What part of 1 day is 3*hr.* 3*m.*? *Ans.* ——

11. What part is 3*hr.* 3*m.* of 2 days? Of 3? Of 4? Of 10? Of 25?

12. Reduce 15*s.* 11*d.* to the fraction of a pound.

13. Reduce 5$\frac{1}{9}$*d.* to the fraction of a shilling.

14. Reduce 1*cwt.* 2*qr.* 6*lb.* 3*oz.* 8$\frac{8}{9}$*dr.* to the fraction of a *cwt.*

15. Reduce 5*oz.* 3$\frac{1}{5}$*gr.* to the fraction of a *lb.* troy.

16. Reduce 3*qr.* 3$\frac{1}{9}$*na.* to the fraction of an English ell.

17. Reduce 147*da.* 15*hr.* to the fraction of a year.

18. What part of a pound is 15*s.* 9$\frac{1}{2}$*d.*?

19. What part of a groat is $\frac{2}{3}$ of three halfpence?

20. Reduce 4*bu.* 2$\frac{2}{7}$*pk.* of corn to the fraction of a quarter.

21. Reduce 1*qr.* 3*na.* to the fraction of a yard.

22. Reduce 2*R.* 15*P.* to the fraction of an acre.

23. Reduce 2℈ 11*gr.* to the fraction of a ℔.

24. Reduce 3*qt.* 1*pt.* 2*gi.* to the fraction of a hogshead.

25. Reduce 184 cubic inches to the fraction of a cubic yard.

26. Reduce 17*bu.* 3*pk.* to the fraction of a London chaldron.

27. Reduce 24′ 33″ to the fraction of a degree.

28. Reduce 27*gal.* 3*qt.* 1*pt.* to the fraction of a hogshead, beer measure

ADDITION OF VULGAR FRACTIONS.

Addition of integer numbers is the process of finding a single number which shall express all the units of the numbers added (ART. 48.)

Addition of fractions is the process of finding a single fraction which shall express the value of all the fractions added.

It is plain that we cannot add fractions so long as they have different units: for, $\frac{1}{2}$ of a £ and $\frac{1}{2}$ of a shilling make neither £1 nor 1 shilling.

Neither can we add parts of the same unit unless they are like parts; for $\frac{1}{3}$ of a £ and $\frac{1}{4}$ of a £ make neither $\frac{2}{3}$ of a £ nor $\frac{2}{4}$ of a £. But $\frac{1}{3}$ of a £ and $\frac{1}{3}$ of a £ may be added: they make $\frac{2}{3}$ of a £. So, $\frac{1}{4}$ of a £ and $\frac{2}{4}$ of a £ make $\frac{3}{4}$ of a £.

Hence, before fractions can be added, two things are necessary.

1st. *That the fractions be reduced to the same denomination; that is, to the same unit:*

2d. *That they be reduced to a common denominator; that is, to the same fractional unit* (ART. 94).

CASE I.

126. When the fractions to be added are of the same denomination and have a common denominator.

Add the numerators together, and place their sum over the common denominator: then reduce the fraction to its lowest terms, or to its equivalent mixed number.

QUEST.—125. What is addition of integer numbers? What is addition of fractions? What two things are necessary before fractions can be added? Can one-half of a £ be added to one-half of a shilling without reduction? Can one-half be added to one-fourth without reduction? 126. When the fractions are of the same denomination and have a common denominator, how do you find their sum? What is the sum of one third and two-thirds? Of three-fourths, one-fourth, and four-fourths? Of three-fifths, six-fifths, and two-fifths? Of three-sixths, seven-sixths, and nine-sixths? Of one-eighth, three-eighths, and four-eighths?

EXAMPLES.

1. Add $\frac{1}{2}$, $\frac{3}{2}$, $\frac{6}{2}$, and $\frac{3}{2}$ together.

It is evident, since all the parts are halves, that the true sum will be expressed by the number of halves; that is, by thirteen halves.	OPERATION. $1+3+6+3=13$; hence, $\frac{13}{2}=$ sum.

2. Add $\frac{1}{7}$ of a £, $\frac{5}{7}$ of a £, and $\frac{9}{7}$ of a £ together.
3. What is the sum of $\frac{3}{9}+\frac{4}{9}+\frac{6}{9}+\frac{13}{9}+\frac{16}{9}$?
4. What is the sum of $\frac{3}{14}+\frac{8}{14}+\frac{9}{14}+\frac{5}{14}+\frac{3}{14}$?
5. What is the sum of $\frac{9}{7}+\frac{6}{7}+\frac{14}{7}+\frac{11}{7}+\frac{15}{7}$?
6. What is the sum of $\frac{3}{11}+\frac{8}{11}+\frac{9}{11}+\frac{10}{11}+\frac{13}{11}+\frac{17}{11}$?

CASE II.

127. When the fractions are of the same denomination but have different denominators.

Reduce complex and compound fractions to simple ones, mixea numbers to improper fractions, and all the fractions to a common denominator. Then add them as in Case I.

EXAMPLES.

1. Add $\frac{6}{2}$, $\frac{4}{3}$, and $\frac{2}{5}$ together.

By reducing to a common denominator, the new fractions are

$\frac{90}{30}+\frac{40}{30}+\frac{12}{30}=\frac{142}{30}$,

which, by reducing to the lowest terms becomes $4\frac{11}{15}$.

OPERATION.

$6\times3\times5=90$ 1st numerator

$4\times2\times5=40$ 2d numerator.

$2\times3\times2=12$ 3d numerator.

$2\times3\times5=30$ the denominator

2. Add $\frac{1}{7}$ of a £, $\frac{2}{9}$ of a £, and $\frac{5}{6}$ of a £ together.
3. Add together $\frac{1}{7}$, $\frac{1}{9}$, $4\frac{1}{3}$, and $6\frac{1}{5}$. *Ans.* ——
4. Find the least common denominator (Art. **119**) and add the fractions $\frac{1}{16}$, $\frac{3}{7}$, $\frac{2}{8}$, and $\frac{4}{9}$. *Ans.* ——

QUEST.—127. How do you add fractions which have different denominators? How do you reduce fractions of different denominators to equivalent fractions having a common denominator?

5. Find the least common denominator and add $\frac{6}{12}$, $\frac{3}{5}$, $\frac{4}{8}$, and $\frac{6}{30}$. *Ans.* ——

6. Find the least common denominator and add $\frac{1}{9}$ of $\frac{3}{5}$, $\frac{3}{4}$ of 19, and $\frac{5}{8}$ of 12 together. *Ans.* ——

7. Add $\frac{3}{5}$, $\frac{9}{10}$ of $\frac{5}{11}$ of $\frac{6}{8}$, and $\frac{6}{7}$ of $\frac{3}{2}$ of 11 together.

128. When there are mixed numbers, instead of reducing them to improper fractions we may add the whole numbers and the fractional parts separately and then add their sums.

EXAMPLES.

1. Add $19\frac{1}{7}$, $6\frac{2}{3}$, and $4\frac{4}{5}$ together.

OPERATION.

Whole numbers.	Fractional parts.
$19 + 6 + 4 = 29.$	$\frac{1}{7} + \frac{2}{3} + \frac{4}{5} = \frac{169}{105} = 1\frac{64}{105}.$

Hence, $29 + 1\frac{64}{105} = 30\frac{64}{105}$, the sum.

2. Add together $3\frac{1}{4}$, $6\frac{5}{7}$, $8\frac{9}{15}$, and $65\frac{9}{8}$. *Ans.* ——
3. Add together $\frac{3}{7}$, $\frac{3}{9}$, 13, and $18\frac{3}{15}$. *Ans.* ——
4. Add together $\frac{4}{16}$, $\frac{12}{7}$, 1, and $\frac{16}{9}$. *Ans.* ——
5. Add together $38\frac{2}{7}$, $13\frac{1}{3}$, and $9\frac{3}{5}$. *Ans.* ——
6. Add together $6\frac{2}{4}$, $13\frac{3}{7}$, $18\frac{3}{15}$, and $132\frac{2}{8}$. *Ans.* ——
7. Add $3\frac{5}{7}$, $4\frac{5}{8}$, and $\frac{5}{11}$ together. *Ans.* ——
8. Add $\frac{3}{5}$ and $\frac{9}{10}$ of $\frac{5}{11}$ of $15\frac{3}{8}$ together. *Ans.* ——
9. Add $\frac{1}{9}$, $7\frac{5}{8}$, $\dfrac{45}{94\frac{7}{11}}$, and $\dfrac{47\frac{5}{9}}{314\frac{3}{5}}$ together. *Ans.* ——

CASE III.

129. When the fractions are of different denominations.

Reduce the fractions to the same denomination. Then reduce all the fractions to a common denominator, and then add them as in Case I.

QUEST.—128. How may you proceed when there are mixed numbers? 129. When the fractions are of different denominations, how are they added? What is the second method?

EXAMPLES.

1. Add $\frac{2}{3}$ of a £ to $\frac{5}{6}$ of a shilling

$\frac{2}{3}$ of a £ $= \frac{2}{3}$ of $\frac{20}{1} = \frac{40}{3}$ of a shilling.

Then, $\frac{40}{3} + \frac{5}{6} = \frac{240}{18} + \frac{15}{18} = \frac{255}{18}s. = \frac{85}{6}s. = 14s.\ 2d.$

Or, the $\frac{5}{6}$ of a shilling might have been reduced to the fraction of a £ thus,

$\frac{5}{6}$ of $\frac{1}{20} = \frac{5}{120}$ of a £ $= \frac{1}{24}$ of a £.

Then, $\frac{2}{3} + \frac{1}{24} = \frac{48}{72} + \frac{3}{72} = \frac{51}{72}$ of a £: which being reduced gives 14*s.* 2*d.*

2. Add $\frac{3}{8}$ of a yard to $\frac{5}{9}$ of an inch.

3. Add $\frac{1}{3}$ of a week, $\frac{1}{4}$ of a day, and $\frac{1}{2}$ of an hour together.

4. Add $\frac{4}{7}$ of a *cwt.*, $8\frac{5}{6}$*lb.*, and $3\frac{9}{10}$*oz.* together.

5. Add $1\frac{1}{4}$ miles, $\frac{7}{10}$ furlongs, and 30 rods together.

NOTE.—The value of each of the fractions may be found separately, and their several values then added.

6. Add $\frac{3}{5}$ of a year, $\frac{1}{3}$ of a week, and $\frac{1}{8}$ of a day together.

$\frac{3}{5}$ of a year $= \frac{3}{5}$ of $\frac{365}{1}$ days $=$ 219 days.
$\frac{1}{3}$ of a week $= \frac{1}{3}$ of 7 days $=$ 2 days 8 hours.
$\frac{1}{8}$ of a day $=$ - - - - 3 hours.
Ans. 221*da.* 11*hr.*

7. Add $\frac{2}{3}$ of a yard, $\frac{3}{4}$ of a foot, and $\frac{7}{8}$ of a mile together.

8. Add $\frac{3}{4}$ of a *cwt.*, $\frac{42}{2}$ of a *lb.* 13*oz.*, and $\frac{1}{2}$ of a *cwt.* 6*lb.* together.

9. Add $\frac{5}{9}$ of a pound, $\frac{3}{8}$ of a shilling, and $\frac{5}{7}$ of a penny together.

10. What is the sum of $\frac{3}{4}$ of £1 10*s.*, $\frac{1}{4}$ of £1 10*s.*, and $\frac{1}{50}$ of a hundred guineas?

11. Add $\frac{1}{5}$ of a *lb.* troy to $\frac{1}{8}$ of an ounce. *Ans.* ——

12. Add $\frac{4}{9}$ of a ton to $\frac{5}{12}$ of a *cwt.* *Ans.* ——

13. Add $\frac{5}{6}$ of 3 ells English to $\frac{5}{12}$ of a yard. *Ans.* ——

14. Add $\frac{2}{3}$ of a yard, $\frac{3}{7}$ of a foot, and $\frac{5}{11}$ of a mile together.

15. Add $\frac{5}{8}$ of an acre, $\frac{3}{5}$ of 19 square feet, and $\frac{3}{7}$ of a square inch together.

16. What is the sum of $\frac{3}{4}$ of a tun of wine and $\frac{3}{5}$ of a *hhd.*?

17. Add $\frac{5}{9}$ of a chaldron to $\frac{3}{7}$ of a bushel.

18 Add $\frac{1}{4}$ of a week, $\frac{1}{3}$ of a day, and $\frac{1}{5}$ of an hour together.

19. Add $\frac{1}{5}$ of $\frac{3}{4}$ of a year, $\frac{3}{8}$ of $\frac{5}{9}$ of a day, and $\frac{7}{8}$ of $\frac{3}{5}$ of $19\frac{1}{2}$ hours together.

SUBTRACTION OF VULGAR FRACTIONS.

130. It has been shown (Art. **125**), that before fractions can be added together they must be reduced to the same unit and to a common denominator. The same reductions must be made before subtraction.

SUBTRACTION *of Vulgar Fractions is the process of finding the difference between two fractional numbers.*

CASE I.

131. When the fractions are of the same denomination and have a common denominator.

Subtract the less numerator from the greater, and place the difference over the common denominator.

EXAMPLES.

1. What is the difference between $\frac{5}{8}$ and $\frac{3}{8}$?

Here we have $5 - 3 = 2$: hence, $\frac{2}{8} =$ the difference.

2. From $\frac{335}{105}$ take $\frac{169}{105}$. *Ans.* ——

3. From $\frac{4978}{9765}$ take $\frac{1697}{9765}$. *Ans.* ——

4. From $\frac{18906}{327}$ take $\frac{909}{327}$. *Ans.* ——

CASE II.

132. When the fractions are of the same denomination, but have different denominators.

QUEST.—130. Can one-third of a shilling be subtracted from one-third of a £ without reduction? Can one-fourth of a shilling be subtracted from one-fifth of a shilling? What reductions are necessary before subtraction? What is subtraction? 131. How do you subtract fractions of the same denomination and denominator?

Reduce mixed numbers to improper fractions, compound and complex fractions to simple ones, and all the fractions to a common denominator: then subtract them as in Case I.

EXAMPLES.

1. What is the difference between $\frac{5}{6}$ and $\frac{1}{3}$?

Here, $\frac{5}{6} - \frac{1}{3} = \frac{5}{6} - \frac{2}{6} = \frac{3}{6} = \frac{1}{2}$ answer.

2. What is the difference between $12\frac{1}{2}$ of $\frac{1}{6}$ and 2?

3. What is the difference between $2\frac{1}{2}$ of a £, and $\frac{3}{12}$ of a £?

4. From $\frac{1}{8}$ of 6, take $\frac{16}{17}$ of $\frac{1}{2}$. *Ans.* ——

5. From $\frac{1}{7}$ of $\frac{3}{6}$ of 7, take $\frac{3}{8}$ of $\frac{5}{4}$. *Ans.* ——

6. From $37\frac{11}{15}$, take $3\frac{5}{7}$ of $\frac{1}{3}$. *Ans.* ——

7. What is the difference between $\frac{3}{5}$ and $\frac{9}{16}$? *Ans.* ——

8. What is the difference between $3\frac{5}{8}$ and $\frac{2}{3}$ of $\frac{5}{8}$?

9. What is the difference between $\dfrac{49\frac{5}{8}}{97}$ and $\dfrac{34\frac{3}{5}}{146\frac{3}{11}}$?

10. From $115\frac{5}{8}$ take $39\frac{7}{8}$. *Ans.* ——

11. Subtract $\frac{54}{713}$ from a unit. *Ans.* ——

12. Subtract $\frac{11}{15}$ from 365. *Ans.* ——

13. What is the difference between $\frac{2}{3}$ of 15 and $\frac{4}{5}$ of 72?

14. To what fraction must I add $\frac{3}{5}$ that the sum may be $\frac{5}{8}$?

15. What number is that to which if $7\frac{2}{3}$ be added, the sum will be $17\frac{3}{5}$?

16. What number is that from which if you subtract $\frac{1}{11}$ of $\frac{5}{9}$ of a unit, and to the remainder add $\frac{3}{5}$ of $\frac{7}{8}$ of a unit, the sum will be 9?

CASE III.

133. When the fractions are of different denominations.

Reduce the fractions to the same denomination. Then reduce them to a common denominator; after which subtract as in Case I.

QUEST.—132. How do you subtract fractions of different denominators? What is the difference between one-half and one-third? 133. How do you subtract fractions which are of different denominations?

EXAMPLES.

1. What is the difference between $\frac{1}{2}$ of a £ and $\frac{1}{3}$ of a shilling?

$\frac{1}{3}$ of a shilling $= \frac{1}{3}$ of $\frac{1}{20} = \frac{1}{60}$ of a £.

Then, $\frac{1}{2} - \frac{1}{60} = \frac{30}{60} - \frac{1}{60} = \frac{29}{60}$ of a £ $= 9s.\ 8d.$

2. What is the difference between $\frac{1}{2}$ of a day and $\frac{2}{3}$ of a second?

3. From $1\frac{3}{4}$ of a *lb.*, troy weight, take $\frac{1}{6}$ of an ounce.

4. What is the difference between $\frac{4}{15}$ of a hogshead and $\frac{6}{19}$ of a quart?

5. From $\frac{1}{2}$ of a £ take $\frac{3}{4}$ of a shilling. *Ans.* ——

6. From $\frac{3}{8}$*oz.* take $\frac{7}{8}$*pwt.* *Ans.* ——

7. From $4\frac{3}{7}$*cwt.* take $4\frac{9}{10}$*lb.* *Ans.* ——

8. What is the difference between $\frac{3}{4}$ of a pound and $\frac{5}{7}$ of a shilling?

9. From $\frac{5}{6}$ of a *lb.* troy take $\frac{5}{8}$ of an ounce. *Ans.* ——

10. From $\frac{3}{8}$ of a ton take $\frac{2}{3}$ of $\frac{3}{4}$ of a *lb.* *Ans.* ——

11. From $\frac{2}{3}$ of $\frac{3}{5}$ of a *hhd.* of wine take $\frac{2}{5}$ of $\frac{1}{2}$ of a pint.

12. From $\frac{3}{5}$ of a league take $\frac{5}{8}$ of a mile. *Ans.* ——

13. From $\frac{5}{9}$ of $365\frac{1}{4}$ days take $\frac{3}{7}$ of $\frac{9}{16}$ of an hour.

14. A pound avoirdupois is equal to 14*oz.* 11*pwt.* 16*gr.* troy; what is the difference, in troy weight, between the ounce avoirdupois and the ounce troy?

MULTIPLICATION OF VULGAR FRACTIONS.

134. MULTIPLICATION is a short method of taking one number, called the multiplicand, as many times as there are units in another number, called the multiplier.

Hence, *when the multiplier is less than* 1 *we do not take the whole of the multiplicand, but only such a part of it as the mul-*

QUEST.—134. What is multiplication? What is required when the multiplier is less than 1? Does multiplication then imply increase? What is the product of 8 multiplied by one-half? By one-fourth? By one-eighth? By three-halves? By six-halves? What is the product of 9 multiplied by one-half? By one-third? By one-sixth? By one-ninth?

tiplier is of unity. For example, if the multiplier be one-half of unity, the product will be half the multiplicand; as, for example, the product of 8 multiplied by $\frac{1}{2}$ is 4. If the multiplier be $\frac{1}{3}$ of unity, the product will be one-third of the multiplicand. Hence, *to multiply by a proper fraction does not imply increase, as in the multiplication of whole numbers.*

CASE I.

135. To multiply a fraction by a whole number.

Multiply the numerator, or divide the denominator by the whole number.

EXAMPLES

1. Multiply the fraction $\frac{5}{8}$ by 4.

OPERATION.

$\frac{5}{8} \times 4 = \frac{20}{8} = \frac{5}{2} = 2\frac{1}{2}$;

or by dividing the denominator by 4, we have

$\frac{5}{8} \times 4 = \frac{5}{4)8} = \frac{5}{2} = 2\frac{1}{2}$.

When it is required to multiply a fraction by a whole number, it is required to increase the fraction as many times as there are units in the multiplier, which may be done by multiplying the numerator (Art. 98), or by dividing the denominator (Art. 101).

2. Multiply $\frac{37}{144}$ by 12. *Ans.* ——
3. Multiply $\frac{47}{49}$ by 7. *Ans.* ——
4. Multiply $\frac{175}{47}$ by 9. *Ans.* ——
5. Multiply $\frac{127}{15}$ by 5. *Ans.* ——
6. Multiply $\frac{369}{145}$ by 49. *Ans.* ——
7. Multiply $\frac{271}{196}$ by 357. *Ans.* ——
8. Multiply $\frac{1823}{9372}$ by 198. *Ans.* ——
9. Multiply $\frac{4291}{87146}$ by 2433. *Ans.* ——

QUEST.—When the multiplier is less than 1, how much of the multiplicand is taken? Does the multiplication by a proper fraction imply increase? 135. How do you multiply a fraction by a whole number? 136. What is the product of one-sixth by one-seventh? Of three-fourths by one-half? Of six-ninths by three-fifths? Give the general rule for the multiplication of fractions.

CASE II.

136. To multiply one fraction by another.

Reduce all the mixed numbers to improper fractions, and all compound and complex fractions to simple ones: then multiply the numerators together for a numerator, and the denominators together for a denominator.

EXAMPLES.

1. Multiply $\frac{3}{4}$ by $\frac{5}{7}$.

In this example $\frac{3}{4}$ is to be taken $\frac{5}{7}$ times; that is, $\frac{3}{4}$ is first to be multiplied by 5 and the product divided by 7, a result which is obtained by multiplying the numerators and denominators together.

OPERATION.

$$\frac{3}{4} \times \frac{5}{7} = \frac{3}{4} \times 5 \times \frac{1}{7} = \frac{15}{28}$$

2. Multiply $\frac{1}{6}$ of $\frac{3}{7}$ by $8\frac{1}{3}$.

We first reduce the compound fraction to the simple one $\frac{3}{42}$, and the mixed number to the equivalent fraction $\frac{25}{3}$; after which, we multiply the numerators and denominators together.

OPERATION.

$$\frac{1}{6} \text{ of } \frac{3}{7} = \frac{3}{42},$$
$$8\frac{1}{3} = \frac{25}{3}.$$
$$\text{Hence, } \frac{3}{42} \times \frac{25}{3} = \frac{75}{126} = \frac{25}{42}.$$

3. Multiply $5\frac{1}{4}$ by $\frac{1}{6}$. *Ans.* ——
4. Multiply $\frac{12}{10}$ by $\frac{3}{4}$ of 9. *Ans.* ——
5. Multiply $\frac{1}{8}$ of 3 of $\frac{1}{6}$ by $15\frac{1}{7}$. *Ans.* ——
6. Multiply $\frac{5}{6}$ by $\frac{2}{3}$ of $\frac{6}{7}$. *Ans.* ——
7. Required the product of 6 by $\frac{2}{3}$ of 5. *Ans.* ——
8. Required the product of $\frac{2}{9}$ of $\frac{3}{5}$ by $\frac{5}{8}$ of $3\frac{2}{7}$.
9. Required the product of $3\frac{2}{7}$ by $4\frac{14}{33}$. *Ans.* ——
10. Required the product of 5, $\frac{2}{3}$, $\frac{2}{7}$ of $\frac{3}{5}$, and $4\frac{1}{6}$.
11. Required the product of $4\frac{1}{2}$, $\frac{3}{4}$ of $\frac{1}{7}$, and $18\frac{4}{5}$.
12. Required the product of 14, $\frac{5}{6}$, $\frac{4}{5}$ of 9, and $6\frac{3}{7}$.
13. What is the product of $16\frac{2}{9}$, $\frac{3}{6\frac{1}{3}}$, $\frac{\frac{7}{8}}{9\frac{1}{2}}$, $\frac{\frac{1}{8}}{19}$, and $\frac{5\frac{3}{4}}{9\frac{5}{16}}$?

137. In multiplying by a mixed number, we may first multiply by the integer, then multiply by the fraction, and then add the two products together. This is the best method when the numerator of the fraction is 1.

EXAMPLES.

1. Multiply 26 by $3\frac{1}{2}$.

We first multiply 26 by 3: the product is 78. Afterwards we multiply 26 by $\frac{1}{2}$: the product is 13: hence the entire product is 91.

OPERATION.

$$\begin{array}{r} 26 \\ 3 \\ \hline 78 \\ 26 \times \frac{1}{2} = 13 \\ \hline 91 \; \textit{Ans.} \end{array}$$

2. Multiply 48 by $8\frac{1}{6}$.

We first multiply by 8, and then add a sixth.

OPERATION.

$$\begin{array}{r} 48 \times 8 = 384 \\ 48 \times \frac{1}{6} = \quad 8 \\ \hline 392 \; \textit{Ans.} \end{array}$$

3. Multiply 67 by $9\frac{1}{12}$. *Ans.* ——
4. Multiply 842 by $7\frac{1}{9}$. *Ans.* ——
5. Multiply 3756 by $3\frac{1}{5}$. *Ans.* ——
6. Multiply 2056 by $5\frac{1}{6}$. *Ans.* ——

GENERAL EXAMPLES.

1. What is the product of $\frac{5}{8}$ of $\frac{3}{7}$, $\frac{2}{3}$ of $15\frac{1}{2}$, and $\frac{3}{11}$ of 2?

2. What is the continued product of $14\frac{3}{7}$, $\frac{4}{5\frac{1}{3}}$, $\frac{7\frac{1}{5}}{15}$, $\frac{\frac{3}{4}}{\frac{5}{7}}$, and $\frac{\frac{5}{7}}{9}$? *Ans.* ——

3. What is the product of $\frac{\frac{1}{7}}{8}$, $\frac{\frac{7}{8}}{9\frac{1}{2}}$, $\frac{7\frac{1}{9}}{\frac{8}{9}}$, $\frac{4\frac{3}{4}}{7\frac{3}{14}}$, $\frac{3}{27}$, $\frac{8}{9}$, and 20? *Ans.* ——

4. What is the product of $\frac{3}{5}$ of $\frac{7}{11}$ of 15, and $\frac{14}{15}$ of $11\frac{5}{8}$? *Ans.* ——

5. What will 7 yards of cloth cost, at $\$\frac{3}{4}$ per yard?

QUEST.—137. How may you multiply by a mixed number? When is this the best method?

6. What will 32 gallons of brandy cost, at \$1$\frac{1}{8}$ per gallon?

7. If 1*lb.* of tea cost \$1$\frac{1}{4}$, what will 6$\frac{1}{4}$*lb.* cost?

8. What will be the cost of 17$\frac{1}{2}$ yards of cambric, at 2$\frac{1}{2}$ shillings per yard?

9. What will 15$\frac{1}{16}$ barrels of cider come to, at \$3 per barrel?

10. What will 3$\frac{3}{8}$ boxes of raisins cost, at \$2$\frac{1}{2}$ per box?

11. What will 15$\frac{1}{2}$ barrels of sugar cost, at 17$\frac{1}{4}$ dollars per barrel?

DIVISION OF VULGAR FRACTIONS.

138. We have seen that division of entire numbers explains the manner of finding how many times a less number is contained in a greater.

In division of fractions the divisor may be larger than the dividend, in which case the quotient will be less than 1.

For example, divide 1 apple into 4 equal parts.

Here it is plain that each part will be $\frac{1}{4}$; or that the dividend will contain the divisor but $\frac{1}{4}$ times.

Again, divide $\frac{1}{2}$ of a pear into 6 equal parts.

If a whole pear were divided into 6 equal parts, each part would be expressed by $\frac{1}{6}$. But since the half of the pear was divided, each part will be expressed by $\frac{1}{2}$ of $\frac{1}{6}$, or $\frac{1}{12}$.

In the division of fractions we should note the following principles:

1st. When the dividend is just equal to the divisor, the quotient will be 1.

2d. When the dividend is greater than the divisor, the quotient will be greater than 1.

QUEST.—138. What does division of whole numbers explain? In division of fractions, may the divisor exceed the dividend? How will the quotient then compare with 1? If an apple be divided in 2 equal parts, what will express each part? If half an apple be divided into 4 equal parts, what will express one of the parts? What is one-half of one-half? What is one-sixth of one-half? What principles do you note in the division of fractions? When will the quotient be 1? When greater than 1?

3d. When the dividend is less than the divisor, the quotient will be less than 1.

4th. The quotient will be just so many times greater than 1, as the dividend is greater than the divisor.

5th. The quotient will be just as many times less than 1 as the dividend is less than the divisor.

CASE I.

139. To divide a fraction by a whole number.

Divide the numerator or multiply the denominator by the whole number.

EXAMPLES.

1. Divide $\frac{4}{3}$ by 2.

OPERATION.

$$\frac{4}{3} \div 2 = \frac{4}{3 \times 2} = \frac{4}{6} = \frac{2}{3}.$$

$$\text{or } \frac{4}{3} \div 2 = \frac{2)4}{3} = \frac{2}{3}.$$

In the first operation we divide the fraction by multiplying the denominator (Art. **100**): in the second we divide the numerator (Art. **99**), giving the same result in both cases.

2. Divide $\frac{18}{37}$ by 9. *Ans.* ——
3. Divide $\frac{406}{19}$ by 15. *Ans.* ——
4. Divide $\frac{2755}{3758}$ by 19. *Ans.* ——
5. Divide $\frac{379}{1267}$ by 15. *Ans.* ——
6. Divide $\frac{37}{19}$ by 8. *Ans.* ——
7. Divide $\frac{61}{21}$ by 37. *Ans.* ——

CASE II.

140. To divide one fraction by another.

EXAMPLES.

1. Let it be required to divide $\frac{10}{24}$ by $\frac{5}{8}$.

The true quotient will be expressed by the complex fraction $\frac{\frac{10}{24}}{\frac{5}{8}}$.

QUEST.—When will the quotient be less than 1? When greater than 1, how many times greater? When less than 1, how many times less? 139. In how many ways may a fraction be divided by a whole number? 140 How do you divide one fraction by another?

Let the terms of this fraction be now multiplied by the denominator with its terms inverted: this will not alter the value of the fraction (Art. 102), and we shall then have.

$$\frac{\frac{10}{24}}{\frac{5}{8}} = \frac{\frac{10}{24} \times \frac{8}{5}}{\frac{5}{8} \times \frac{8}{5}} = \frac{\frac{10}{24} \times \frac{8}{5}}{1} = \frac{10}{24} \times \frac{8}{5} = \text{quotient.}$$

It will be seen that the quotient is obtained by simply multiplying the numerator by the denominator with its terms inverted. This quotient may be further simplified by cancelling the common factors 5 and 8, giving $\frac{2}{3}$ for the true quotient

SECOND METHOD OF PROOF.

OPERATION.
$\frac{10}{24} \div 5 = \frac{10}{120}$
$\frac{10}{120} \times 8 = \frac{80}{120}.$

Let us first divide the dividend by 5. This is done by multiplying the denominator (Art. 100), which gives $\frac{10}{120}$. But the divisor being but $\frac{1}{8}$ of 5, this quotient is 8 times too small, since the eighth of a number will be contained in the dividend 8 times more than the number itself. Therefore, by multiplying $\frac{10}{120}$ by 8, we obtain $\frac{80}{120}$ for the true quotient.

Hence, to divide one fraction by another,

Reduce compound and complex fractions to simple ones, also whole and mixed numbers to improper fractions: then multiply the dividend by the divisor with its terms inverted, and the product reduced to its simplest terms will be the quotient sought.

EXAMPLES.

1. Divide $\frac{1}{8}$ by $\frac{1}{7}$. *Ans.* ——
2. Divide $3\frac{1}{4}$ by $\frac{1}{9}$. *Ans.* ——
3. Divide $16\frac{1}{2}$ of $\frac{1}{3}$ by $4\frac{1}{7}$. *Ans.* ——
4. Divide $44\frac{1}{33}$ by $\frac{213}{333}$. *Ans.* ——
5. Divide $371\frac{1}{2}$ by $\frac{1}{414}$. *Ans.* ——
6. Divide $\frac{64}{111}$ by $\frac{23}{1213}$. *Ans.* ——
7. Divide $\frac{1}{2}$ of $\frac{2}{3}$ by $\frac{2}{3}$ of $\frac{3}{4}$. *Ans.* ——
8. Divide 5 by $\frac{7}{10}$. *Ans.* ——
9. Divide $5205\frac{1}{5}$ by $\frac{4}{5}$ of 91. *Ans.* ——

10. Divide 100 by $4\frac{7}{8}$. *Ans.* ——

11. Divide $\frac{3}{4}$ of $\frac{7}{8}$ by $\frac{2}{3}$. *Ans.* ——

12. Divide $\frac{5}{6}$ of 50 by $4\frac{1}{3}$. *Ans.* ——

13. Divide $14\frac{1}{8}$ of $\frac{1}{9}$ by $3\frac{1}{3}$ of 6. *Ans.* ——

14. Divide $34\frac{1}{7}$ by $\frac{54\frac{1}{8}}{93\frac{5}{11}}$. *Ans.* ——

15. Divide $\frac{51\frac{1}{11}}{95}$ by $\frac{71}{149\frac{3}{8}}$. *Ans.* ——

16. What number multiplied by $\frac{3}{4}$ will give $15\frac{3}{4}$ for the product?

17. What part of 108 is $\frac{5}{12}$? *Ans.* ——

18. What number is that which, if multiplied by $\frac{5}{8}$ of $\frac{3}{7}$ of $15\frac{1}{8}$, will produce $\frac{5}{8}$?

19. If 7*lb.* of sugar cost $\frac{44}{75}$ of a dollar, what is the price per pound?

20. If $\frac{3}{7}$ of a dollar will pay for $10\frac{1}{2}$*lb.* of nails, how much is the price per pound?

21. If $\frac{4}{7}$ of a yard of cloth cost \$3, what is the price per yard?

22. If \$$21\frac{1}{2}$ will buy $7\frac{13}{16}$ barrels of apples, how much are they per barrel?

23. If $4\frac{1}{7}$ gallons of molasses cost \$$2\frac{5}{9}$, how much is it per quart?

24. If $1\frac{1}{8}$*hhd.* of wine cost \$$250\frac{1}{3}$, how much is the wine per quart?

25. If 8 pounds of tea cost $7\frac{5}{8}$ of a dollar, how much is it per pound?

26. In $8\frac{1}{2}$ weeks a family consumes $165\frac{3}{9}$ pounds of butter: how much do they consume a week?

27. If a piece of cloth containing $176\frac{3}{4}$ yards costs \$$375\frac{6}{5}$, what does it cost per yard?

28. Divide $15\frac{1}{2}$ of $\frac{\frac{2}{7}}{\frac{7}{9}}$ of $\frac{7}{\frac{2}{3}}$ of $\frac{\frac{7}{10}}{3}$ by $\frac{4\frac{5}{9}}{7}$ of $\frac{3}{4\frac{3}{4}}$ of $\frac{\frac{7}{8}}{3\frac{1}{2}}$ of $\frac{2\frac{3}{4}}{\frac{4}{7}}$.

DECIMAL FRACTIONS.

141. If the unit 1 be divided into 10 equal parts, the parts are called *tenths*, because each part is one-tenth of unity.

If the unit 1 be divided into one hundred equal parts, the parts are called *hundredths*, because each part is one-hundredth of unity.

If the unit 1 be divided into one thousand equal parts, the parts are called *thousandths*, because each part is one-thousandth of unity: and we have similar expressions for the parts when the unit is divided into ten thousand, one hundred thousand, &c., equal parts.

The division of the unit into tenths, hundredths, thousandths, &c., forms a system of numbers called *Decimal Fractions.* They may be written,

Four-tenths, - - - - - - $\frac{4}{10}$

Six-tenths, - - - - - - $\frac{6}{10}$.

Forty-five hundredths, - - - - - $\frac{45}{100}$.

125 thousandths, - - - - - $\frac{125}{1000}$.

1047 ten thousandths, - - - - - $\frac{1047}{10000}$.

From which we see, that in each case the denominator gives denomination or name to the fraction; that is, determines whether the parts are tenths, hundredths, thousandths, &c.

142. The denominators of decimal fractions are seldom set down. The fractions are usually expressed by means of

---

Quest.—141. When the unit 1 is divided into 10 equal parts, what is each part called? What is each part called when it is divided into 100 equal parts? When into 1000? Into 10,000, &c.? How are decimal fractions formed? What gives denomination to the fraction? 142. Are the denominators of decimal fractions generally set down? How are the fractions expressed?

a comma, or period, which is called the decimal point, and is placed at the left of the numerator.

Thus, $\frac{4}{10}$	- -	is written	- -	.4
$\frac{45}{100}$	- -	" "	- -	.45
$\frac{125}{1000}$	- -	" "	- -	.125
$\frac{1047}{10000}$	- -	" "	- -	.1047.
&c.,		&c.,		&c.

This manner of expressing decimal fractions is a mere language, and is used to avoid the inconvenience of writing the denominators. The denominator, however, of every decimal fraction is always understood. *It is a unit* 1, *with as many ciphers annexed as there are places of figures in the numerator.*

The place next to the decimal point, is called the place of tenths, and its unit is 1 tenth. The next place to the right is the place of hundredths, and its unit is 1 hundredth; the next is the place of thousandths, and its unit is 1 thousandth; and similarly for places still to the right.

DECIMAL NUMERATION TABLE.

Tenths.	Hundredths.	Thousandths.	Tenths of thousandths.	Hundredths of thousandths.	Millionths.	Tenths of millionths.	&c., &c.		
.4								is read	4 tenths.
.6	4							" "	64 hundredths.
.0	6	4						" "	64 thousandths.
.6	7	5	4					" "	6754 ten-thousandths.
.0	1	2	3	4				" "	1234 hundred-thousandths.
.0	0	7	6	5	4			" "	7654 millionths.
.0	0	4	3	6	0	4		" "	43604 ten-millionths.

QUEST.—Is the denominator understood? What is it? What is the place next the decimal point called? What is its unit? What is the next place called? What is its unit? The next? Its unit? Which way are decimals numerated?

Decimal fractions are numerated from the left hand to the right, beginning with the tenths, hundredths, &c., as in the table.

143. Let us now write and numerate the following decimals.

Four-tenths, - - - - -	.4.
Four hundredths, - - -	.0 4.
Four thousandths, - - -	.0 0 4.
Four ten-thousandths, - -	.0 0 0 4.
Four hundred thousandths, - -	.0 0 0 0 4.
Four millionths, - - -	.0 0 0 0 0 4.
Four ten-millionths, - - -	.0 0 0 0 0 0 4.

Here we see, that the same figure expresses units of different values, according to the place which it occupies.

But $\frac{1}{10}$ of $\frac{4}{10}$ is equal to $\frac{4}{100} = .04.$
$\frac{1}{10}$ of $\frac{4}{100}$ " " $\frac{4}{1000} = .004.$
$\frac{1}{10}$ of $\frac{4}{1000}$ " " $\frac{4}{10000} = .0004.$
$\frac{1}{10}$ of $\frac{4}{10000}$ " " $\frac{4}{100000} = .00004.$
$\frac{1}{10}$ of $\frac{4}{100000}$ " " $\frac{4}{1000000} = .000004.$
$\frac{1}{10}$ of $\frac{4}{1000000}$ " " $\frac{4}{10000000} = .0000004.$

Therefore the value of the units of the different places, in passing from the left to the right, diminishes according to the scale of tens.

Hence, ten of the parts in any one of the places are equal to one of the parts in the place next to the left; that is, ten thousandths make one hundredth, ten hundredths make one-tenth, and ten tenths a unit 1.

This law of increase from the right hand towards the left, is the same as in whole numbers. *Therefore, whole numbers and decimal fractions may be united by placing the decimal point between them.* Thus,

---

QUEST.—143. Does the value of the unit of a figure depend upon the place which it occupies? How does the value of the unit change from the left towards the right? What do ten parts of any one place make? How do the units of place increase from the right towards the left? How may whole numbers be joined with decimals?

Whole numbers.									Decimals.						
Tens of millions.	Millions.	Hundreds of thousands.	Tens of thousands.	Thousands.	Hundreds.	Tens.	UNITS.		Tenths.	Hundredths.	Thousandths.	Tenths of thousandths.	Hundredths of thousandths.	Millionths.	Tenths of millionths.
8	3	6	3	0	6	4	1	.	0	4	7	8	9	7	6.

A number composed partly of a whole number and partly of a decimal, is called a mixed number.

Write the following numbers in figures, and numerate them.

1. Forty-one, and three tenths. 41.3.
2. Sixteen, and three millionths. 16.000003
3. Five, and nine hundredths. 5.09
4. Sixty-five, and fifteen thousandths.
5. Eighty, and three millionths.
6. Two, and three hundred millionths.
7. Four hundred and ninety-two thousandths.
8. Three thousand, and twenty-one ten-thousandths.
9. Forty-seven, and twenty-one ten-thousandths.
10. Fifteen hundred and three millionths.
11. Thirty-nine, and six hundred and forty thousandths.
12. Three thousand, eight hundred and forty millionths.
13. Six hundred and fifty thousandths.
14. Fifty thousand, and four hundredths.
15. Six hundred, and eighteen ten-thousandths.
16. Three millionths.
17. Thirty-nine hundred-thousandths.

144. The denominations of Federal Money will correspond to the decimal division, if we regard 1 dollar as the unit.

QUEST.—What is a number called when composed partly of whole numbers and partly of decimals? 144. If the denominations of Federal Money be expressed decimally, what is the unit?

For, the dimes are tenths of the dollar, the cents are hundredths of the dollar, and the mills, being tenths of the cent, are thousandths of the dollar.

EXAMPLES.

1. Express $17, 3 dimes 8 cents and 9 mills decimally.
2. Express $92, 8 dimes 9 cents 5 mills decimally.
3. Express $107, 9 dimes 6 cents 8 mills decimally.
4. Express $47 and 25 cents decimally.
5. Express $39, 39 cents and 7 mills decimally.
6. Express $12 and 3 mills decimally. *Ans.* ——
7. Express $147 and 4 cents decimally. *Ans.* ——
8. Express $148, 4 mills decimally. *Ans.* ——
9. Express four dollars, six mills decimally. *Ans.* ——
10. Express $14, 3 cents 9 mills decimally. *Ans.* ——
11. Express $149, 33 cents 2 mills decimally.
12. Express $1328, 5 mills decimally. *Ans.* ——
13. Express 9 dimes 4 mills decimally. *Ans.* ——
14. Express 5 cents 8 mills decimally. *Ans.* ——
15. Express $3856, 2 cents decimally. *Ans.* ——

145. A cipher is annexed to a number when it is placed on the right of it. If ciphers be annexed to the numerator of a decimal fraction, the same number of ciphers must also be annexed to the denominator; for there must always be as many ciphers in the denominator as there are places of figures in the numerator (Art. **142**). The numerator and denominator will therefore be multiplied by the same number, and consequently the value of the fraction will not be changed (Art. **102**). Hence,

Annexing ciphers to a decimal fraction does not alter its value.

QUEST.—What part of a dollar is one dime? What part of a dime is a cent? What part of a cent is a mill? What part of a dollar is 1 cent? 1 mill? 145. When is a cipher annexed to a number? Does the annexing of ciphers to a decimal alter its value? Why not? What do three-tenths become by annexing a cipher? What by annexing two ciphers?

We may take as an example the decimal $.3 = \frac{3}{10}$. If, now, we annex a cipher to the numerator, we must, at the same time, annex one to the denominator, which gives

$.30 = \frac{30}{100}$ by annexing one cipher,
$.300 = \frac{300}{1000}$ by annexing two ciphers,
$.3000 = \frac{3000}{10000}$ all of which are equal to $\frac{3}{10} = .3$.

Also, $.5 = \frac{5}{10} = .50 = \frac{50}{100} = .500 = \frac{500}{1000}$.

Also, $.8 = .80 = .800 = .8000 = .80000$.

146. Prefixing a cipher is placing it on the left of a number. If ciphers be prefixed to the numerator of a decimal fraction, that is, placed at the left hand of the significant figures, the same number of ciphers must be annexed to the denominator. Now, the numerator will remain unchanged while the denominator will be increased ten times for every cipher which is annexed, and the value of the fraction will be decreased in the same proportion (Art. 100). Hence,

Prefixing ciphers to a decimal fraction diminishes its value ten times for every cipher prefixed.

Take as an example the fraction $.2 = \frac{2}{10}$.

$.02 = \frac{02}{100}$ by prefixing one cipher,
$.002 = \frac{002}{1000}$ by prefixing two ciphers,
$.0002 = \frac{0002}{10000}$ by prefixing three ciphers:

in which the fraction is diminished ten times for every cipher prefixed.

Also, .03 becomes .003 by prefixing one cipher; and .0003 by prefixing two.

QUEST.—What does .8 become by annexing a cipher? By annexing two ciphers? By annexing three ciphers? 146. When is a cipher prefixed to a number? When prefixed to a decimal, does it increase the numerator? Does it increase the denominator? What effect then has it on the value of the fraction? What does .5 become by prefixing a cipher? By prefixing two ciphers? By prefixing three? What does .07 become by prefixing a cipher? By prefixing two? By prefixing three? By prefixing four?

ADDITION OF DECIMAL FRACTIONS.

147. It must be recollected that only like parts of unity can be added together, and therefore in setting down the numbers for addition, the figures occupying places of the same value must be placed in the same column.

The addition of decimal fractions is then made in the same manner as that of whole numbers.

Add 37.04, 704.3, and .0376 together.

In this example, we place the tenths under tenths, the hundredths under hundredths, and this brings the decimal points and the like parts of the unit directly under each other. We then add as in whole numbers.

OPERATION

```
  37.04
 704.3
    .0376
 --------
 741.3776
```

Hence, for addition of decimals,

I. *Set down the numbers to be added so that tenths shall fall under tenths, hundredths under hundredths, &c. This will bring all the decimal points under each other.*

II. *Then add as in simple numbers and point off in the sum, from the right hand, so many places for decimals as are equal to the greatest number of places in any of the added numbers.*

EXAMPLES.

1. Add 6.035, 763.196, 445.3741, and 91.5754 together.
2. Add 465.103113, .78012, 1.34976, .3549, and 61.11.
3. Add 57.406 + 97.004 + 4 + .6 + .06 + .3.
4. Add .0009 + 1.0436 + .4 + .05 + .047.
5. Add .0049 + 49.0426 + 37.0410 + 360.0039.
6. Add 5.714, 3.456, .543, 17.4957 together.

QUEST.—147. What parts of unity may be added together? How do you set down the numbers for addition? How will the decimal points fall? How do you then add? How many decimal places do you point off in the sum?

7. Add 3.754, 47.5, .00857, 37.5 together.

8. Add 54.34, .375, 14.795, 1.5 together.

9. Add 71.25, 1.749, 1759.5, 3.1 together.

10. Add 375.94, 5.732, 14.375, 1.5 together.

11. Add .005, .0057, 31.008, .00594 together.

12. Required the sum of 9 tenths, 19 hundredths, 18 thousandths, 211 hundred-thousandths, and 19 millionths.

13. Required the sum of twenty-nine and 3 tenths, four hundred and sixty-five, and two hundred and twenty-one thousandths.

14. Required the sum of two hundred dollars one dime three cents and nine mills, four hundred and forty dollars nine mills, and one dollar one dime and one mill.

15. What is the sum of one tenth, one hundredth, and one thousandth?

16. What is the sum of 4, and 6 ten-thousandths?

17. What is the sum of 3 thousandths, 9 millionths, 5 hundredths, 6 hundredths, 3 tenths, and 2 units?

18. Required, in dollars and decimals, the sum of one dollar one dime one cent one mill, six dollars three mills, four dollars eight cents, nine dollars six mills, one hundred dollars six dimes, nine dimes one mill, and eight dollars six cents.

19. What is the sum of 4 dollars 6 cents, 9 dollars 3 mills, 14 dollars 3 dimes 9 cents 1 mill, 104 dollars 9 dimes 9 cents 9 mills, 999 dollars 9 dimes 1 mill, 4 mills, 6 mills, and 1 mill?

SUBTRACTION OF DECIMAL FRACTIONS.

148. Subtraction of Decimal Fractions is the process of finding the difference between two decimal numbers.

1. From 3.275 take .0879.

In this example a cipher is annexed to the minuend to make the number of decimal places equal to the number in the subtrahend. This does not alter the value of the minuend (Art. 145).

OPERATION.

$$\begin{array}{r} 3.2750 \\ .0879 \\ \hline 3.1871 \end{array}$$

Hence, for the subtraction of decimal numbers,

I. *Set down the less number under the greater, so that figures occupying places of the same value shall fall in the same column.*

II. ***Then subtract as in simple numbers, and point off in the remainder, from the right hand, as many places for decimals as are equal to the greatest number of places in either of the given numbers***

EXAMPLES.

2. From 3278 take .0879. *Ans.* ——
3. From 291.10001 take 41.496. *Ans.* ——
4. From 10.00001 take .111111. *Ans.* ——
5. Required the difference between 57.49 and 5.768.
6. What is the difference between .3054 and 3.075?
7. Required the difference between 1745.3 and 173.45.
8. What is the difference between seven-tenths and 54 ten-thousandths?
9. What is the difference between .105 and 1.00075?
10. What is the difference between 150.43 and 754.355?
11. From 1754.754 take 375.49478. *Ans.* ——
12. Take 75.304 from 175.01. *Ans.* ——
13. Required the difference between 17.541 and 35.49.
14. Required the difference between 7 tenths and 7 millionths.
15. From 396 take 8 ten-thousandths. *Ans.* ——
16. From 1 take one-thousandth. *Ans.* ——
17. From 6374 take one-tenth. *Ans.* ——
18. From 365.0075 take 5 millionths. *Ans.* ——
19. From 21.004 take 98 ten-thousandths. *Ans.* ——
20. From 260.3609 take 47 ten-millionths. *Ans.* ——
21. From 10.0302 take 19 millionths. *Ans.* ——
22. From 2.03 take 6 ten-thousandths *Ans.* ——

QUEST.—148. What does subtraction teach? How do you set down the numbers for subtraction? How do you then subtract? How many decimal places do you point off in the remainder?

MULTIPLICATION OF DECIMAL FRACTIONS.

149.—1. Multiply .37 by .8.

We may first write $.37 = \frac{37}{100}$, and $.8 = \frac{8}{10}$.

If, now, we multiply the fraction $\frac{37}{100}$ by $\frac{8}{10}$, we find the product to be $\frac{296}{1000}$; the number of ciphers in the denominator of this product is equal to the number of decimal places in the two factors, and the same will be true for any two factors whatever.

OPERATION.

$$\begin{aligned} .37 &= \tfrac{37}{100}. \\ .8 &= \tfrac{8}{10}. \\ \hline .296 &= \tfrac{296}{1000}. \\ &= .296. \end{aligned}$$

2. Multiply .3 by .02.

OPERATION.

$$.3 \times .02 = \tfrac{3}{10} \times \tfrac{02}{100} = \tfrac{6}{1000} = .006 \text{ answer.}$$

Now, to express the 6 thousandths decimally, we have to prefix two ciphers to the 6, and this makes as many decimal places in the product as there are in both multiplicand and multiplier.

Therefore, to multiply one decimal by another,

Multiply as in simple numbers, and point off in the product, from the right hand, as many figures for decimals as are equal to the number of decimal places in the multiplicand and multiplier; and if there be not so many in the product, supply the deficiency by prefixing ciphers.

EXAMPLES.

1. Multiply 3.049 by .012. *Ans.* .036588.

(2.)		(3.)	
Multiply	365.491	Multiply	496.0135
by	.901	by	1.496
Ans.		*Ans.*	

QUEST.—149. After multiplying, how many decimal places will you point off in the product? When there are not so many in the product, what do you do? Give the rule for the multiplication of decimals.

4. Multiply one and one millionth by one thousandth.

5. Multiply 473.54 by .057. *Ans.* ——

6. Multiply 137.549 by 75.437. *Ans.* ——

7. Multiply 3.7495 by 73487. *Ans.* ——

8. Multiply .04375 by .47134. *Ans.* ——

9. Multiply .371343 by 75493. *Ans.* ——

10. Multiply 49.0754 by 3.5714. *Ans.* ——

11. Multiply .573005 by .000754. *Ans.* ——

12. Multiply .375494 by 574.375. *Ans.* ——

13. Multiply two hundred and ninety-four millionths, by one millionth.

14. Multiply three hundred, and twenty-seven hundredths by 62. *Ans.* ——

15. Multiply 93.01401 by 10.03962. *Ans.* ——

16. What is the product of five-tenths by five-tenths?

17. What is the product of five-tenths by five thousandths?

18. Multiply 596.04 by 0.000012. *Ans.* ——

19. Multiply 38049.079 by 0.000016. *Ans.* ——

20. Multiply 1192.08 by 0.000024. *Ans.* ——

21. Multiply 76098.158 by 0.000032. *Ans.* ——

CONTRACTION IN MULTIPLICATION.

150. CONTRACTION in the multiplication of decimals is a short method of finding the product of two decimal numbers in such a manner, that it shall contain but a given number of decimal places.

1. Let it be required to find the product of 2.38645 multiplied by 38.2175, in such a manner that it shall contain but four decimal places.

In this example it is proposed to take the multiplicand 2.38645, 38 times, then 2 tenths times, then 1 hundredth times, then 7 thousandth times, then 5 ten-thousandth times,

---

QUEST.—150. What is contraction in the multiplication of decimals? What is proposed in the example? How are the numbers written down for multiplication?

and the sum of these several products will be the product sought.

Write the unit figure of the multiplier directly under that place of the multiplicand which is to be retained in the product, and the remaining places of integer numbers, if any, to the right, and then write the decimal places to the left in their order, tenths, hundredths, &c.

OPERATION.

```
  2.38645
  5712.83
  -------
   715935
   190916
     4773
      239
      167
       12
  -------
  91.2042
```

When the numbers are so written, the product of any figure in the multiplier by the figure of the multiplicand directly over it, will be of the same order of value as the last figure to be retained in the product. Therefore, the first figure of each product is always to be arranged directly under the last retained figure of the multiplicand. But it is the whole of the multiplicand which should be multiplied by each figure of the multiplier, and not a part of it only. Hence, to compensate for the part omitted, we begin with the figure to the right of the one directly over any multiplier, and carry one when the product is greater than 5 and less than 15, 2 when it falls between 15 and 25, 3 when it falls between 25 and 35, and so on for the higher numbers.

For example, when we multiply by the 8, instead of saying 8 times 4 are 32, and writing down the 2, we say first, 8 times 5 are 40, and then carry 4 to the product 32, which gives 36. So, when we multiply by the last figure 5, we first say, 5 times 3 are 15, then 5 times 2 are 10 and 2 to carry make 12, which is written down.

EXAMPLES.

1. Multiply 36.74637 by 127.0463, retaining three decimal places in the product.

QUEST.—When the numbers are so written, what will be the order of value of the product of any figure of the multiplier by the figure directly over it? Where then is the first figure by each product to be written? How do you compensate for the part omitted?

CONTRACTION.	COMMON WAY.
36.74637	36.74637
3640.721	127.0463
3674637	11023911
734927	22047822
257224	14698548
1470	25722459
220	7349274
11	3674637
4668.489	4668.490346931

2. Multiply 54.7494367 by 4.714753, reserving five places of decimals in the product.

3. Multiply 475.710564 by .3416494, retaining three decimal places in the product.

4. Multiply 3754.4078 by .734576, retaining five decimal places in the product.

5. Multiply 4745.679 by 751.4549, and reserve only whole numbers in the product.

151. Note.—When a decimal number is to be multiplied by 10, 100, 1000, &c., the multiplication may be made by removing the decimal point as many places to the right hand as there are ciphers in the multiplier; and if there be not so many figures on the right of the decimal point, supply the deficiency by annexing ciphers.

Thus, 6.79 multiplied by	=
10	67.9
100	679.
1000	6790.
10000	67900.
100000	679000.

Also, 370.036 multiplied by	=
10	3700.36
100	37003.6
1000	370036.
10000	3700360.
100000	37003600.

Quest.—151. How do you multiply a decimal number by 10, 100, 1000, &c.? If there are not as many decimal figures as there are ciphers in the multiplier, what do you do?

DIVISION OF DECIMAL FRACTIONS.

152. Division of Decimal Fractions is similar to that of simple numbers.

We have just seen that, if one decimal fraction be multiplied by another, the product will contain as many places of decimals as there were in both the factors. Now, if this product be divided by one of the factors, the quotient will be the other factor (Art. 79). Hence, in division, the dividend must contain just as many decimal places as the divisor and quotient together. *The quotient, therefore, will contain as many places as the dividend, less the number in the divisor.*

EXAMPLES.

1. Divide 1.38483 by 60.21.

There are five decimal places in the dividend, and two in the divisor: there must therefore be three places in the quotient: hence one 0 must be prefixed to the 23, and the decimal point placed before it.

OPERATION.

```
60.21)1.38483(23
      1.2042
      ------
       18063
       18063
      ------
  Ans. .023.
```

Hence, for the division of decimals,

Divide as in simple numbers, and point off in the quotient, from the right hand, so many places for decimals as the decimal places in the dividend exceed those in the divisor; and if there are not so many, supply the deficiency by prefixing ciphers.

2. Divide 4.6842 by 2.11. *Ans.* ——
3. Divide 12.82561 by 1.505. *Ans.* ——
4. Divide 33.66431 by 1.01. *Ans.* ——

Quest.—152. If one decimal fraction be multiplied by another, how many decimal places will there be in the product? How does the number of decimal places in the dividend compare with those in the divisor and quotient? How do you determine the number of decimal places in the quotient? If the divisor contains four places and the dividend six, how many in the quotient? If the divisor contains three places and the dividend five, how many in the quotient? Give the rule for the division of decimals.

5. Divide .010001 by .01. *Ans.* ——

6. Divide 24.8410 by .002. *Ans.* ——

7. What is the quotient of 75.15204, divided by 3? By .3? By .03? By .003? By .0003?

8. What is the quotient of 389.27688, divided by 8? By .08? By .008? By .0008? By .00008?

9. What is the quotient of 374.598, divided by 9? By .9? By .09? By .009? By .0009? By .00009?

10. What is the quotient of 1528.4086488, divided by 6? By .06? By .006? By .0006? By .00006? By .000006?

11. Divide 17.543275 by 125.7. *Ans.* ——

12. Divide 1437.5435 by .7493. *Ans.* ——

13. Divide .000177089 by .0374. *Ans.* ——

14. Divide 1674.35520 by 960. *Ans.* ——

15. Divide 120463.2000 by 1728. *Ans.* ——

16. Divide 47.54936 by 34.75. *Ans.* ——

17. Divide 74.35716 by .00573. *Ans.* ——

18. Divide .37545987 by 75.714. *Ans.* ——

153. Note 1.—When any decimal number is to be divided by 10, 100, 1000, &c., the division is made by removing the decimal point as many places to the left as there are 0's in the divisor; and if there be not so many figures on the left of the decimal point, the deficiency must be supplied by prefixing ciphers.

$$27.69 \text{ divided by } \left\{\begin{array}{l}10\\100\\1000\\10000\end{array}\right\} = \left\{\begin{array}{r}2.769\\.2769\\.02769\\.002769\end{array}\right.$$

$$642.89 \text{ divided by } \left\{\begin{array}{l}10\\100\\1000\\10000\\100000\end{array}\right\} = \left\{\begin{array}{r}64.289\\6.4289\\.64289\\.064289\\.0064289\end{array}\right.$$

Quest.—153. How do you divide a decimal number by 10, 100, 1000, &c.? If there be not as many figures to the left of the decimal point as there are ciphers in the divisor, what do you do?

154. NOTE 2.—When there are more decimal places in the divisor than in the dividend, annex as many ciphers to the dividend as are necessary to make its decimal places equal to those of the divisor; *all the figures of the quotient will then be whole numbers.* Always bear in mind that *the quotient is as many times greater than unity, as the dividend is greater than the divisor.*

EXAMPLES.

1. Divide 4397.4 by 3.49.

We annex one 0 to the dividend. Had it contained no decimal place we should have annexed two.

OPERATION.

```
3.49)4397.40(1260
     349
     ---
     907
     698
     ----
     2094
     2094
     ----
     Ans. 1260.
```

2. Divide 1097.01097 by .100001. *Ans.* ——
3. Divide 9811.0047 by .1629735. *Ans.* ——
4. Divide .1 by .0001. *Ans.* ——
5. Divide 10 by .1. *Ans.* ——
6. Divide 6 by .6. By .06. By .006. By .2. By .3 By .003. By .5. By .005. By .000012.

155. NOTE 3.—When it is necessary to continue the division farther than the figures of the dividend will allow, we may annex ciphers to it, and consider them as decimal places.

EXAMPLES.

1. Divide 4.25 by 1.25.

In this example, after having exhausted the decimals of the dividend, we annex an 0, and then the decimal places used in the dividend will exceed those in the divisor by 1.

OPERATION.

```
1.25)4.25(3.4
     3.75
     ----
      500
      500
     ----
Ans.  3.4
```

QUEST.—154. If there are more decimal places in the divisor than in the dividend, what do you do? What will the figures of the quotient then be? 155. How do you continue the division after you have brought down all the figures of the dividend?

2. Divide .2 by .06.

We see in this example that the division will never terminate. In such cases the division should be carried to the third or fourth place, which will give the answer true enough for all practical purposes, and the sign + should then be written, to show that the division may still be continued.

OPERATION.

```
.06).20(3.333 +
     18
     --
      20
      18
      --
       20
       18
       --
        20
Ans. 3.333+.
```

3. Divide 37.4 by 4.5. *Ans.* ——

4. Divide 586.4 by 375. *Ans.* ——

5. Divide 94.0369 by 81.032. *Ans.* ——

REMARKS.

156. The unit of Federal Money, the currency of the United States, is one dollar, and all the lower denominations, dimes, cents, and mills, are decimals of the dollar. Hence, all the operations upon Federal Money are the same as the corresponding operations on decimal fractions.

APPLICATIONS IN THE FOUR PRECEDING RULES.

1. A merchant sold 4 parcels of cloth; the 1st contained 239 and 3 thousandths yards; the 2d, 6 and 5 tenths yards; the 3d, 4 and one hundredth yards; the 4th, 90 and one millionth yards: how many yards did he sell in all?

2. A merchant buys three chests of tea; the first contains 70 and one thousandth *lb.*; the second, 49 and one ten-thousandth *lb.*; the third, 36 and one-tenth *lb.*: how much did he buy in all?

3. What is the sum of $20 and three hundredths; $44 and one-tenth, $6 and one thousandth, and $18 and one hundredth?

QUEST.—When the division does not terminate, what sign do you place after the quotient? What does it show? 156. What is the unit of the currency of the United States? What parts of this unit are the inferior denominations, dimes, cents, and mills?

4. A puts in trade $1504.342; B puts in $350.1965; C puts in $100.11; D puts in $99.334; and E puts in $9001.31: what is the whole amount put in?

5. B has $936, and A has $5, 3 dimes, and 1 mill: how much more money has B than A?

6. A merchant buys 112.5 yards of cloth, at one dollar twenty-five cents per yard: how much does the whole come to?

7. A farmer sells to a merchant 13.12 cords of wood at $4.25 per cord, and 17 bushels of wheat at $1.06 per bushel: he is to take in payment 13 yards of broadcloth at $4.07 per yard, and the remainder in cash: how much money did he receive?

8. If 11 men had each $339 1 dime 9 cents and 3 mills, what would be the total amount of their money?

9. If one man can remove 5.91 cubic yards of earth in a day, how much could 38 men remove?

10. What is the cost of 24.9 yards of cloth, at $5.47 per yard?

11. If a man earns one dollar and one mill per day, how much will he earn in a year?

12. What will be the cost of 675 thousandths of a cord of wood, at $2 per cord?

13. A farmer purchased a farm containing 56 acres of woodland, for which he paid $46.347 per acre; 176 acres of meadow land at the rate of $59.465 per acre; besides which there was a swamp on the farm that covered 37 acres, for which he was charged $13.836 per acre. What was the area of the land; what its cost; and what the average price per acre?

14. A person dying has $8345 in cash, and 6 houses valued at $4379.837 each; he ordered his debts to be paid, amounting to $3976.480, and $120 to be expended at his funeral; the residue was to be divided among his five sons in the following manner: the eldest was to have a fourth part, and each of the other sons to have equal shares. What was the share of each son?

CONTRACTION IN DIVISION.

157. Contraction in division is a short method of obtaining the quotient of one decimal number divided by another.

EXAMPLES

1. Divide 754.347385 by 61.34775, and let the quotient contain three places of decimals.

COMMON METHOD.

61.34775)754.34738500(12.296

61347	75
14086	988
12269	550
1817	4385
1226	9550
590	48350
552	12975
38	353750
36	808650
1	545100

CONTRACTED METHOD.

61.34775)754.347385(12.296

61348
14086
12269
1817
1227
590
552
38
37
1

It is plain that all the work by the common method, which stands on the right of the vertical line, does not affect the quotient figures. On what principle is the work omitted in the contracted method?

In every division, the first figure of the quotient will always be of the same order of value as that figure of the dividend under which is written the product of the first figure of the quotient by the unit's figure of the divisor.

Having determined the order of value of the first quotient figure, make use of as many figures of the divisor as you wish places of figures in the quotient.

Let each remainder be a new dividend, and in each following division omit one figure to the right hand of the divisor,

---

QUEST—157. What is contraction in division? In every division, what will be the order of the first quotient figure? How many figures of the divisor will you use? How will you then make the division?

observing to carry for the increase of the figures cut off, as in contraction of multiplication.

In the example above, the order of the first quotient figure was obviously tens; hence, as there were three decimal places required in the quotient, five figures of the divisor must be used.

2. Divide 59 by .74571345, and let the quotient contain four places of decimals.

3. Divide 17493.407704962 by 495.783269, and let the quotient contain four places of decimals.

4. Divide 98.1874З7 by 8.4765618, and let the quotient contain ten places of decimals.

5. Divide 47194.379457 by 14.73495, and let the quotient contain as many decimal places as there will be integers in it.

REDUCTION OF VULGAR FRACTIONS TO DECIMALS.

158. The value of every vulgar fraction is equal to the quotient arising from dividing the numerator by the denominator (Art. 94).

EXAMPLES.

1. What is the value in decimals of $\frac{9}{2}$?

We first divide 9 by 2, which gives a quotient 4, and 1 for a remainder. Now, 1 is equal to 10 tenths. If, then, we add a cipher, 2 will divide 10, giving the quotient 5 tenths. Hence, the true quotient is 4.5.

OPERATION.

$\frac{9}{2} = 4\frac{1}{2}$; but

$4\frac{1}{2} = 4\frac{10}{2} = 4.5.$

2. What is the value of $\frac{13}{4}$?

We first divide by 4, which gives a quotient 3 and a remainder 1. But 1 is equal to 100 hundredths. If, then, we add two ciphers, 4 will divide the 100, giving a quotient of 25 hundredths.

OPERATION.

$\frac{13}{4} = 3\frac{1}{4}$; but

$3\frac{1}{4} = 3\frac{100}{4} = 3.25.$

QUEST.—What is the order of the first quotient figure in Ex. 2? In 3? In 4? 158. What is the value of a fraction equal to? What is the value of four-halves?

Hence, to reduce a vulgar fraction to a decimal,

I. *Annex one or more ciphers to the numerator and then divide by the denominator.*

II. *If there is a remainder, annex a cipher or ciphers, and divide again, and continue to annex ciphers and to divide until there is no remainder, or until the quotient is sufficiently exact: the number of decimal places to be pointed off in the quotient is the same as the number of ciphers used; and when there are not so many, ciphers must be prefixed to supply the deficiency.*

EXAMPLES.

1. Reduce $\frac{635}{125}$ to its equivalent decimal.

We here use two ciphers, and therefore point off two decimal places in the quotient.

OPERATION.

```
125)635(5.08
    625
    ----
    1000
    1000
    ----
```

2. Reduce $\frac{1}{8}$ and $\frac{18}{1129}$ to decimals. *Ans.* ——
3. Reduce $\frac{6}{480}$, $\frac{27}{39}$, $\frac{3}{10000}$, and $\frac{11}{60006}$ to decimals.
4. Reduce $\frac{1}{2}$ and $\frac{5}{1785}$ to decimals. *Ans.* ——
5. Reduce $\frac{314957123}{210456891}$ to a decimal. *Ans.* ——
6. Reduce $\frac{8}{6}$, $\frac{1375}{8436}$, $\frac{3265}{4121}$, $\frac{574}{123}$ to decimals. *Ans.* ——
7. Reduce $\frac{30}{1280}$ to decimals. *Ans.* ——
8. Reduce $\frac{347}{2560}$ to decimals. *Ans.* ——
9. Reduce $\frac{3}{10000}$ to decimals. *Ans.* ——
10. Reduce $\frac{3476}{15625}$ to deeimals. *Ans.* ——
11. Reduce $\frac{1}{2048000}$ to decimals. *Ans.* ——
12. Reduce $\frac{6}{7}$ to decimals. *Ans.* ——
13. Reduce $\frac{15}{34}$ to decimals. *Ans.* ——
14. Reduce $\frac{9}{35}$ to decimals. *Ans.* ——
15. Reduce $\frac{17}{19}$ to decimals. *Ans.* ——
16. Reduce $\frac{19}{2365}$ to decimals. *Ans.* ——
17. Reduce $\frac{1412}{5907}$ to decimals. *Ans.* ——

QUEST.—What is the decimal value of one-half? Of three-fourths? Of six-fourths? Of nine-halves? Of seven-halves? Of five-fourths? Of one-fourth? Give the rule for reducing a vulgar fraction to a decimal

18. What is the decimal value of $\frac{2}{3}$ of $\frac{3}{5}$ multiplied by $\frac{5}{12}$?

19. What is the value in decimals of $\frac{1}{2}$ of $\frac{2}{3}$ of $\frac{7}{8}$ divided by $\frac{3}{8}$ of $\frac{2}{4}$?

20. A man owns $\frac{7}{8}$ of a ship; he sells $\frac{4}{22}$ of his share what part is that of the whole, expressed in decimals?

21. Bought $\frac{11}{30}$ of $87\frac{3}{11}$ bushels of wheat for $\frac{9}{20}$ of 7 dollars a bushel: how much did it come to, expressed in decimals?

22. If a man receives $\frac{8}{5}$ of a dollar at one time, $7\frac{1}{2}$ at another, and $8\frac{3}{4}$ at a third: how many in all, expressed in decimals?

23. What decimal is equal to $\frac{5}{4}$ of 18, and $\frac{8}{11}$ of $\frac{11}{12}$ of $7\frac{4}{11}$ added together?

24. What decimal is equal to $\frac{2}{3}$ of 6 taken from $\frac{3}{5}$ of $8\frac{3}{4}$?

25. What decimal is equal to $\frac{21}{22}$, $\frac{13}{7}$, $\frac{2}{9}$, added together?

REDUCTION OF DENOMINATE DECIMALS.

159. We have seen that a denominate number is one in which the *kind* of unit is denominated or expressed (Art. 14).

A denominate decimal is a decimal fraction in which the kind of unit that has been divided is expressed. Thus, .5 of a £, and .6 of a shilling are denominate decimals: the unit that was divided in the first fraction being £1, and that in the second 1 shilling.

CASE I.

160. To find the value of a denominate number in decimals of a higher denomination.

1. Reduce 9*d*. to the decimal of a £.

We first find that there are 240 pence in £1. We then divide 9*d*. by 240, which gives the quotient .0375 of a £. This is the true value of 9*d*. in the decimal of a £.

OPERATION.

240*d*.=£1

240)9(.0375

Ans. £.0375.

QUEST.—159. What is a denominate number? What is a denominate decimal? In the decimal five-tenths of a £, what is the unit? In the decimal six-tenths of a shilling, what is the unit?

Hence, to make the reduction,

I. *Consider how many units of the given denomination make one unit of the denomination to which you would reduce.*

II. *Divide the given denominate number by the number so found, and the quotient will be the value in the required denomination.*

EXAMPLES.

1. Reduce 14 drams to the decimal of a *lb.* avoirdupois.
2. Reduce 78*d.* to the decimal of a £.
3. Reduce .056 poles to the decimal of an acre.
4. Reduce 42 minutes to the decimal of a day.
5. Reduce 63 pints to the decimal of a peck.
6. Reduce 9 hours to the decimal of a day.
7. Reduce 375678 feet to the decimal of a mile.
8. Reduce 72 yards to the decimal of a rod.
9. Reduce .5 quarts to the decimal of a barrel.
10. Reduce 4*ft.* 6*in.* to the decimal of a yard.
11. Reduce 7*oz.* 19*pwt.* of silver to the decimal of a pound.
12. Reduce 9½ months to the decimal of a year.
13. Reduce 62 days to the decimal of a year of 365¼ days.
14. Reduce £25 19*s.* 6½*d.* to the decimal of a pound.
15. Reduce 3*qr.* 21*lb.* to the decimal of a *cwt.*
16. Reduce 5*fur.* 36*rd.* 2*yd.* 2*ft.* 9*in.* to the decimal of a mile.
17. Reduce 4*cwt.* 2⅗*qr.* to the decimal of a ton.
18. Reduce 3*cwt.* 7*lb.* 8*oz.* to the decimal of a ton.
19. Reduce 17*hr.* 6*m.* 43*sec.* to the decimal of a day.

CASE II.

161. To reduce denominate numbers of different denominations to an equivalent decimal of a given denomination.

QUEST.—160. How do you find the value of a denominate number in a decimal of a higher denomination?

1. Reduce £1 4*s.* $9\frac{3}{4}$*d.* to the denomination of pounds.

We first reduce 3 farthings to the decimal of a penny, by dividing by 4. We then annex the quotient .75 to the 9 pence. We next divide by 12, giving .8125, which is the decimal of a shilling. This we annex to the shillings, and then divide by 20.

OPERATION.

$\frac{3}{4}$*d.* = .75*d.*; hence,
$9\frac{3}{4}$*d.* = 9.75*d.*
12)9.75*d.*
.8125*s.*, and
20)4.8125*s.*
£.240625; therefore,
£1 4*s.* $9\frac{3}{4}$*d.* = £1.240625.

Hence, to make the reduction,

Divide the lowest denomination named, by that number which makes one of the denomination next higher, annexing ciphers if necessary; then annex this quotient to the next higher denomination, and divide as before: proceed in the same manner through all the denominations to the last: the last result will be the answer sought.

EXAMPLES.

1. Reduce £19 17*s.* $3\frac{1}{4}$*d.* to the decimal of a £.
2. Reduce 46*s.* 6*d.* to the denomination of pounds.
3. Reduce $7\frac{1}{2}$*d.* to the decimal of a shilling.
4. Reduce 2*lb.* 5*oz.* 12*pwt.* 16*gr.* troy to the decimal of a *lb.*
5. Reduce 7 feet 6 inches to the denomination of yards.
6. Reduce 1*lb.* 12*dr.* avoirdupois to the denomination of pounds.
7. Reduce 10 leagues 4 furlongs to the denomination of leagues.
8. Reduce 7*s.* $5\frac{1}{2}$*d.* to the decimal of a pound.
9. What decimal part of a pound is three halfpence?
10. Reduce 4*s.* $7\frac{9}{11}$*d.* to the decimal of a pound.
11. Reduce 1*oz.* 11*pwt.* 3*gr.* to the decimal of a pound troy.

QUEST.—161. How do you reduce denominate numbers of different denominations to equivalent decimals of a given denomination?

12. Reduce 24 grains to the decimal of an ounce troy.

13. Reduce 5*oz.* 4*dr.* avoirdupois to the decimal of a pound troy.

14. Reduce 3*cwt.* 1*qr.* 14*lb.* to the decimal of a ton.

15. Reduce 2*qr.* 15*lb.* to the decimal of a hundred-weight.

16. Reduce 5*lb.* 10*oz.* 3*pwt.* 13*gr.* troy to the decimal of a hundred-weight avoirdupois.

17. Reduce 1*qr.* 1*na.* to the decimal of a yard.

18. Reduce 2*qr.* 3*na.* to the decimal of an English ell.

19. Reduce 2*yds.* 2*ft.* 6½*in.* to the decimal of a mile.

20. What decimal part of an acre is 1*R.* 37*P* ?

21. What decimal part of a hogshead of wine is 2 quarts 1 pint?

22. Reduce 3 bushels 3 pecks to the decimal of a chaldron of 36 bushels.

23. What decimal part of a year is 3*wk.* 6*da.* 7*hr.*, reckoning 365*da.* 6*hr.* a year?

24. Reduce 2.45 shillings to the decimal of a £.

25. Reduce 1.047 roods to the decimal of an acre.

26. Reduce 176.9 yards to the decimal of a mile.

CASE III.

162. To find the value of a denominate decimal in terms of integers of inferior denominations.

1. What is the value of .832296 of a £?

We first multiply the decimal by 20, which brings it to shillings, and after cutting off from the right as many places for decimals as in the given number, we have 16*s.* and the decimal .645920 over. This we reduce to pence by multiplying by 12, and then reduce to farthings by multiplying by 4.

OPERATION.

```
     .832296
          20
   16.645920
          12
    7.751040
           4
    3.004160
Ans. 16s. 7d. 3far.
```

Hence, to make the reduction,

I. *Consider how many in the next less denomination make one of the given denomination, and multiply the decimal by this number. Then cut off from the right hand as many places as there are in the given decimal.*

II. *Multiply the figures so cut off by the number which it takes in the next less denomination to make one of a higher, and cut off as before. Proceed in the same way to the lowest denomination: the figures to the left will form the answer sought*

EXAMPLES.

1. What is the value of .625 of a *cwt.*? *Ans.* ——
2. What is the value of .625 of a gallon? *Ans.* ——
3. What is the value of .004168*lb.* troy? *Ans.* ——
4. What is the value of .375 hogshead of beer?
5. What is the value of .375 of a year of 365 days?
6. What is the value of .085 of a £? *Ans.* ——
7. What is the value of .258 of a *cwt.*? *Ans.* ——
8. What is the difference between .82 of a day and .64 of an hour?
9. What is the value of 2.078 miles? *Ans.* ——
10. What is the value of £.3375? *Ans.* ——
11. What is the value of .3375 of a ton? *Ans.* ——
12. What is the value of .05 of an acre? *Ans.* ——
13. What is the value of .875 pipes of wine?
14. What is the value of .046875 of a pound, avoirdupois?
15. What is the value of .56986 of a year of $365\frac{1}{4}$ days?
16. What is the value of £2.092? *Ans.* ——
17. What is the value of £5.64? *Ans.* ——
18. What is the value of .36974 of a last, wool weight?
19. What is the value of .827364*qr.*, corn measure?
20. What is the value of .093765℔.? *Ans.* ——

QUEST.—162. How do you find the value of a denominate decimal in integers of inferior denominations? What is the value in shillings of one-half of a £? In pence of one-half of a shilling?

CIRCULATING OR REPEATING DECIMALS.

163. We have seen that in changing a vulgar into a decimal fraction, cases will arise in which the division does not terminate, and then the vulgar fraction cannot be exactly expressed by a decimal (Art. **158**).

Let it be required to reduce $\frac{5}{12}$ to its equivalent decimal.

We find the equivalent decimal to be .4166 + &c., giving 6's, as far as we choose to continue the division.

OPERATION.

$$12)\underline{50000}$$
$$.4166 +$$

The further the division is continued the nearer the decimal will approach to the true value of the vulgar fraction; and we express this approach to equality of value by saying, that if the division be continued *without limit*, that is, to *infinity*, the value of the decimal will then be equal to that of the vulgar fraction. Thus, we also say,

$$.999 +, \text{ continued to infinity} = 1,$$

because every annexation of a 9 brings the value nearer to 1.

164. Let us now examine the circumstances under which, in the reduction of a vulgar to a decimal fraction, the division will not terminate.

If the vulgar fraction be first reduced to its lowest terms, (which we suppose to be done in all the cases which follow,) there will be no factor common to its numerator and denominator. Now, by the addition of 0's to the numerator we may increase its value ten times for every 0 annexed; that is, we introduce into the numerator the two factors 2 and 5 for every

---

QUEST.—163. Can a vulgar fraction always be exactly expressed by a decimal? Can five-twelfths? If we continue the division, does the quotient approach to the true value? By what language do we express this fact? 164. In annexing a 0 to the numerator, what factors do we introduce into it?

additional 0. But the numerator can never be exactly divided by the denominator, if the denominator contains any prime factor not found in the numerator (Art. 107): hence it can never be so divided, if the denominator contains any prime factor other than 2 or 5. Hence, to determine whether a vulgar fraction in its lowest terms can be expressed by an exact decimal,

Decompose the denominator into its prime factors, and if there are any factors other than 2 or 5, the exact division cannot be made.

EXAMPLES.

1. Can $\frac{7}{25}$ be exactly expressed by decimals?

$25 = 5 \times 5$; hence, the fraction can be exactly expressed by a decimal.

OPERATION.

```
25) 70 (.28
    50
    200
    200
```

2. Can $\frac{5}{36}$ be exactly expressed by decimals?

$36 = 18 \times 2 = 9 \times 2 \times 2 = 3 \times 3 \times 2 \times 2$; in which we see that the denominator contains other factors than 2 and 5, and hence the fraction cannot be exactly expressed by decimals.

OPERATION.

```
36) 50 (.1388 +
    36
    140
    108
     320
     288
      320
      288
```

3. Can $\frac{9}{150}$ be exactly expressed by decimals?
4. Can $\frac{13}{148}$ be exactly expressed by decimals?
5. Can $\frac{11}{320}$ be exactly expressed by decimals?
6. Can $\frac{17}{1280}$ be exactly expressed by decimals?

QUEST.—Under what circumstances will the numerator be exactly divisible by the denominator? When not so? How do you determine whether a vulgar fraction can be exactly divisible by a decimal?

NOTE.—**165.** When there are no prime factors in the denominator other than 2 or 5, the division will always be exact, and the number of decimal places in the quotient will be equal to the greatest number of factors among the 2's or 5's.

7. What is the decimal corresponding to the fraction $\frac{7}{250}$?
8. What is the decimal corresponding to $\frac{11}{625}$?
9. What is the decimal corresponding to $\frac{17}{128}$?

166. The decimals which arise from vulgar fractions, where the division does not terminate, are called *circulating decimals*, because of the continual repetition of the same figures. The set of figures which repeats, is called a *repetend.*

167. A SINGLE REPETEND is one in which only a single figure repeats, as $\frac{2}{9}=.2222+$, or $\frac{3}{9}=.3333+$. Such repetends are expressed by simply putting a mark over the first figure; thus, .2222+ is denoted by .2̀ +, and .3333 + by .3̀ +.

168. A COMPOUND REPETEND has the same figures circulating alternately: thus $\frac{19}{33}=.5757+$ and $\frac{5723}{9999}=.57235723+$ are compound repetends, and are distinguished by marking the first and last figures of the circulating period. Thus .5757 + is written .5̀7′ +, and .57235723 + is written .5̀723′ +.

169. A PURE REPETEND is an expression in which there are no figures except the repeating figures which make up the repetend; as .3̀ +, .5̀ +, .4̀73′ +, &c.

170. A MIXED REPETEND is one which has significant figures or ciphers between the repetend and the decimal

---

QUEST.—165. If there are no prime factors in the denominator other than 2 and 5, will the division be exact? How many decimal places will there be in the quotient? 166. What are the decimals called when the division does not terminate? What is the set of figures which repeats called? 167. What is a single repetend? How is it expressed? 168. What is a compound repetend? How is it expressed? 169. What is a pure repetend? 170. What is a mixed repetend?

point, or which has whole numbers at the left hand of the decimal point: such figures are called finite figures. Thus, .0\`4+, .0\`733′+, .4\`73′+, .3\`573′+, 6\`.5, and 4.\`375′+ are all mixed repetends, .0, .4, .3, and 6 are the finite figures.

171. SIMILAR REPETENDS are such as begin at an equal distance from the decimal points; as .3\`54′+, 2.7\`534′+.

172. DISSIMILAR REPETENDS are such as begin at different places from the decimal point; as .\`253′+, .47\`52′+.

173. CONTERMINOUS REPETENDS are such as end at the same distance from the decimal points; as .1\`25′+, .\`354′+, &c.

174. SIMILAR AND CONTERMINOUS REPETENDS are such as begin and end at the same distance from the decimal point: thus, 53.2\`753′+, 4.6\`325′+, and .4\`632′+, are similar and conterminous repetends.

REDUCTION OF CIRCULATING DECIMALS.

CASE I.

175. To reduce a pure repetend to its equivalent vulgar fraction.

Since $\frac{1}{9}$ = .\`1 +, and $\frac{3}{9}$ = .\`3 +, and $\frac{54}{99}$ = .\`54′+; and since all repetends may be placed under similar forms; therefore, to find the finite value of a pure repetend,

Make the given repetend the numerator, and write a denominator containing as many 9's as there are places in the repetend, and this fraction reduced to its lowest terms will be the equivalent fraction sought.

QUEST.—What are such figures called? 171. What are similar repetends? 172. What are dissimilar repetends? 173. What are conterminous repetends? 174. What are similar and conterminous repetends? 175 How do you reduce a pure repetend to its equivalent vulgar fraction?

EXAMPLES.

1. What is the equivalent vulgar fraction of the repetend $0.\grave{3}+$?

Now, $\frac{3}{9}=\frac{1}{3}=0.33333+. = 0.\grave{3}+.$

2. What is the equivalent vulgar fraction of the repetend $.\grave{1}62'+$?

We have, $\frac{162}{999}=\frac{18}{111}$ *Ans.*

3. What are the simplest equivalent vulgar fractions of the repetends $.\grave{6}+$, $.\grave{1}62'+$, $0.\grave{7}69230'+$, $.\grave{9}45'+$, and $.\grave{0}9'+$?

4. What are the least equivalent vulgar fractions of the repetends $.\grave{5}94405'+$, $.\grave{3}6'+$, and $.\grave{1}42857'+$?

CASE II.

176. To reduce a mixed repetend to its equivalent vulgar fraction.

A mixed repetend is composed of the finite figures which precede, and of the repetend itself; and hence its value must be equal to such finite figures plus the repetend. Hence, to find such value,

To the finite figures add the repetend divided by as many 9's as it contains places of figures, with as many 0's annexed to them as there are places of decimal figures which precede the repetend; the sum reduced to its simplest form will be the equivalent fraction sought.

EXAMPLES.

1. Required the least equivalent vulgar fraction of the mixed repetend $2.4\grave{1}8'+$.

Now,

$2.4\grave{1}8'+=2+\frac{4}{10}+\frac{18}{1000}+=2+\frac{4}{10}+\frac{18}{990}=2\frac{23}{55}$ *Ans.*

QUEST.—176. How do you reduce a mixed repetend to its equivalent vulgar fraction?

2. Required the least equivalent vulgar fraction of the mixed repetend $.5\grave{9}25'+$.

We have, $.5\grave{9}25'+ = \frac{5}{10} + \frac{925}{9990} = \frac{16}{27}$ *Ans.*

3. What is the least equivalent vulgar fraction of the repetend $.008\grave{4}97133'+$?

We have, $.008\grave{4}97133'+ = \frac{8}{1000} + \frac{497133}{999999000} = \frac{83}{9768}$.

4. Required the least equivalent vulgar fractions of the mixed repetends $.13\grave{8}+$, $7.5\grave{4}3'+$, $.04\grave{3}54'+$, $37.5\grave{4}+$, $.6\grave{7}5'+$, and $.7\grave{5}4347'+$.

5. Required the least equivalent vulgar fractions of the mixed repetends $0.7\grave{5}+$, $0.4\grave{3}8'+$, $.09\grave{3}+$, $4.7\grave{5}43'+$, $.009\grave{8}7'+$, and $.4\grave{5}+$.

177. There are some properties of the repetends which it is important to remark.

1. Any finite decimal may be considered as a circulating decimal by making ciphers to recur: thus,

$.35 = .35\grave{0}+ = .35\grave{0}0'+ = .35\grave{0}00'+ = .35\grave{0}000'+$, &c.

2. If any circulating decimal have a repetend of any number of figures, it may be reduced to one having twice or thrice that number of figures, or any multiple of that number.

Thus, a repetend $2.3\grave{5}7'+$, having two figures, may be reduced to one having 4, 6, 8, or 10 places of figures. For, $2.3\grave{5}7'+ = 2.3\grave{5}757'+ = 2.3\grave{5}75757'+ = 2.3\grave{5}7575757'+$ &c.; so, the repetend $4.16\grave{3}16'+$ may be written $4.16\grave{3}16'+ = 4.16\grave{3}16316'+ = 4.16\grave{3}16316316'+$ &c. &c., and the same may be shown of any other. Hence, two or more repetends, having a different number of places in each, may be reduced to repetends having the same number of places, in the following manner:

QUEST.—177. What is the first property of the circulating decimals? How do you reduce several repetends having different places in each, to repetends having the same number of places?

Find the least common multiple of the number of places in each repetend, and reduce each repetend to such number of places.

1. Reduce $.13\dot{8}+$, $7.5\dot{4}\dot{3}+$, $.04\dot{3}5\dot{4}+$ to repetends having the same number of places.

Since the number of places are now 1, 2, and 3, the common multiple will be 6, and hence each new repetend will contain 6 places. Hence,

$$.13\dot{8}+=.13\dot{8}8888\dot{8}+;\quad 7.5\dot{4}\dot{3}+=7.5\dot{4}3434\dot{3}+;$$
$$0.4\dot{3}5\dot{4}+=0.4\dot{3}5435\dot{4}+.$$

2. Reduce $2.4\dot{1}\dot{8}+$, $.5\dot{9}2\dot{5}+$, $.008\dot{4}9713\dot{3}+$ to repetends having the same number of places.

3. Any circulating decimal may be transformed into another having finite decimals and a repetend of the *same* number of figures as the first. Thus,

$.\dot{5}\dot{7}+=.5\dot{7}\dot{5}+=.57\dot{5}\dot{7}+=.575\dot{7}\dot{5}+=.5757\dot{5}\dot{7}$; and
$3.4\dot{7}8\dot{5}+=3.47\dot{8}5\dot{7}+=3.478\dot{5}7\dot{8}+=3.4785\dot{7}8\dot{5}+$;
and hence, any two repetends may be made similar. These properties may be proved by changing the repetends into their equivalent vulgar fractions.

4. Having made two or more repetends similar by the last article, they may be rendered conterminous by the previous one: thus, *two or more repetends may always be made similar and conterminous.*

1. Reduce the circulating decimals $165.\dot{1}6\dot{4}+$, $.\dot{0}\dot{4}+$, $.03\dot{7}+$ to such as are similar and conterminous.

2. Reduce the circulating decimals $.5\dot{3}+$, $.4\dot{7}\dot{5}+$, and $1.\dot{7}5\dot{7}+$, to such as are similar and conterminous.

5. If two or more circulating decimals, having several

QUEST.—When a repetend has more than one figure, may it be transformed into a circulating decimal having finite decimals? How many places must there be in the repetend? What are similar and conterminous repetends? May all circulating decimals be made similar and conterminous?

repetends of equal places, be added together, their sum will have a repetend of the same number of places; for every two sets of repetends will give the same sum.

6. If any circulating decimal be multiplied by any number, the product will be a circulating decimal having the same number of places in the repetend; for, each repetend will be taken the same number of times, and consequently must produce the same product.

CASE III.

178. To find the number of places in the repetend corresponding to any vulgar fraction which cannot be expressed by a finite decimal.

Let the fraction be first reduced to its lowest terms, after which find all the prime factors 2 and 5 of the denominator. Then separate the fraction into two factors, viz.,

1st. The numerator divided by the product of all the prime factors 2 and 5; and

2d. Unity divided by the remaining factor of the denominator.

As an example, let us decompose the fraction $\frac{3}{280}$ into the two factors named above. They are,

$$\frac{3}{280} = \frac{3}{2 \times 2 \times 2 \times 5} \times \frac{1}{7}.$$

If, now, we add a 0 to the 1 and proceed to make the division, every remainder will be less than the divisor, and hence we cannot make more divisions than there are units in the divisor less 1, without reducing the remainder to unity, when the first quotient figures will repeat. And observe carefully when any remainder becomes the same as a remainder previously used, *for at this point the repeating figures begin.*

QUEST.—What is the fifth property named? What is the sixth? 178. What is the first operation in finding the form of the decimal corresponding to a given vulgar fraction? Into how many factors is it then divided? What are these factors? How many divisions may be performed in the second factor?

If, now, we suppose the remainder 1 to be subtracted from the dividend so used, there would remain as many 9's as there were divisions. Hence,

If, after having taken out the 2's and 5's from the denominator, we divide a succession of 9's by the result until there is no remainder, the number of 9's so used will be equal to the number of places of the repetend, which can never exceed the number of units in the denominator less one.

Having found the number of finite decimals which precede the repetend, and the number of places in the repetend, as above,

Divide the numerator of the vulgar fraction, reduced to its lowest terms, by the denominator, and point off in the quotient the finite decimals, if any, and the repetend.

EXAMPLES.

1. Required to find whether the decimal equivalent to $\frac{249}{29304}$ is finite or circulating; the number of places in the repetend and the place at which the repetend begins; and, also, the equivalent circulating decimal.

We first reduce the fraction to its lowest terms, giving $\frac{83}{9768}$. We then search for the factors 2 and 5 in the denominator, and find that 2 is a factor 3 times: hence we know that there are three finite decimals preceding the repetend. We next divide 99999, &c., by the factor 1221 of the denominator, and find that we use six nines before the remainder becomes 0: hence, we know that there are six places of figures in the

OPERATION.

$$\frac{249}{29304} = \frac{83}{9768}$$

```
2)9768
2)4884
2)2442
  1221

1221)999999(819
```

$$\frac{83}{9768} = .008\grave{4}97133' +$$

QUEST.—What will determine the highest limit of the number of figures in the repetend? What will determine the number of finite decimals? How then will you find the equivalent decimal?

repetend. We then divide 83 by 9768, and point off the proper places in the quotient.

2. Find the finite decimals, if any, and the repetend, if any, of the fraction $\frac{210}{1120}$.

3. Find the finite decimals, if any, and the repetend, if any, of the fraction $\frac{4}{1160}$.

4. Find the finite decimals, if any, and the repetend, if any, of the fractions $\frac{12}{123}$, $\frac{80}{135}$, $\frac{72}{135}$.

ADDITION OF CIRCULATING DECIMALS.

179. To add circulating decimals.

Make the repetends similar and conterminous, and then write the places of like value under each other, and so many figures of the second repetends to the right as shall indicate with certainty how many are to be carried from one repetend to the other; after which add them as in whole numbers. If all the figures of a repetend be 9's, omit them and add one to the figure next to the left.

EXAMPLES

1. Add $.12\grave{5}+$, $4.\grave{1}6\acute{3}+$, $1.\grave{7}14\acute{3}$, and $2.\grave{5}\acute{4}$ together.

Dissimilar.	*Similar.*	*Similar and conterminous.*	
$.12\grave{5}+$	$= .12\grave{5}+$	$= .12\grave{5}5555555555\acute{5}+$	. . 5555
$4.\grave{1}6\acute{3}+$	$= 4.16\grave{3}1\acute{6}+$	$= 4.16\grave{3}1631631631\acute{6}+$	. . 3163
$1.\grave{7}14\acute{3}+$	$= 1.71\grave{4}37\acute{1}+$	$= 1.71\grave{4}3714371437\acute{1}+$	. . 4371
$2.\grave{5}\acute{4}+$	$= 2.54\grave{5}\acute{4}+$	$= 2.54\grave{5}4545454545\acute{4}+$	. . 5454
	The true sum	$= 8.54\grave{8}5447013169\acute{7}+$, 1 to carry.	

2. Add $67.3\grave{4}\acute{5}+$, $9.\grave{6}5\acute{1}+$, $.\grave{2}\acute{5}+$, $17.4\grave{7}+$, $.\grave{5}+$ together.

3. Add $.\grave{4}7\acute{5}+$, $3.75\grave{4}\acute{3}+$, $64.\grave{7}\acute{5}+$, $.\grave{5}\acute{7}+$, $.1\grave{7}8\acute{8}+$ together.

QUEST—179. How do you add circulating decimals?

4. Add .`5 +, 4.3`7 +, 49.4`57′+, .4`954′+, .`7345′+ together.

5. Add .`175′+, 42.`57′+, .3`753′+, .5`945′+. 3.7`54′+ together.

6. Add 165, .`164′ +, 147.`04′ +, 4.`95′ +, 94.3`7 +, 4.`7123456′+ together.

SUBTRACTION OF CIRCULATING DECIMALS.

180. To subtract one finite decimal from another.

Make the repetends similar and conterminous, and subtract as in finite decimals, observing that when the repetend of the lower line is the largest its first right hand figure must be increased by unity.

EXAMPLES.

1. From 11.4`75′+ take 3.45`735′+.

Dissimilar.	*Similar.*	*Similar and conterminous.*	
11.4`75′+	= 11.47`57′+	= 11.47`575757′+	. . . 575
3.45`735′+ =	3.45`735′+	= 3.45`735735′+	. . . 735
	The true difference	= 8.01`840021′+, 1 to carry.	

2. From 47.5`3 + take 1.`757′+. *Ans.* ——
3. From 17.`573′+ take 14.5`7 +. *Ans.* ——
4. From 17.4`3 + take 12.34`3 +. *Ans.* ——
5. From 1.12`754′+ take .4`7384′+. *Ans.* ——
6. From 4.75 take .37`5 +. *Ans.* ——
7. From 4.794 take .1`744′+. *Ans.* ——
8. From 1.45`7 + take .3654. *Ans.* ——
9. From 1.4`937′+ take .1475. *Ans.* ——

QUEST.—180. How do you subtract circulating decimals?

MULTIPLICATION OF CIRCULATING DECIMALS.

181. To multiply one circulating decimal by another.

Change the circulating decimals into their equivalent vulgar fractions, and then multiply them together; after which reduce the product to its equivalent circulating decimal, as in Art. 178.

EXAMPLES.

1. Multiply $4.25\grave{3}+$ by .257.

OPERATION.

$$4.25\grave{3}+ = 4.\tfrac{25}{100} + \tfrac{3}{900} = 4 + \tfrac{225}{900} + \tfrac{3}{900} = \tfrac{228}{900} = \tfrac{3828}{900}$$
$$= \tfrac{1914}{450} = \tfrac{957}{225}.$$

Also, $.257 = \tfrac{257}{1000}$: hence,

$$\tfrac{957}{225} \times \tfrac{257}{1000} = \tfrac{245949}{225000} = 1.09310\grave{6}+;$$

and since $225000 = 5 \times 5 \times 5 \times 5 \times 5 \times 2 \times 2 \times 2 \times 9$, there will be five places of finite decimals, and one figure in the repetend (Art. 178).

NOTE. *Much labor will be saved in this and the next rule by keeping every fraction in its lowest terms, and when two fractions are to be multiplied together, cancelling all the factors common to both terms before making the multiplication.*

2. Multiply $.375\grave{4}+$ by 14.75. *Ans.* ——
3. Multiply $.4\grave{2}53'+$ by 2.57. *Ans.* ——
4. Multiply .437 by $3.7\grave{5}+$. *Ans.* ——
5. Multiply 4.573 by $.3\grave{7}5'+$. *Ans.* ——
6. Multiply $3.45\grave{6}+$ by $.42\grave{5}+$. *Ans.* ——
7. Multiply $1.\grave{4}56'+$ by $4.2\grave{3}+$. *Ans.* ——
8. Multiply $45.1\grave{3}+$ by $.\grave{2}45'+$. *Ans.* ——
9. Multiply $.4705\grave{3}+$ by $1.7\grave{3}5'+$. *Ans.* ——
10. Multiply $3.457\grave{3}+$ by $54.\grave{7}53'+$. *Ans.* ——

QUEST.—181. How do you multiply circulating decimals? What is to be observed in regard to keeping fractions in their lowest terms?

DIVISION OF CIRCULATING DECIMALS.

182. To divide one circulating decimal by another.

Change the decimals into their equivalent vulgar fractions, and find the quotient of these fractions. Then change the quotient into its equivalent decimal, as in Art. **178.**

EXAMPLES.

1. Divide $56.\grave{6}+$ by 137.

OPERATION.

$$56.\grave{6}+ = 56 + \tfrac{6}{9} = \tfrac{510}{9} = \tfrac{170}{3}.$$

Then, $\tfrac{170}{3} \div 137 = \tfrac{170}{3} \times \tfrac{1}{137} = \tfrac{170}{411} = .\grave{4}136253\acute{0}+.$

2. Divide $24.3\grave{1}\acute{8}+$ by 1.792. *Ans.* ——
3. Divide 8.5968 by $.2\grave{4}\acute{5}+$. *Ans.* ——
4. Divide 2.295 by $.\grave{2}9\acute{7}+$. *Ans.* ——
5. Divide 47.345 by $1.\grave{7}\acute{6}+$. *Ans.* ——
6. Divide $13.5\grave{1}6953\acute{3}+$ by $4.\grave{2}9\acute{7}+$. *Ans.* ——
7. Divide $.\grave{4}\acute{5}+$ by $.\grave{1}1888\acute{1}+$. *Ans.* ——
8. Divide $.\grave{4}7\acute{5}+$ by $.3\grave{7}5\acute{3}+$. *Ans.* ——
9. Divide $3.\grave{7}5\acute{3}$ by $.\grave{2}\acute{4}+$. *Ans.* ——

QUEST.—182. How do you divide circulating decimals?

OF THE RATIO AND PROPORTION OF NUMBERS

183. Two numbers having the same unit may be compared together in two ways.

1st. By considering *how much* one is greater or less than the other, which is shown by their difference; and

2d. By considering *how many times* one is greater or less than the other, which is shown by their quotient.

Thus, in comparing the numbers 3 and 12 together with respect to their difference, we find that 12 *exceeds* 3 by 9; and in comparing them together with respect to their quotient, we find that 12 contains 3 four times, or that 12 is four times as great as 3.

The quotient which arises from dividing the second number by the first, is called the *ratio* of the numbers, and shows how many times the second number is greater than the first, or how many times it is less.

Thus, the ratio of 3 to 9 is 3, since $9 \div 3 = 3$. The ratio of 2 to 4 is 2, since $4 \div 2 = 2$.

We may also compare a larger number with a less. For example, the ratio of 4 to 2 is $\frac{1}{2}$; for, $2 \div 4 = \frac{1}{2}$. The ratio of 9 to 3 is $\frac{1}{3}$, since $3 \div 9 = \frac{1}{3}$.

EXAMPLES.

1. What is the ratio of 9 to 18? *Ans.* ——
2. What is the ratio of 6 to 24? *Ans.* ——

QUEST.—183. In how many ways may two numbers having the same unit be compared? How do you determine how much one number is greater than another? How do you determine how many times it is greater or less? How much does 12 exceed 3? How many times is 12 greater than 3? What is the quotient called which arises from dividing the second number by the first? What does it show? When the second number is the least, what does it show?

3. What is the ratio of 12 to 48? *Ans.* ——
4. What is the ratio of 11 to 13? *Ans.* ——
5. What part of 20 is 4? Or what is the ratio of 20 to 4?
6. What part of 100 is 30? Or what is the ratio of 100 to 30?
7. What part of 6 is 3? *Ans.* ——
8. What part of 9 is 3? *Ans.* ——
9. What part of 12 is 4? *Ans.* ——
10. What part of 50 is 5? *Ans.* ——
11. What part of 75 is 3? *Ans.* ——

NOTE.—In determining *what part* one number is of another, it is plain that the number to *be measured*, must be written in the numerator; while the *standard*, or number with which it is compared, and of which it forms a part, is written in the denominator. *This fraction, reduced to its lowest terms, will express the part.*

184. If one yard of cloth cost $2, how many dollars will 6 yards of cloth cost at the same rate?

It is plain that 6 yards of cloth will cost 6 times as much as one yard; that is, the cost will contain $2 as many times as 6 contains 1. Hence the cost will be $12.

In this example there are four numbers considered, viz., 1 yard of cloth, 6 yards of cloth, $2, and $12: these numbers are called *terms*.

1 yard of cloth is the	1st term,
6 yards of cloth is the	2d term,
$2 is the - - -	3d term,
$12 is the - - -	4th term.

Now the ratio of the first term to the second is the same as the ratio of the third to the fourth.

This relation between four numbers is called a *proportion*; and generally

Four numbers are said to be in proportion when the ratio of

QUEST.—How do you determine what part one number is of another? 184. If one yard of cloth cost $2, what will 6 yards cost? How many numbers are here considered? What are they called? What is the ratio of the first to the second equal to? What is this relation between numbers called? When are four numbers said to be in proportion?

the first to the second is the same as that of the third to the fourth. Hence,

A PROPORTION *is an equality of ratios between numbers compared together two and two.*

185. We express that four numbers are in proportion thus:

1 : 6 : : 2 : 12.

That is, we write the numbers in the same line and place two dots between the 1st and 2d terms, four between the 2d and 3d, and two between the 3d and 4th terms. We read the proportion thus,

as 1 is to 6, so is 2 to 12.

The 1st and 2d terms of a proportion always express quantities of the same kind, and so likewise do the 3d and 4th terms. As in the example,

yd.		*yd.*		$		$
1	:	6	: :	2	:	12.

This is implied by the definition of a ratio; for, it is only quantities of the same kind which can be divided the one by the other. The ratio of the first term to the second, or of the third to the fourth, is called the ratio of the proportion.

1. What are the ratios of the proportions

3	:	9	: :	12	:	36?	*Ans.* ——
2	:	10	: :	12	:	60?	*Ans.* ——
4	:	2	: :	8	:	4?	*Ans.* ——
9	:	1	: :	90	:	10?	*Ans.* ——
16	:	15	: :	48	:	45?	*Ans.* ——

186. When two numbers are compared together, the first is called the *antecedent*, and the second the *consequent;* and when four numbers are compared, the first antecedent and consequent are called the *first couplet*, and the second antecedent and consequent the *second couplet.* Thus, in the last

---

QUEST.—How do you define proportion? 185. How do you indicate that four numbers are in proportion? How is the proportion read? What do you remark of the first and second terms? Also of the third and fourth? 186. When two numbers are compared together, what is the first called? What the second? When four numbers are compared, what are the two first called? What the two second?

proportion, 16 and 48 are the antecedents, and 15 and 45 the consequents; also, 16 and 15 make the first couplet, and 48 and 45 the second.

187. We have said that proportion is an equality of ratios (Art. 184). Besides the method above, we may express that equality thus:

$$\frac{4}{2}=\frac{6}{3};$$

and we may then write the proportion thus:

$$2 : 4 :: 3 : 6.$$

Put the following equal ratios into proportion.

1. $\frac{8}{9}=\frac{16}{18}$.
2. $\frac{17}{51}=\frac{19}{57}$.
3. $\frac{9}{16}=\frac{27}{48}$.
4. $\frac{19}{13}=\frac{76}{52}$.
5. $\frac{21}{16}=\frac{105}{80}$.
6. $\frac{42}{35}=\frac{252}{210}$.
7. $\frac{29}{37}=\frac{232}{296}$.
8. $\frac{45}{23}=\frac{405}{207}$.

188. If 4*lb.* of tea cost $8, what will 12*lb.* cost at the same rate?

OPERATION.

lb. *lb.* $ $
As 4 : 12 : : 8 : *Ans.*

```
 12
4)96
$24 the cost of 12lb. of tea.
```

$$\frac{12}{4}\times 8 = 3\times 8 = 24.$$

Ans. $24.

It is evident that the 4th term, or cost of 12*lb.* of tea, must be as many times greater than $8, as 12*lb.* is greater than 4*lb.* But the ratio of 4*lb.* to 12*lb.* is 3; hence, 3 is the number of times which the cost exceeds $8: that is, the cost is

---

QUEST.—187. What has proportion been called? By what second method may this equality be expressed? 188. Explain this example mentally.

equal to \$8 × 3 = \$24. But instead of writing the numbers

$$\frac{12}{4} \times 8 = 24,$$

we may write them

$$(12 \times 8) \div 4 = 24:$$

and as the same may be shown for every proportion, we conclude,

That the 4th term of every proportion may be found by multiplying the 2d and 3d terms together, and dividing their product by the 1st term.

EXAMPLES.

1. The first three terms of a proportion are 1, 2, and 3: what is the fourth? *Ans.* ——
2. The first three terms are 6, 2, and 1: what is the 4th? *Ans.* ——
3. The first three terms are 10, 3, and 1: what is the 4th? *Ans.* ——

189. The 1st and 4th terms of a proportion are called the two extremes, and the 2d and 3d terms are called the two means.

Now, since the 4th term is obtained by dividing the product of the 2d and 3d terms by the 1st term, and since the product of the divisor by the quotient is equal to the dividend, it follows,

That in every proportion the product of the two extremes is equal to the product of the two means.

Thus, in the example, Art. **184** we have

	1 : 6 : : 2 : 12;	and	1 × 12 = 2 × 6;
also,	4 : 12 : : 8 : 24;	and	4 × 24 = 12 × 8;
"	6 : 9 : : 10 : 15;	and	6 × 15 = 9 × 10;
"	7 : 15 : : 14 : 30;	and	7 × 30 = 15 × 14.

QUEST.—How may the fourth term of every proportion be found? 189. What are the first and fourth terms of a proportion called? What are the second and third terms called? In every proportion, what is the product of the extremes equal to?

OF CANCELLING.

190. When one number is to be divided by another, the operation may often be shortened by striking out or cancelling the factors common to both, before the division is made.

1. For example, suppose it were required to divide 360 by 120.

We first write the dividend above a horizontal line, and the divisor beneath it, after the form of a fraction. We next separate both of them into factors, and then cancel the factors which are alike.

OPERATION.

$$\frac{360}{120} = \frac{12 \times 30}{12 \times 10} = \frac{\not{12} \times 3 \times \not{10}}{\not{12} \times \not{10}} = 3.$$

2. Divide 630 by 35.

We separate the dividend and divisor into like factors, and then cancel those which are common in both.

OPERATION.

$$\frac{630}{35} = \frac{3 \times \not{5} \times 6 \times \not{7}}{\not{5} \times \not{7}} = 18.$$

3. Divide 1860 by 36.
4 Divide 7920 by 720.
5. Divide 1890 by 210.
6. Divide 1260 by 504.
7. Divide 1768 by 221.
8. Divide 2856 by 238.
9. Divide 3420 by 285.
10. Divide 9072 by 1512.

191. If two or more numbers are to be multiplied together and their product divided by the product of other numbers, the operation may be abridged by *striking out* or *cancelling* any factor which is common to the dividend and divisor. For example, if 6 is to be multiplied by 8 and the product divided by 4, we have

$$\frac{6 \times 8}{4} = \frac{48}{4} = 12; \text{ or, } \frac{6 \times 8}{4} = 6 \times 2 = 12:$$

QUEST.—190. How may the division of two numbers be often abridged? Explain the example mentally. Also the second example. 191. When two numbers are multiplied together and their product divided by a third how may the operation be abridged?

in the latter case we cancelled the factor 4 in the numerator and denominator, and multiplied 6 by the quotient 2.

1. Let it be required to multiply 24 by 16 and divide the product by 12.

OPERATION.

$$\frac{\overset{2}{\not{24}} \times 16}{\underset{1}{\not{12}}} = 32.$$

Having written the product of the figures, which form the dividend, above the line, and the product of the figures which form the divisor below it, then

We cancel the common factors in the numerator and denominator, and write the quotients over and under the numbers in which such common factors are found, and if the quotients still have a common factor, they may be again divided.

2. Reduce the compound fraction $\frac{4}{5}$ of $\frac{6}{9}$ of $\frac{3}{12}$ of $\frac{5}{16}$ to a simple fraction.

OPERATION.

$$\frac{\overset{1}{\not{4}}}{\underset{1}{\not{5}}} \times \frac{\overset{1}{\not{6}}}{\underset{3}{\not{9}}} \times \frac{\overset{1}{\not{3}}}{\underset{2}{\not{12}}} \times \frac{\overset{1}{\not{5}}}{\underset{4}{\not{16}}} = \frac{1}{24}$$

Beginning with the first numerator, we find it is once a factor of itself and 4 times in 16; 6 is twice a factor in 12; 3 three times a factor in 9; and 5, once a factor in the denominator 5.

3. What is the product of 3 × 8 × 9 × 7 × 15 divided by 63 × 24 × 3 × 5?

OPERATION.

$$\frac{\not{3} \times \not{8} \times \not{9} \times \not{7} \times \not{15}}{\not{63} \times \not{24} \times \not{3} \times \not{5}} = 1.$$

This example presents a case that often arises, in which the *product* of two factors may be cancelled. Thus, 3 × 8 is 24: then cancel the 3 and 8 in the numerator and the 24 in the denominator. Again, 9 times 7 are 63; therefore cancel the 9 and 7 in the numerator and the 63 in the denominator. Also, 3 × 5 in the denominator cancels the 15 remaining in the numerator: hence, the quotient is unity.

4. What is the product of 126 × 16 × 3 divided by 7 × 12?

We see that 7 is a factor of 126, giving a quotient 18, which we place over 126, crossing at the same time 126 and the 7 below. We then divide 18 and 12 by 6, crossing them both and writing down the quotients 3 and 2. We next divide 16 and 2 by 2, giving the quotients 8 and 1. Hence, the result is 72.

OPERATION.

$$\frac{\overset{\overset{3}{\cancel{18}}}{\cancel{126}} \times \overset{8}{\cancel{16}} \times 3}{\cancel{7} \times \underset{\underset{1}{\cancel{2}}}{\cancel{12}}} = 72.$$

EXAMPLES.

1. What is the product of $1 \times 6 \times 9 \times 14 \times 15 \times 7 \times 8$ divided by $36 \times 128 \times 56 \times 20$?

2. What is the value of $18 \times 36 \times 72 \times 144$ divided by $6 \times 6 \times 8 \times 9 \times 12 \times 8$?

3. What is the product of $3 \times 9 \times 7 \times 3 \times 14 \times 36$ divided by $252 \times 81 \times 2 \times 7$?

4. What is the product of $19 \times 17 \times 16 \times 8 \times 9 \times 6$ divided by $32 \times 4 \times 27 \times 2$?

5. What is the product of $4 \times 12 \times 16 \times 30 \times 16 \times 48 \times 48$ divided by $9 \times 10 \times 14 \times 24 \times 44 \times 40$?

192. The process of cancelling may be applied to the terms of a proportion.

If we have any proportion, as

$$6 : 15 :: 28 : 70,$$

We may always cancel like factors in either couplet. Thus,

$$\overset{2}{\cancel{6}} : \overset{5}{\cancel{15}} :: \overset{14}{\cancel{28}} : \overset{35}{\cancel{70}},$$

in which we divide the terms of the first couplet by 3, and those of the second by 2, and write the quotients above.

EXAMPLES.

1. What is the simplest form of 18 : 72 : : 100 : 400?
2. What is the simplest form of 14 : 49 : : 42 : 147?
3. What is the simplest form of 365 : 876 : : 140 : 336?

QUEST.—192. How else may the process of cancelling be applied? What may be cancelled in each couplet?

RULE OF THREE.

193. THE Rule of Three takes its name from the circumstance that three numbers are always given to find a fourth, which shall bear the same proportion to one of the given numbers as exists between the other two.

The following is the manner of finding the fourth term:

I. *Reduce the two numbers which have different names from the answer sought, to the lowest denomination named in either of them.*

II. *Set the number which is of the same kind with the answer sought in the third place, and then consider from the nature of the question whether the answer will be greater or less than the third term.*

III. *When the answer is greater than the third term, write the least of the remaining numbers in the first place, but when it is less place the greater there.*

IV. *Then multiply the second and third terms together, and divide the product by the first term: the quotient will be the fourth term or answer sought, and will be of the same denomination as the third term.*

EXAMPLES.

1. If 48 yards of cloth cost $67,25 what will 144 yards cost at the same rate?

QUEST.—193. From what does the Rule of Three take its name? What is the first thing to be done in stating the question? Which number do you make the third term? How do you determine which to put in the first? After stating the question, how do you find the fourth term? What will be its denomination?

In this example, as the answer is to be dollars, we place the $67,25 in the third place. Then, as 144 yards of cloth will cost more than 48 yards, the fourth term must be greater than the third, and therefore, we write the least of the two remaining numbers in the first place. The product of the second and third terms is $9684,00: dividing this by the first term, we obtain $201,75 for the cost of 144 yards of cloth.

OPERATION.

```
yd.   yd.        $       $
48 : 144 : : 67,25 : Ans
               144
             26900
            26900
            6725
        48)9684,00($201,75
           96
             84
             48
             360
             336
              240
              240
```

2. If 6 men can dig a certain ditch in 40 days, how many days would 30 men be employed in digging it?

As the answer must be days, the 40 days are written in the third place. Then, as it is evident that 30 men will do the same work in a shorter time than 6 men, it is plain that the fourth term must be less than the third; therefore, 30 men, the greater of the remaining numbers, is taken as the first term. Besides, it is plain that the fourth term must be just so many times less than 40, as 6 is less than 30.

OPERATION.

```
men  men.  days   days
30 : 6 : : 40  :  Ans.
            6
       3|0)24|0 days.
       Ans.  8  days.
```

3. If 25 yards of cloth cost £2 3*s.* 4*d.*, what will 5 yards cost at the same rate?

QUEST.—In the first example which is greater, the third or fourth term? Which number must then be in the first term? How many times will the fourth term be greater or less than the third?

When we come to divide the product of the second and third terms by the first, it is found the £10 does not contain 25. We then reduce to the next lower denomination, and divide as in division of denominate numbers.

OPERATION.

```
yd.   yd.       £  s.  d.
25  :  5  : :  2  3   4  :  Ans
                      5
          25)£10 16s. 8d.
               20
          25)216(8s.
             200
              16
              12
          25)200(8d.
             200
                        Ans. 8s. 8d.
```

4. If 3*cwt.* of sugar cost £9 2*s.* 0*d.*, what will 4*cwt.* 3*qr* 26*lb.* cost at the same rate?

```
          3cwt.    4cwt. 3qr. 26lb.
            4        4                   £9 2s. 0d.
           12       19                      20
         (  7        7                     182s.
4×7 = 28 { 84      133                      12
         (  4        4                     2184
         336lb. : 558lb. : : 2184d. : Ans.
```

We first reduce the first and second terms to pounds, then the third term to pence. The answer comes out in pence, and is afterwards reduced to pounds, shillings, and pence.

```
           558
         17472
        10920
       10920
336)1218672(3627d.
    1008
     2106             12)3627
     2016             20) 302s. 3d.
      907                 £15 2s.
      672
     2352
     2352
                    Ans. £15 2s. 3d.
```

PROOF.

194. The product of the two means is equal to the product of the extremes (Art. **189**). Hence, if either of these equal products be divided by one of the mean terms the quotient will be the other. Therefore,

Divide the product of the extremes by one of the mean terms, and if the work is right the quotient will be the other mean term.

EXAMPLES.

1. The first term is 4, the second 8, the third 12, and the answer 24 : is the answer true ?

The product of the extremes is 96. If this be divided by 8 the quotient is 12 ; if by 12 the quotient is 8 : hence, the answer is right.

OPERATION OF PROOF.

$24 \times 4 = 96$

8)96(12 ;

or, 12)96(8

APPLICATIONS.

1. If 8 hats cost $24, what will 110 cost at the same rate ?

2. What is the value of 4*cwt.* of sugar at 5*d.* per pound ?

3. If 80 yards of cloth cost $340, what will 650 yards cost ?

4. If 120 sheep yield 330 pounds of wool, how many pounds will be obtained from 1200 ?

5. If 6 gallons of molasses cost $1,95, what will 6 hogsheads cost ?

6. If 16 men perform a piece of work in 24 days, how many men would it take to perform the work in 12 days ?

7. Suppose a cistern has two pipes, and that one can fill it in $8\frac{1}{2}$ hours, the other in $4\frac{3}{4}$: in what time can both fill it together ?

8. If a man travels at the rate of 630 miles in 12 days, how far will he travel in a year, supposing him not to travel on Sundays ?

QUEST.—194. How do you prove the Rule of Three ?

9. If 2 yards of cloth cost $3,25, what will be the cost of 3 pieces, each containing 25 yards?

10. If 30 barrels of flour will support 100 men for 40 days, how long would it subsist 25 men?

11. If 30 barrels of flour will support 100 men for 40 days, how long would it subsist 200 men?

12. If 50 persons consume 600 bushels of wheat in a year, how much will they consume in 7 years?

13. What will be the cost of a piece of silver weighing 73*lb.* 5*oz.* 15*pwt.*, at 5*s.* 9*d.* per ounce?

14. If the penny loaf weighs 8 ounces when the bushel of wheat costs 7*s.* 3*d.*, what ought it to weigh when the wheat is 8*s.* 4*d.* per bushel?

15. If one acre of land costs £2 15*s.* 4*d.*, what will be the cost of 173*A.* 2*R.* 14*P.* at the same rate?

16. A gentleman's estate is worth £4215 4*s.* a year: what may he spend per day and yet save £1000 per annum?

17. A father left his son a fortune, $\frac{1}{4}$ of which he ran through in 8 months, $\frac{3}{7}$ of the remainder lasted him 12 months longer, when he had barely £820 left: what sum did his father leave him?

18. There are 1000 men besieged in a town with provisions for 5 weeks, allowing each man 16 ounces a day. If they are reinforced by 500 more and no relief can be afforded till the end of 8 weeks, how many ounces must be given daily to each man?

19. A father gave $\frac{7}{18}$ of his estate to one son, and $\frac{7}{18}$ of the remainder to another, leaving the rest to his widow. The difference of the children's legacies was £514 6*s.* 8*d.*: what was the widow's portion?

20. If 14*cwt.* 2*qr.* of sugar cost $129,92, what will be the price of 9*cwt.*?

21. The clothing of a regiment of foot of 750 men amounts to £2831 5*s.*: what will it cost to clothe a body of 10500 men?

22. How many yards of carpeting, that is 3 feet wide, will cover a floor that is 40 feet long and 27 feet broad?

23. After laying out $\frac{1}{4}$ of my money, and $\frac{1}{5}$ of the remainder, I had 114 guineas left: how much had I at first?

24. A reservoir has three pipes, the first can fill it in 24 days, the second in 22 days, and the third can empty it in 28 days: in what time will it be filled if they are all running together?

25. If the freight of 80 tierces of sugar, each weighing $3\frac{1}{3}$*cwt.*, 150 miles, cost $84, what must be paid for the freight of 30*hhd.* of sugar, each weighing 12*cwt.*, 50 miles?

26. If 1500 men require 45000 rations of bread for a month, how many rations will a garrison of 3600 men require?

27. The quick step in marching is 2 paces per second, at 28 inches each: at what rate is that per hour, and how long will a troop be in reaching a place 60 miles distant, allowing a halt of an hour and a half for refreshment?

28. Two persons A and B are on the opposite sides of a wood which is 536 yards in circumference; they begin to travel in the same direction at the same moment; A goes at the rate of 11 yards per minute, and B at the rate of 34 yards in 3 minutes: how many times must the quicker one go round the wood before he overtakes the slower?

29. Two men and a boy were engaged to do a piece of work, one of the men could do it in 10 days, the other in 16 days, and the boy could do it in 20 days: how long would it take the three together to do it?

30. A certain amount of provisions will subsist an army of 9000 men for 90 days. If the army be increased by 6000, how long will the same provisions subsist it?

31. Four thousand soldiers were supplied with bread for 24 weeks, each man to receive 14*oz.* per day; but by some accident 210 barrels containing 200*lb.* each were spoiled: what must each man receive in order that the remainder may last the same time?

32. Let us suppose the 4000 soldiers having one-fourteenth of their bread spoiled, to be put on an allowance of 13*oz.* of bread per day for 24 weeks: required the weight of their bread, good and spoiled, and the amount spoiled.

33. If 56 yards of cloth cost 40 guineas, how many ells Flemish can be bought for £1135 10*s.*?

34. If a pack of wool weighs 2*cwt.* 2*qr.* 14*lb.*, what is it worth at 17*s.* 6*d.* per tod?

35. A merchant bought a quantity of broadcloth and baize for £124; there was $117\frac{1}{2}$ yards of broadcloth at 17*s.* 9*d.* per yard; for every 5 yards of broadcloth he had $1\frac{1}{2}$ yards of baize; how many yards of baize did he buy, and what did it cost him per yard?

36. If $\frac{1}{9}$ of a pole stands in the mud, 1 foot in the water, and $\frac{5}{6}$ in the air, or above the water, what is the length of the pole?

37. A bankrupt's effects amount to $1000\frac{1}{2}$ guineas. His debts amount to £2547 14*s.* 9*d.*: what will his creditors receive in the pound?

38. If 12 dozen copies of a certain book cost $54,72, what will 297 copies cost at the same rate?

39. Suppose 4000 soldiers after losing 210 barrels of bread, each containing 200*lb.*, were to subsist on 13*oz.* each a day for 24 weeks; had none been lost they might have received 14*oz.* a day: what was the whole weight, and how much did they receive?

40. Let us now suppose 4000 soldiers to lose one-fourteenth of their bread, then to receive 13*oz.* each a day for 24 weeks: what was the whole weight of their bread including the lost, and how much would each have received per day had none been spoiled?

41. Provisions in a garrison were sufficient for 1800 men for 12 months; but at the end of 3 months it was reinforced by 600 men, and 4 months after that a second reinforcement of 400 men was sent in. How long did the provisions last?

RULE OF THREE BY ANALYSIS.

195. The solution of questions in the Rule of Three by analysis consists in finding the ratio of two of the given terms, and multiplying this ratio by the other term.

The ratio of two of the terms will generally express the value or cost of a single thing.

EXAMPLES.

1. If 3 barrels of flour cost $24, what will 7 barrels cost?

By dividing the $24 by 3 we get the cost of 1 barrel. For, if $24 will buy 3 barrels, it is plain that $\frac{1}{3}$ of it will buy 1 barrel. This, multiplied by 7, gives $56 the cost of 7 barrels.

OPERATION

3)24
8
$8 \times 7 = 56$
Ans. $56.

2. If in 20 days a man travels 58 miles, how far will he travel in 30 days?

3. If 6 men consume 1 barrel of flour in 30 days, how much would 48 men consume?

It is evident that $\frac{1}{6}$ of a barrel would be the amount consumed by 1 man: hence, 48 times $\frac{1}{6}$ is the amount consumed by 48 men.

OPERATION.

$\frac{1}{6} \times 48 = 8.$
Ans. 8.

4. If 2 barrels of flour cost $13, what will 12 barrels cost?

5. If I walk 168 miles in 6 days, how far should I walk at the same rate in 18?

6. If 8*lb.* of sugar cost $1,28, how much will 13*lb.* cost? What is 16 × 13?

7. If $\frac{3}{4}$ of a piece of cloth cost $8,25, what will $\frac{9}{4}$ cost?

8. If 300 barrels of flour cost $570, what will 200 cost? What is $\frac{2}{3} \times 570$?

9. If $\frac{8}{7}$ of a barrel of cider cost $\frac{9}{11}$ of a dollar, what will $\frac{3}{5}$ cost? What is $\frac{21}{40} \times \frac{9}{11}$?

QUEST.—195. In what does the solution of questions by analysis consist? What does the ratio of the two terms express? If this ratio be multiplied by the other term, what will be the product?

10 If 90 bushels of oats will feed 40 horses for 6 days, how long would 450 bushels feed them?

11. If 5 oxen, or 7 colts, eat up a certain quantity of grass, in 87 days, in what time will 2 oxen and 3 colts eat up the same quantity of grass?

12. A person's income is £146 per annum: how much is that each day?

13. If 3 paces of common steps be equal to 2 yards, how many yards will 160 paces make?

14. A certain work can be done in 12 days by working 4 hours each day: how long would it require to do the work by working 9 hours a day?

15. A pasture of a certain extent having supplied a body of horse, consisting of 3000, with forage for 18 days, how many days would the same pasture have supplied a body of 2000 horse?

16. The governor of a besieged city has provisions for 54 days, at the rate of 2*lb.* of bread per day, but is desirous of prolonging the siege to 80 days, in expectation of succor: in that case what must be the allowance of bread per day?

17. If a person pays half a guinea a week for his board, how long can he be boarded for £21?

18. If a person drinks 80 bottles of wine per month, when it costs 2*s.* per bottle, how much can he drink, without increasing the expense, when it costs 2*s.* 6*d.* per bottle?

19. How long will a person be in saving £100, if he saves 1*s.* 6*d.* per week?

20. A merchant bought 21 pieces of cloth, each containing 41 yards, for which he paid $1260; he sold the cloth at $1,75 per yard: did he gain or lose by the bargain?

21. A grocer bought a puncheon of rum for £41 14*s.* 6*d.*, to which he added as much water as reduced its cost to 5*s.* 6*d.* per gallon; how much water did he put in?

22. If one pound of tea be equal in value to 50 oranges, and 70 oranges be worth 84 lemons, what is the value of a pound of tea when a lemon is worth two cents?

RULE OF THREE BY CANCELLING.

196. The cancelling process may be applied to all ques tions in the Rule of Three, where the second or third terms have a factor common to the first. Let the second and third terms be written above the line, with the sign of multiplication between them, and the first term below it, and then cancel the common factors.

EXAMPLES.

1. If 48 yards of cloth cost $67,25, what will 144 yards cost?

OPERATION.

The process here is obvious, being entirely similar to that explained in Art. **191**.

$$\frac{67{,}25 \times \overset{3}{\not{1}\not{4}\not{4}}}{48} = \$201{,}75.$$

2. If 25 yards of cloth cost £2 3*s*. 4*d*., what will 5 yards cost?

3. If 24 hats cost $120, how much will 80 cost?

4. If 90 barrels of flour will subsist 100 men for 120 days, how long will it subsist 75?

5. If 60 sheep yield 180*lb*. of wool, how many pounds will be obtained from 900?

6. If a man travel 210 miles in 6 days, how far will he travel in 120 days?

7. If the freight of 40 tierces of sugar, each weighing $3\frac{1}{2}$*cwt*., 150 miles, cost $42, what must be paid for the freight of 10 hogsheads, each weighing 12*cwt*., 50 miles?

8. A certain amount of provisions will subsist an army of 9000 men for 90 days: if the army be increased by 4500, how long would the same provisions subsist it?

9. If 50 persons consume 600 bushels of wheat in one year, how much will 278 persons consume in 7 years?

10. If 3 yards of cloth cost 18*s*., what will 24 yards cost?

QUEST.—196. To what questions may the cancelling process be applied? How are the numbers written? What factors do you cancel?

11. If 112 pounds of sugar cost 56*s*., what will 1 pound cost?

12. If 4 men can do a piece of work in 80 days, how many days of the same length will 16 men require to do the same work?

13. If 21 pioneers make a trench in 18 days, how many days of the same length will 7 men require to make a similar trench?

14. If a field of grass be mowed by 10 men in 12 days, in how many days would it be mowed by 20 men?

15. A certain piece of grass was to be mowed by 20 men in 6 days; an extraordinary occasion calls off half the workmen. It is required to find in what time the rest will finish it.

16. If the penny loaf weighs 5*oz.* when flour is 2*s.* a peck, what should it weigh when flour is sold for 2*s.* 6*d.* a peck?

17. Provisions in a garrison are found sufficient to last 1800 soldiers for three months; but a reinforcement being wanted that the provisions may last for one month only, what number of soldiers must be added to the garrison?

18. If 3*yd.* 2*qr.* of cloth of 1*yd.* 3*qr.* wide will make a suit of clothes, how many yards of stuff of 3*qr.* wide will make a suit for the same person?

19. If I lend my friend £200 for 12 months on condition of his returning the favor, how long ought he to lend me £150 to requite my kindness?

20. If an acre be 220 yards long, the breadth will be 22 yards; but if the breadth of an acre be 40 yards, what then will be the length?

21. How many pounds of sugar at 12*d.* per pound are equal in value to 24*lb.* of tea, worth 9*s.* 6*d.* per pound?

22. A tax of £225 10*s.* was laid upon four villages A, B, C, and D; it had been the custom with these villages, that whenever any taxes were to be levied, as often as A, B, and C paid each 3*d.*, D paid only 2*d.*: how much did each village pay?

EXAMPLES INVOLVING FRACTIONS.

1. If $\frac{3}{8}$ of a yard of cloth cost $3,20, what will $2\frac{1}{2}$ yards cost?

We state the question exactly as in whole numbers. In multiplying the second and third terms together, we observe the rules for multiplying fractions, and in dividing by the first term, the rules for division. Thus, in this example, we invert the terms of the divisor and multiply.

OPERATION.

$$\frac{3}{8} : 2\frac{1}{2} :: 3,20 : Ans.$$

$$\begin{array}{lr} & 2\frac{1}{2} \\ \hline & 6,40 \\ \text{by multiplying by } \frac{1}{2} & 1,60 \\ \hline & 8,00 \end{array}$$

$$8,00 \div \frac{3}{8} = 8,00 \times \frac{8}{3} = \frac{64,00}{3} = \$21,33+.$$

2. If $\frac{5}{7}$*oz.* cost £$\frac{11}{12}$, what will $1\frac{1}{2}$*oz.* cost? *Ans.* ——

3. If $\frac{3}{16}$ of a ship cost £273 3*s.* 6*d.*, what will $\frac{15}{32}$ of her cost?

4. If 375 yards of cloth cost £164 9*d.* what will $257\frac{1}{2}$ yards cost?

5. If 14 yards of cloth can be bought for 10 guineas, how many ells Flemish can be bought for £283.875?

6. If $1\frac{3}{4}$*oz.* of plate cost 10*s.* $11\frac{1}{4}$*d.*, what will a service weighing 327.61875*oz.* cost?

7. If $14\frac{2}{3}$*lb.* of sugar cost $$1\frac{3}{4}$, what will 12*lb.* cost?

8. If $\frac{4}{5}$ of a yard of cloth cost $$1\frac{5}{9}$, what will $7\frac{1}{2}$ yards cost?

9. If 2 men can do 125 rods of ditching in 65 days, in how many days can 18 men do $242\frac{4}{13}$ rods?

10. If a wedge of gold weighing $17\frac{3}{7}$*lb.* troy, be worth £$679\frac{5}{7}$, what is the value of $1\frac{3}{13}$*qr.* of that gold?

11. If the carriage of 2.5 tons of goods 2.9 miles cost 0.75 guinea, what is that per *cwt.* for a mile?

12. If $\frac{1}{2}$*cwt.* of tobacco cost £4 18*s.*, how much may be bought for £7?

13. If $14\frac{1}{2}$ yards of cloth cost $\$19\frac{1}{3}$, how much will $39\frac{3}{8}$ yards cost?

14. If .3 of a house cost $201.5, what would .95 cost?

15. A man receives $\frac{3}{5}$ of his income, and finds it equal to $3724.16: how much is his whole income?

16. If 3.5 yards of cloth cost £2 14*s.* 3*d.*, what will 27.75 yards cost at the same rate?

17. If 12 men and a boy can perform a piece of work in $100\frac{3}{8}$ days, the boy doing $\frac{1}{2}$ as much work as one man, in how many days will 20 men perform the same?

18. A mercer bought $10\frac{1}{2}$ pieces of silk, each containing $24\frac{1}{3}$ yards; he paid 6*s.* $\frac{1}{2}$*d.* per yard: what does the whole come to?

19. If 4*lb.* of beef cost $\frac{3}{8}$ of a dollar, what will 60*lb.* cost?

20. If a traveller perform a journey in 35.5 days when the days are 13.625 hours long; in how many days of 11.9 hours would he perform the same journey?

21. If 5400 bricks be required to pave a yard, when the bricks are .5 foot long and .25 broad, how many will be required of .75 foot long and $\frac{1}{3}$ foot broad?

22. A man with his family, consisting of 5 persons, usually drink 7.8 gallons of beer in a week: how much would they drink in 23.5 weeks, if the family was to be increased by 3 persons?

23. If 248 men in $60\frac{1}{2}$ hours dig a trench containing $13924\frac{1}{6}$ solid yards of earth, how long would it take the same number of men to dig a similar trench containing 26460 solid yards of earth?

24. The earth turns round on its axis from west to east in 23 hours 56 minutes, and the circumference of every circle on its surface is supposed to be divided into 360 degrees. At the equator a degree is 69.07 miles; at Madras, Barbadoes, &c., 67.21 miles; at Madrid, Philadelphia, &c., 52.85 miles: and at Petersburg, 34.53 miles. How many miles per hour are the inhabitants in each of these places carried from west to east by the revolution of the earth on its axis?

OF QUESTIONS REQUIRING TWO STATEMENTS.

197. The answer to each of the questions heretofore considered, has been found by a single statement. Questions, however, frequently occur in which two or more statements will be necessary, if the question be resolved by the principles above explained.

EXAMPLES.

1. If a family of 6 persons expend $300 in 8 months, how much will serve a family of 15 persons for 20 months?

First question. If $300 will support a family of 6 persons for 8 months, how many dollars will support 15 persons for the same time?

1ST OPERATION.

```
persons. persons.     $        $
BY CANCELLING.
   2   :   5   : :   300   :  Ans.
                       5
                  2)1500
            Ans.  $750.
```

Second question. If $750 will support a family of 15 persons for 8 months, how much will serve them for 20 months?

2D OPERATION.

```
months. months.      $        $
   2   :   5   : :   750   :  Ans.
                       5
                  2)3750
            Ans.  $1875
```

2. If 16 men build 18 feet of wall in 12 days, how many men must be employed to build 72 feet in 8 days, working at the same rate?

The first question is, how long would it take the 16 men to build the 72 feet of wall?

OPERATION.

```
feet.   feet.      days.    days.
  1   :   4   : :   12   :  Ans.
                     4
                    48 days.
```

It is evident that 18 feet of wall, is to 72 feet, as 12 days,

---

QUEST.—197. How many statements have been necessary in the questions heretofore considered? What other questions frequently occur?

the time necessary to build 18 feet, is to 48 days, the time necessary to build 72 feet.

The second question is, if 16 men can build 72 feet of wall in 48 days, how many men are necessary to build it in 8 days?

Make 16 men the third term. Then as the same work is to be done in less time, more men will be necessary; therefore, the fourth term will be greater than the third, and hence 8 days are placed in the first term (Art. **193**).

OPERATION.

days.		*days.*		*men.*		*men.*
8	:	48	: :	16	:	*Ans.*
				48		
				128		
				64		
				8)768		
			Ans.	96 *men.*		

3. If a man travel 217 miles in 7 days, travelling 6 hours a day, how far would he travel in 9 days, if he travelled 11 hours a day?

1ST OPERATION.

days.		*days.*		*miles.*		*miles.*
7	:	9	: :	217	:	279
				9		
				7)1953		
				279		

2D OPERATION.

hours.		*hours.*		*miles.*		*miles.*
6	:	11	: :	279	:	$511\frac{3}{6}$
				11		
				6)3069		
				$511\frac{3}{6}$		
				Ans. $511\frac{3}{6}$ *miles*		

DOUBLE RULE OF THREE.

198. The last three questions, and all similar ones involving five, seven, or even nine terms, have generally been classed under a separate rule, called the DOUBLE RULE OF THREE, or COMPOUND PROPORTION. They may be thus stated and resolved:

I. *Make the first statement as though the question were to be solved by two or more statements by the Single Rule of Three, and suppose the fourth term to be found.*

QUEST.—198. Under what rule have questions similar to the last three been classed? How may they be stated and resolved? Give the rule.

II. *If it is of the same name with the answer sought, mark its place blank under the third term; if not, mark its place under the second term, and in either case arrange the two remaining terms as though it were a question in the Single Rule of Three. If there are more than five terms in the question, suppose the fourth term of the second proportion to be found, and make the third statement in the same manner as the second was made.*

III. *Then multiply the second and third terms together, and divide their product by the product of the first terms, and the quotient will be the answer sought.*

EXAMPLES.

Let us first resolve each of the last three questions by this rule.

1. If a family of 6 persons expend $300 in 8 months, how much will serve a family of 15 persons for 20 months?

OPERATION.

persons.		*persons.*		*$*		
6	:	15	: :	300	:	1*st answer*
months.		*months.*				
8	:	20	: :	1*st ans.*	:	*true answer*

$$\frac{15 \times \overset{5}{\cancel{20}} \times \overset{\overset{25}{\cancel{50}}}{\cancel{300}}}{\cancel{6} \times \cancel{8}} = 15 \times 5 \times 25 = \$1875.$$

Having made the first statement, we see that the fourth term is of the same name with the answer sought, and that if this term be placed in the second proportion, the true answer will be found. But since the first answer is equal to the product of the second and third terms divided by the first, it is plain that the true answer will always be equal to the continued product of the second and third terms divided by the product of the first terms, and similarly when there are more than five terms. In the operation we first cancel the 6 in the 300, then the 4 from 20, and then the 2 from the 50 over the 300.

2. If 16 men build 18 feet of wall in 12 days, how many men must be employed to build 72 feet in 8 days, working at the same rate?

OPERATION.

feet.		*feet.*		*days.*		
18	:	72	: :	12	:	1*st answer.*
days.		*days.*		*men.*		
8	:	—	: :	16	:	*true answer.*

Then, $\dfrac{\overset{4}{\cancel{72}} \times 12 \times \overset{2}{\cancel{16}}}{\cancel{18} \times \cancel{8}} = 4 \times 12 \times 2 = 96$ *Ans.*

3. If a man travel 217 miles in 7 days, travelling 6 hours a day, how far would he travel in 9 days, if he travelled 11 hours a day?

OPERATION.

days.		*days.*		*miles.*		*miles.*
7	:	$\overset{3}{\cancel{9}}$	: :	217	:	1*st answer.*
$\underset{2}{\cancel{6}}$	:	11	: :	—	:	*true answer.*

4. If 4 compositors, in 16 days of 12 hours long, can compose 14 sheets of 24 pages each sheet, 44 lines in a page, and 40 letters in a line; in how many days of 10 hours long will 9 compositors compose a volume consisting of 30 sheets, 16 pages in a sheet, 48 lines in a page, and 45 letters in a line?

The number of letters set by the first compositors is expressed by $14 \times 24 \times 44 \times 40$; and the letters to be set by the second by $30 \times 16 \times 48 \times 45$.

OPERATION.

com.		*com.*		*days.*		*Ans.*
9	:	4	: :	16	:	1*st answer.*
hours.		*hours.*				
10	:	12	: :	—	:	2*d answer.*
14×24×44×40	:	30×16×48×45	: :	—	:	*true answer.*

$$\frac{4\times12\times16\times30\times16\times48\times45}{9\times10\times14\times24\times44\times40}=\frac{4\times3\times16\times3\times2}{7\times11}=$$

$$=\frac{1152}{77}=14\tfrac{74}{77}\text{ days}=\textit{Ans.}$$

Let us now analyze this statement. Had the compositors worked the *same number of hours* per day, and had *the same work to do*, the first would have been the true answer; and the second would have been the true answer had the time only been different and the work to be done been the same. The third proportion accounts for the inequality of the work done, and gives the answer under all the suppositions. It is evident the same answer would have been obtained, had the first answer been substituted in the second proportion, and the second answer in the third proportion. Hence, the reason of the rule is obvious.

5. If a pasture of 16 acres will feed 6 horses for 4 months, how many acres will feed 12 horses for 9 months?

6. If 25 persons consume 300 bushels of corn in 1 year, how much will 139 persons consume in 7 years at the same rate?

7. If 32 men build a wall 36 feet long, 8 feet high, and 4 feet wide in 4 days; in what time will 48 men build a wall 864 feet long, 6 feet high, and 3 feet wide?

8. If a regiment of 1878 soldiers consume 702 quarters of wheat in 336 days, how many quarters will an army of 22536 soldiers consume in 112 days?

9. If 12 tailors in 7 days can finish 13 suits of clothes, how many tailors in 19 days of the same length, can finish the clothes of a regiment of soldiers consisting of 494 men?

10. An ordinary of 100 men drank £20 worth of wine at 2*s.* 6*d.* per bottle; how many men, at the same rate of drinking, will £7 worth suffice, when wine is rated at 1*s.* 9*d.* per bottle?

11. If 60 bushels of oats will serve 24 horses for 40 days, how long will 30 bushels serve 48 horses at the same rate?

12. If a garrison of 3600 men, in 35 days, at 24*oz.* per

day each man, eat a certain quantity of bread, how many men in 45 days, at the rate of 14*oz.* per day each man, will eat double the quantity?

13. A garrison of 3600 men has just bread enough to allow 24*oz.* a day to each man for 35 days; but a siege coming on, the garrison was reinforced to the number of 4800 men. How many ounces of bread a day must each man be allowed, to hold out 45 days against the enemy?

14. If 336 men, in 5 days of 10 hours each, dig a trench of 5 degrees of hardness, 70 yards long, 3 wide, and 2 deep, what length of trench of 6 degrees of hardness, 5 yards wide, and 3 deep, may be dug by 240 men in 9 days of 12 hours each?

15. If 12 pieces of cannon, eighteen-pounders, can batter down a castle in an hour, in what time would nine twenty-four-pounders batter down the same castle, both pieces of cannon being fired the same number of times, and their balls flying with the same degree of velocity?

16. If 15 weavers by working 10 hours a day for 10 days, can make 250 yards of cloth, how many must work 9 hours a day for 15 days, to make $607\frac{1}{2}$ yards?

17. If £$3\frac{1}{2}$ be the wages of 13 men for $7\frac{1}{2}$ days, what will be the wages of 20 men for $15\frac{1}{3}$ days?

18. If a footman travel 294 miles in $7\frac{2}{3}$ days, of $12\frac{1}{2}$ hours long, in how many days, of $10\frac{2}{3}$ hours long each, will he travel $147\frac{1}{8}$ miles?

19. Bought 5000 planks, of 15 feet long and $2\frac{1}{2}$ inches thick; how many planks are they equivalent to, of $12\frac{1}{2}$ feet long and $1\frac{3}{4}$ inches thick?

20. If 248 men, in $5\frac{1}{2}$ days of 11 hours each, dig a trench of 7 degrees of hardness, $232\frac{1}{2}$ yards long, $3\frac{2}{3}$ wide, and $2\frac{1}{3}$ deep; in how many days, of 9 hours long, will 24 men dig a trench of 4 degrees of hardness, $337\frac{1}{2}$ yards long, $5\frac{3}{8}$ wide, and $3\frac{1}{2}$ deep?

PRACTICE.

199. PRACTICE is an easy and concise method of applying the rules of arithmetic to questions which occur in trade and business. It is only a contraction of the RULE OF THREE when the first term is unity.

For example, if 1 yard of cloth cost half a dollar, what will 60 yards cost? This is a question which may be answered by the rule called Practice. The cost is obviously $30.

200. One number is said to be an aliquot part of another, when it forms an exact part of it: that is, when it is contained in that other an exact number of times. Hence, an aliquot part is an exact or even part.

For example, 25 cents is an aliquot part of a dollar. It is an exact fourth part, and is contained in the dollar four times. So also, 2 months, 3 months, 4 months, and 6 months, are all aliquot parts of a year.

TABLE OF ALIQUOT PARTS.

Cts.	Parts of $1.	Mo.	Parts of a year.	Days.	Parts of 1 mo.	Parts of £1.	Parts of 1 shilling
50 =	$\frac{1}{2}$	6=	$\frac{1}{2}$	15 =	$\frac{1}{2}$	10*s.* = $\frac{1}{2}$	6 *d.* = $\frac{1}{2}$
$33\frac{1}{3}$=	$\frac{1}{3}$	4=	$\frac{1}{3}$	10 =	$\frac{1}{3}$	6*s.* 8*d.*= $\frac{1}{3}$	4 *d.* = $\frac{1}{3}$
25 =	$\frac{1}{4}$	3=	$\frac{1}{4}$	$7\frac{1}{2}$=	$\frac{1}{4}$	5*s.* = $\frac{1}{4}$	3 *d.* = $\frac{1}{4}$
20 =	$\frac{1}{5}$	2=	$\frac{1}{6}$	6 =	$\frac{1}{5}$	4*s.* = $\frac{1}{5}$	2 *d.* = $\frac{1}{6}$
$12\frac{1}{2}$=	$\frac{1}{8}$	1=	$\frac{1}{12}$	5 =	$\frac{1}{6}$	3*s.* 4*d.*= $\frac{1}{6}$	$1\frac{1}{2}$*d.* = $\frac{1}{8}$
$6\frac{1}{4}$=	$\frac{1}{16}$		or $\frac{1}{3}$ of	3 =	$\frac{1}{10}$	2*s.* 6*d.*= $\frac{1}{8}$	1 *d.* = $\frac{1}{12}$
5 =	$\frac{1}{20}$		3 *mo.*			1*s.* 8*d.*= $\frac{1}{12}$	

QUEST.—199. What is Practice? If one yard of cloth cost $8, what will half a yard cost? What will one quarter of a yard cost? 200. When is one number said to be an aliquot part of another? What is an aliquot part? What are the aliquot parts of a dollar expressed in the table? What the aliquot parts of a year? What the aliquot parts of a month? What the aliquot parts of a pound? What are the aliquot parts of a shilling?

EXAMPLES.

1. What is the cost of 376 yards of cloth at $0,75, or $\frac{3}{4}$ of a dollar per yard?

Had the cloth cost $1 per yard, the cost of the 376 yards would have been $376. Had it cost 50*cts.* per yard, the cost would have been $\frac{1}{2}$ of $376, or $188: had it been 25*cts.* per yard, the cost would have been $\frac{1}{4}$ of $376, or $94; but the price being 75*cts.* per yard, the cost is 188 + 94 = $282.

OPERATION.

cts.		$	
50	$\frac{1}{2}$	376	
		188	cost at 50*cts.*
25	$\frac{1}{4}$	94	cost at 25*cts.*
75	$\frac{3}{4}$	$282	cost at $\frac{3}{4}$*doll.*

2. What is the cost of 196 yards of cotton, at 9*d.* per yard?

196*yd.* at 6*d.* or $\frac{1}{2}$*s.* = 98*s.*
196*yd.* at 3*d.* or $\frac{1}{4}$*s.* = 49*s.*
Therefore, 196*yd.* at 9*d.* or $\frac{3}{4}$*s.* = 147*s.* = £7 7*s.* *Ans.*

3. What is the cost of 4715 yards of tape, at $\frac{1}{4}$*d.* per yard?

$\frac{1}{4}$*d.* - - 4)4715
12)1178$\frac{3}{4}$*d.* = *cost.*
20)98*s.* 2$\frac{3}{4}$*d.*
Ans. = £4 18*s.* 2$\frac{3}{4}$*d.*

4. What is the cost of 425 yards at 1 penny per yard?

1*d.* = $\frac{1}{12}$*s.* - 12)425
20)35*s.* 5*d.*
Ans. £1 15*s.* 5*d.*

5. What will be the cost of 354 yards at 1$\frac{1}{4}$ per yard?

1*d.* = $\frac{1}{12}$*s.* - 12)354
4)29*s.* 6*d.*
$\frac{1}{4}$*d.* - - - - 7*s.* 4$\frac{1}{2}$*d.*
cost 36*s.* 10$\frac{1}{2}$*d.*
= £1 16*s.* 10$\frac{1}{2}$*d.*

6. What will be the cost of 4756 yards of cotton shirting, at 12$\frac{1}{2}$ cents per yard?

12$\frac{1}{2}$*cts.* = $\frac{1}{8}$ of 1$. 8)4756
594$\frac{1}{2}$
Ans. $594,50.

7. What will be the cost of 3754 pairs of gloves, at 2*s*. 6*d*. per pair?

2*s*. 6*d*.$=\frac{1}{8}$£. - - 8)3754

$469\frac{2}{8}$

Ans. £469 5*s*.

8. What will be the cost of 5320 bushels of wheat, at 3*s*. 6*d*. per bushel?

	5320
	3
at 3*s*. - - - -	15960
at $\frac{1}{2}$*s*. - - - -	2660
at 3*s*. 6*d*. - -	18620*s*.
Ans.	£9310

9. What will be the cost of 435 yards of cloth, at £2 7*s*. per yard?

	435
	2
Cost at £2 - -	£870
5*s*.$=\frac{1}{4}$ of £.	108 15*s*.
2*s*.$=\frac{1}{10}$ of £.	43 10*s*.
Total cost	£1022 5*s*.

10. What will be the cost of 660 yards at 2*s*. 6*d*?

2*s*.$=\frac{1}{10}$£. 10)660

6*d*.$=\frac{1}{4}$ of 2*s*. 4)66 cost at 2*s*.

16 10*s*.

Ans. £82 10*s*.

201. When the price in shillings is less than 20,

Multiply by half the number of shillings, and the figures to the left of the right hand figure will express the pounds, and this figure doubled will be the shillings.

11. What is the cost of 56 yards of cloth, at 16*s*. per yard?

$$16s. = \tfrac{16}{20} \text{ of a £: Hence}$$

$$56 \times \tfrac{16}{20} = \text{the amount in pounds.}$$

But $56 \times \frac{16}{20} = 56 \times \frac{8}{10} =$ £44.8, in which the right hand figure 8 expresses tenths of pounds, and by doubling it, we obtain twentieths of pounds, or shillings: therefore, the reason of the rule is manifest.

QUEST.—201. When the price is in shillings and less than 20, how will you find the cost? What is the reason of the rule?

12. What will be the cost of 4514 yards of cloth at £2 17*s.* $7\frac{1}{2}$*d.* per yard?

	4514
	2
Cost at £2 - - - - - -	9028
4514 × $8\frac{1}{2}$ gives - - -	3836 18*s.*
at 6*d.* = $\frac{1}{40}$ of £ - - -	112 17*s.*
$1\frac{1}{2}$*d.* = $\frac{1}{4}$ of 6*d.* - - - -	28 4*s.* 3*d.*
Total cost	£13005 19*s.* 3*d.*

GENERAL EXAMPLES.

13. What will 19*cwt.* 3*qr.* 11*lb.* of hops cost, at £4 11*s.* 9*d.* per *cwt.*?

14. 19*cwt.* 3*qr.* 19*lb.* of sugar, at £2 4*s.* 8*d.* per *cwt.*?

15. 11*cwt.* 1*qr.* 16*lb.* of soap, at £3 7*s.* per *cwt.*?

16. 9*cwt.* 3*qr.* 10*lb.* of treacle, at £1 18*s.* 9*d.* per *cwt.*?

17. What is the cost of 40*lb.* of soap, at $6\frac{3}{4}$*cts.* per pound?

18. What is the cost of 70 yards of tape, at $2\frac{1}{4}$*cts.* per yard?

19. What is the cost of 876 bushels of apples, at $62\frac{1}{2}$*cts.* per bushel?

20. What is the cost of 1000 quills, at $\frac{1}{2}$ cent per quill?

21. What is the cost of 1800 lead pencils, at 6 cents apiece?

22. What is the cost of 9*T.* 13*cwt.* 19*lb.* of pewter, at £14 15*s.* 9*d.* per ton?

23. 3*qr.* 19*lb.* 10*oz.*, at £11 12*s.* $5\frac{1}{2}$*d.* per *cwt.*?

24. 74*oz.* 2*pwt.* 12*gr.* of silver, at 4*s.* $11\frac{1}{2}$*d.* per *oz.*?

25. A pair of chased silver salts, weight 7*oz.* 11*pwt.*, at 8*s.* $11\frac{1}{4}$*d.* per *oz.*?

26. 571*oz.* 14*pwt.* $16\frac{1}{2}$*gr.*, at £3 11*s.* $9\frac{1}{4}$*d.* per *oz.*?

27. What will be the cost of $85\frac{1}{2}$ yards of cloth, at \$$9\frac{1}{2}$ per yard?

28. What will be the cost of 1848 yards of linen, at $87\frac{1}{2}$ cents per yard?

29. What is the cost of $51\frac{1}{2}$ tons of hay, at \$12,50 per ton?

30. What is the cost of 693 yards of linen, at 75*cts.* per yard?

31. What is the rent of 725*A.* 2*R.* 19*P.* of land, at £2 11*s.* 9*d.* per acre?

32. 51*A.* 3*R.* 15*P.* at £4 10*s.* per acre?

33. 97*A.* 14*P.* at £3 11*s.* 10*d.* per acre?

34. What is the cost of $28\frac{1}{2}$ yards of cloth, at \$$4\frac{3}{4}$ per yard?

35. What will be the cost of 2000 quills, at $\frac{3}{4}$ cent per quill?

36. What will $154\frac{1}{2}$ tons of hay come to, at \$12 per ton?

37. What is the cost of 514*yd.* 3*qr.* 2*na.*, at 17*s.* $9\frac{1}{2}$*d.* per yard?

38. 125*E. E.* 1*qr.* 1*na.* at £1 11*s.* $9\frac{1}{2}$ per ell?

39. What will be the cost of 1752 bushels of apples, at $62\frac{1}{2}$ cents per bushel?

40. What is the cost of 280 yards of tape, at $2\frac{1}{2}$ cents per yard?

41. What is the cost of 120 pounds of soap, at $6\frac{3}{4}$ cents per pound?

42. What cost 17*E.Fr.* 1*qr.* 3*na.* of Brussels lace, at £3 19*s.* $11\frac{1}{4}$*d.* per ell?

43. 349*E.Fl.* 1*qr.* 3*na.* of holland, at £1 11*s.* 6*d.* per ell?

44. 475*yd.* 3*qr.* 2*na.* at £1 14*s.* $9\frac{1}{4}$*d.* per ell English?

45. $375\frac{3}{8}$*E.E.* at 18*s.* $11\frac{3}{4}$*d.* per yard?

46. What will be the cost of 2*hhd.* 5*gal.* 3*qt.* 2*gi.* of molasses, at $12\frac{1}{2}$ cents per quart?

47. What will be the cost of 376 yards of cloth, at \$$1\frac{1}{2}$ per yard?

48. What will be the cost of 1*hhd.* 2*gal.* 3*qt.* 1*pt.* 1*gi.* of brandy, at $56\frac{1}{4}$ cents per quart?

49. What will be the cost of 27*bu.* 3*pk.* 6*qt.* 1*pt.* of wheat at \$1,75 per bushel?

TARE AND TRET.

202. *Tare* and *Tret* are allowances made in selling goods by weight.

Draft is an allowance on the gross weight in favor of the buyer or importer: it is always deducted before the *Tare.*

Tare is an allowance made to the buyer for the weight of the hogshead, barrel or bag, &c., containing the commodity sold.

Gross Weight is the whole weight of the goods, together with that of the hogshead, barrel, bag, &c., which contains them.

Suttle is what remains after a *part* of the allowances have been deducted from the gross weight.

Net Weight is what remains after all the deductions are made.

EXAMPLES.

1. What is the net weight of 25 hogsheads of sugar, the gross weight being 66*cwt.* 3*qr.* 14*lb.*; tare 11*lb.* per hogshead?

	cwt.	*qr.*	*lb.*	
	66	3	14	gross.
25 × 11 = 275*lb.* - -	2	1	23	tare.
Ans.				net.

2. If the tare be 4*lb.* per hundred, what will be the tare on 6*T.* 2*cwt.* 3*qr.* 14*lb.*?

Tare for 6*T.* or 120*cwt.*	=	480*lb.*
2*cwt.*	=	8
3*qr.*	=	3
14*lb.*	=	$0\frac{1}{2}$
Tare - - -	-	$491\frac{1}{2}$

QUEST.—202. What are Tare and Tret? What is Draft? What is Tare? What is Gross Weight? What is Suttle? What is Net Weight?

3. What is the tare on 32 boxes of soap, weighing 31550*lb.*, allowing 4*lb.* per box for draft and 12*lb.* in every hundred for tare?

	31550 gross.	31422
32 × 4 =	128 draft.	12
	31422	3770.64

Ans. ——

4. What will be the cost of 3 hogsheads of tobacco at $9,47 per *cwt.* net, the gross weight and tare being of

	cwt.	*qr.*	*lb.*		*lb.*
No. 1 - -	9	3	25	- - tare	146
" 2 - -	10	2	12	- - "	150
" 3 - -	11	1	25	- - "	158

Ans. ——

5. At £1 5*s.* per *cwt.* net; tare 4*lb.* per *cwt.*: what will be the cost of 4 hogsheads of sugar, weighing gross,

	cwt.	*qr.*	*lb.*	
No. 1 - - -	10	3	6	
" 2 - - -	12	5	19	
" 3 - - -	13	1	10	
" 4 - - -	11	2	7	
	49	0	14	gross.
Tare 4*lb.* per cwt	1	3	0	8*oz.*
	47	1	13	8*oz.* net.

Ans. ——

6. At 21 cents per lb., what will be the cost of 5*hhd.* of coffee, the tare and gross weight being as follows:

	cwt.	*qr.*	*lb.*		*lb.*
No. 1 - -	6	2	14	- - tare	94
" 2 - -	9	1	20	- - "	100
" 3 - -	6	2	22	- - "	88
" 4 - -	7	2	25	- - "	89
" 5 - -	8	0	13	- - "	100

Ans. ——

7. At £7 5*s.* per cwt. net, how much will 16*hhd.* of sugar come to, each weighing gross 8*cwt.* 3*qr.* 7*lb.*; tare 12*lb.* per cwt.?

8. What is the net weight of 18*hhd.* of tobacco, each weighing gross 8*cwt.* 3*qr.* 14*lb.*; tare 16*lb.* to the *cwt.*?

9. In 4*T.* 3*cwt.* 3*qr.* gross, tare 20*lb.* to the *cwt.*, what is the net weight?

10. What is the net weight and value of 80 kegs of figs, gross weight 7*T.* 11*cwt.* 3*qr.*, tare 14*lb.* per *cwt.*, at $2,31 per *cwt.*?

11. A merchant bought 19*cwt.* 1*qr.* 27*lb.* gross of tobacco in leaf, at $24,28 per *cwt.*; and 12*cwt.* 3*qr.* 19*lb.* gross in rolls, at $28,56 per *cwt.*; the tare of the former was 149*lb.*, and of the latter 48½*lb.*: what did the tobacco cost him net?

12. A grocer bought 17¼*hhd.* of sugar, each 10*cwt.* 1*qr.* 14*lb.*, draft 7*lb.* per *cwt.*, tare 4*lb.* per 104*lb.* What is the value at $7,30 per *cwt.* net?

13. In 29 parcels, each weighing 3*cwt.* 3*qr.* 14*lb.* gross, draft 8*lb.* per *cwt.*, tare 4*lb.* per 104*lb.*, how much net weight, and what is the value at $7,50 per *cwt.* net?

14. A merchant bought 7 hogsheads of molasses, each weighing 4*cwt.* 3*qr.* 14*lb.* gross, draft 17*lb.* per *cwt.*, tare 8*lb.* per hogshead, and damage in the whole 99¾*lb.* What is the value at $8,45 per *cwt.* net?

15. The net value of a hogshead of Barbadoes sugar was $22,50; the custom and fees $12,49, freight $5,11, factorage $1,31; the gross weight was 11*cwt.* 1*qr.* 15*lb.*, tare 11⅕*lb.* per *cwt.* What was the sugar rated at per *cwt.* net. in the bill of parcels?

16. In 7*hhd.* of oil, each weighing 3*cwt.* 2*qr.* 14*lb.* gross, tare 21*lb.* per *cwt.*, how many gallons net, and what is the value at $1,24 per gallon?

17. I have imported 87 jars of Lucca oil, each containing 47 gallons: what came the freight to at $1,19 per *cwt.* net reckoning 1*lb.* in 11*lb.* for tare, and 9*lb.* of oil to the gallon?

PERCENTAGE.

203. THE term per cent comes from per centum, and means by the hundred. The term is generally used to express the interest on money, but may also be employed to designate hundredth parts of other things. Thus, when we say twenty per cent of a bushel of wheat, we mean twenty hundredths, or one-fifth of it.

204. The rate per cent may always be expressed by a decimal fraction. Thus, five per cent may be expressed by .05, eight per cent by .08, fifteen per cent by .15, &c.

Hence, to find the amount of percentage on any number,

Multiply the number by the rate per cent, expressed in a decimal fraction, and the product will be the percentage.

EXAMPLES.

1. A has $852 deposited in the bank, and wishes to draw out 5 per cent of it: how much must he draw for?

2. A merchant has 1200 barrels of flour; he shipped 64 per cent of it and sold the remainder: how much did he sell?

3. A merchant bought 1200 hogsheads of molasses. On getting it into his store, he found it short $3\frac{1}{2}$ per cent: how many hogsheads were wanting?

4. Two men had each $240. One of them spends 14 per cent, and the other $18\frac{1}{2}$ per cent: how many dollars more did one spend than the other?

5. What is the difference between $5\frac{1}{2}$ per cent of $800 and $6\frac{1}{2}$ per cent of $1050?

6. A trader laid out $1600 as follows: he pays 24 per ct. of his money for broadcloths; 30 per ct. of what is left for linens; 12 per ct. of what is left for calicoes; and then 5 per ct. of the residue for cottons: how much did he pay for cottons?

QUEST.—203. What do you understand by the term per cent? For what is the term generally used? What do you understand by twenty per cent? What by eight per cent? 204. How may the rate per cent be expressed? How do you express five per cent? Eight per cent? How do you find the amount of percentage on any given number?

7. A man purchased 250 boxes of oranges, and found that he had lost in bad ones 18 per cent: to how many full boxes were his good oranges equal?

8. If I buy 895 gallons of molasses and lose 17 per cent by leakage, how much have I left?

205. To find the per cent which one number is of another.

If I buy 6 hogsheads of molasses for $200 and sell them for $220, what do I gain per cent, on the money expended?

It is plain that $20 is the amount made. What per cent is $20 of $200; that is, how many hundredths of $200? If we add two ciphers to the first, and then divide it by the second, the quotient will express the hundredths. Thus,

$$\frac{2000}{200} = 10;$$

that is, 20 is ten per cent of 200.

Hence, to determine the per cent which one number is of another,

I. *Bring the number which determines the per cent to hundredths by annexing two ciphers or removing the decimal point two places to the right.*

II. *Divide the number so formed by the number on which the percentage is estimated, and the quotient will express the per cent.*

EXAMPLES.

1. A man has $550 and purchases goods to the amount of $82,75: what per cent of his money does he expend?

2. A merchant goes to New York with $1500; he first lays out 20 per ct., after which he expends $600: what per ct. was his last purchase of the money that remained after his first?

3. Out of a cask containing 300 gallons, 60 gallons are drawn: what per cent is this?

4. If I pay $698,33 for 3 hhds. of molasses and sell them for $837,996, how much do I gain per ct. on the money laid out?

5. If I pay $698,33 for 3 hhds. of sugar and sell them for $837,996, how much do I make per ct. on the amount received?

QUEST.—205. How do you find the per ct. which one number is of another?

SIMPLE INTEREST.

206. INTEREST is an allowance made for the use of money that is borrowed.

For example, if I borrow $100 of Mr. Wilson for one year, and agree to pay him $6 for the use of it, the $6 is called the *interest* of $100 for one year, and at the end of the time Mr. Wilson should receive back his $100 together with the $6 interest, making the sum of $106.

The money on which interest is paid, is called the *Principal.*

The money paid for the use of the principal, is called the *Interest.*

The principal and interest, taken together, are called the *Amount.*

In the above example,

$100 is the principal,
$ 6 is the interest, and
$106 is the amount.

The interest of $100 for one year, determines the rate of interest, or rate per cent. In the example above, the rate of interest is 6 per cent, or $6 for the use of the hundred. Had $8 been paid for the use of the $100, the rate of interest would have been 8 per cent; or had $3 only been paid, the rate of interest would have been 3 per cent.

Legal interest is the rate of interest established by law. In the New England States, and indeed in most of the other states, the legal interest is 6 per cent per annum, that is, 6 per cent by the year.

QUEST.—206. What is Interest? What is the money called on which interest is paid? What is the money called which is paid for the use of the principal? What is the amount? What determines the rate of interest? What is legal interest? What is meant by per annum? How much is the interest per annum in most of the states? What is it in New York? In Alabama?

In New York, however, it is 7, and in Alabama 8 per cent.

CASE I.

207. To find the interest on any given principal for one or more years.

The interest of each dollar, for a single year, will be so many hundredths of itself as are expressed by the rate of interest. Thus, if the rate of interest be 4 per cent, each dollar will produce annually an interest of .04 of a dollar, or 4 cents: if the rate be 5 per cent, it will produce .05 of a dollar, or 5 cents: if 6 per cent, .06, or 6 cents, &c.

Hence, to find the interest on any given sum for one or more years,

Multiply the principal by the decimal fraction which expresses the rate of interest, and the product so arising by the number of years. Or,

Multiply the decimal fraction which expresses the rate of interest by the number of years, and then multiply the principal by this product.

EXAMPLES.

1. What is the interest on $1960 for four years, at 7 per cent per annum?

The rate of interest being 7 per cent, each dollar will produce .07 of a dollar, or 7 cents, in one year: hence, $137,20 will be the interest on the sum for one year, and $548,80 for 4 years.

OPERATION.

$1960	
.07	
$137,20	int. for 1 year.
4	number of years.
$548,80	*Ans.*

QUEST.—207. What will be the interest of one dollar for one year? What will express decimally the interest on one dollar for one year at 4 per cent? What will express it at 5 per cent? At 6? At 7? At 8? How do you find the interest on any sum for one or more years? What will be the multiplier when the rate of interest is 4 per cent, and the time 3 years? When the rate is 6 per cent and the time 5 years? When the rate is 8 per cent and the time 3 years?

2. What is the interest on $78,457 dollars for three years, at 5 per cent per annum?

Since there are three places of decimals in the multiplicand and two in the multiplier, there will be five in the product (Art. **149**). Observe *that the two first, counting from the comma to the right,* are cents, the third mills, the fourth tenths of mills, &c.

OPERATION.

```
               78,457
.05 × 3 =         .15
               392285
               78457
Ans.      $11,76855
```

3. What is the interest on $365,874 for one year, at $5\frac{1}{2}$ per cent?

We first find the interest at 5 per cent, and then the interest for $\frac{1}{2}$ per cent: the sum is the interest at $5\frac{1}{2}$ per cent.

OPERATION.

```
$365,874
     .05
18,29370
 1,82937  ½ per cent.
$20,12307 Ans.
```

4. What is the interest on $2871,24 for 6 years, at 7 per cent?

5. What is the interest on $535,50 for 25 years, at 6 per cent?

6. What is the interest on $1125,819 for 5 years, at 8 per cent per annum?

7. What is the interest on $8089,74 for 12 years, at 5 per cent?

8. What is the interest on $1226,35 for 7 years, at $7\frac{1}{2}$ per cent?

9. What is the interest on $3153,82 for 9 years, at $4\frac{1}{2}$ per cent?

10. What is the interest on $982,35 for 4 years, at 6 per cent?

11. What is the interest on $1914,16 for 18 years, at $3\frac{1}{2}$ per cent?

12. What is the interest on $2866,28 for 6 years, at 8 per cent?

13. What is the interest on $16199,48 for 16 years, at 5 per cent?

14. What is the interest on $897,50 for 21 years, at 6 per cent?

CASE II.

208. To find the interest for any number of months, at the rate of 6 per cent per annum.

At the rate of 6 per cent per annum, one month produces $\frac{1}{2}$ per cent on the principal; and hence, every two months produces one per cent on the principal. Therefore to find the interest for months,

Divide the number of months by 2 and regard the quotient as hundredths. Then multiply the principal by the decimal so found, and the product will be the interest.

EXAMPLES.

1. What is the interest on $651 for 8 months, at 6 per cent per annum?

The decimal corresponding to 8 months, which gives 4 per cent, is .04: hence, the interest is $26,04.

OPERATION.

$651	
.04	half the number of months.
$26,04	

2. What is the interest on $614,364 for 9 months, at 6 per cent per annum?

The decimal corresponding to 9 months is .04½, and hence the interest is $27,64638.

OPERATION.

$614,364
.04½
2457456
307182
$27,64638

QUEST.—208. At the rate of 6 per cent, what will be the interest on any principal for one month? What time will produce one per cent? How do you find the interest on any principal for any number of months? What is the multiplier for 4 months? What for 6 months? What for 7? What for 8? For 9? What for 10? For 11? What for 12?

3. What is the interest on $17507,30 for 14 months, at 6 per cent?

4. What is the interest on $982,41 for 9 months, at 6 per cent?

5. What is the interest on $75192,84 for 16 months, at 6 per cent?

6. What is the interest on $7953,70 for 9 months, at 6 per cent?

7. What is the interest on $15907,40 for 27 months, at 6 per cent?

8. What is the interest on $4918,50 for 11 months, at 6 per cent?

9. What is the interest on $84377,91 for 7 months, at 6 per cent?

10. What is the interest on $91358,24 for 17 months, at 6 per cent?

11. What is the interest on $31573,25 for 10 months, at 6 per cent?

12. What is the interest on $959875,45 for 18 months, at 6 per cent?

CASE III.

209. To find the interest at 6 per cent per annum, for any number of days.

In computing interest the month is reckoned at 30 days. Hence, 60 days, which make two months, will give an interest of one per cent on the principal, and consequently, 6 days will give an interest of one mill on the dollar, or one-thousandth of the principal. If, therefore, the days be divided by 6, the quotient will show how many thousandths of the principal must be taken on account of the days. Hence, to find the interest for any number of days less than 60,

---

QUEST.—209. In computing interest for days, at what is the month reckoned? How many days give one per cent? What part of the principal is one per cent? How many days will give one-thousandth of the principal? How will you find how many thousandths of the principal must be taken for the days?

Divide the days by 6, and multiply the principal by the quotient, considered as thousandths.

EXAMPLES.

1. What is the interest on $297,047 for 28 days, at 6 per cent per annum?

We find that the 28 days give $4\frac{2}{3}$ thousandths. We multiply the principal by .004, and then add $\frac{2}{3}$ of the principal multiplied by one-thousandth for the fractional part.

OPERATION.

	$297,047
$28 \div 6 = 4\frac{2}{3}$.	$.004\frac{2}{3}$
	1188188
Add $\frac{1}{3}$	99015
" $\frac{1}{3}$	99015
	$1,386218

210. To avoid the fractions which sometimes appear in the multipliers, we may, if we please, first multiply the principal by the number of days, and then divide the product by 6, which will give the same quotient as found above. Hence, to find the interest for any number of days,

Multiply the principal by the number of days, divide the product by 6, and then point off in the quotient three more places for decimals than there are decimals in the given principal.

2. What is the interest on $657,87 for 13 days, at 6 per cent per annum?

We first multiply the given principal by 13; we then divide the product by 6; and since there are two places of decimals in the principal, we point off five in the quotient.

OPERATION.

```
  $657,87
       13
   197361
   65787
 6)855231
 $1,42538
```

NOTE—Let each of the following examples be worked by both methods; though, when the days exceed 60, the second method is preferable.

QUEST.—How do you find the interest for less than 60 days? What is the multiplier for 6 days? For 9 days? For 10 days? For 15 days? For 20 days? For 25 days? 210. How may the interest for days be found by the second method?

3. Find the interest on $785,469 for 25 days. Also, the interest on $8709,27 for 100 days.

4. What is the interest on $2691,12 for 150 days?

5. What is the interest on $1151,44 for 29 days?

6. What is the interest on $136,25 for 19 days?

7. What is the interest on $981,90 for 70 days?

8. What is the interest on $757,06 for 9 days?

9. What is the interest on $864 for 95 days?

10. What is the interest on $11268,75 for 17 days?

11. What is the interest on $4428,10 for 165 days?

12. What is the interest on $975,95 for 14 days?

13. What is the interest on $28793,28 for 127 days?

211. Note.—The above method of computing interest for days, is the one in general use. It, however, considers the year as made up of 360 instead of 365 days; and hence the result is too large by 5 of the 365 parts into which the interest found may be divided. Hence, the interest found will be too large by its $\frac{1}{73}$ part, by which it must be diminished when entire accuracy is desired.

CASE IV.

212. To find the interest at 6 per cent per annum for years, months, and days.

Find the interest for the years by Case I., for the months by Case II., and for the days by Case III.; then add the several results together, and their sum will be the answer sought. Or,

Form a single multiplier for the years, months, and days, and then multiply the principal by it.

EXAMPLES.

1. What is the interest on $1597,27 at 6 per cent, for 3 years 9 months and 11 days?

Quest.—211. How many days does the above method give to the year? Is the result obtained too great or too small? By how much is it too great? How will you find the exact interest? 212. How do you find the interest at 6 per cent per annum for years, months, and days? What is the multiplier for 1 year 4 months and 12 days? What for 2 years 8 months and 18 days? For 3 years 10 months and 24 days?

1ST METHOD.

$1597,27	$1597,27	$1,59727 for 6 days.
.06 × 3 = .18	.04½	,79863 for 3 days.
1277816	638908	,53242 for 2 days.
159727	79863½	$2,92832 for 11 days.
$287,5086	$71,8771½	

Interest for 3 years	$287,5086
" " 9 months	71,8771+
" " 11 days	2,9283+
Total interest	$362,3140+

2D METHOD.

Multiplier for 3 years = .06 × 3	= .18.
" " 9 months	= .045.
" " 11 days = $\frac{11}{6}$	= .001$\frac{5}{6}$.
Entire multiplier	0.226$\frac{5}{6}$.

Then, $1597,27 × 0.226$\frac{5}{6}$ = $362,3140+.

2. What is the interest on $252803,87 for 1 year 1 month and 1 day?

3. What is the interest on $3195,54 for 7 years 6 months and 22 days?

4. What is the interest on $1352,25 for 4 years and 7 months?

5. What is the interest on $23518,20 for 9 years, 11 months, and 16 days?

6. What is the interest on $2420,70 for 1 year and 10 months?

7. What is the interest on $19574 for 12 years and 1 day?

8. What will be the amount of $1947,66 after 21 years and 8 months?

9. What is the interest on $1330,50 for 14 years, 4 months, and 24 days?

10. What is the interest on $3227,60 for 2 years, 8 months, and 20 days?

11. What is the interest on $79265,375 for 8 years 7 months and 6 days?

12. What will be the amount of $9537,15 after 11 years, 2 months, and 18 days?

CASE V.

213. To find the interest when there are months and days, and the rate of interest is greater or less than 6 per cent.

Find the interest at 6 per cent. Then add to it or subtract from it such a part of the interest so found as the given rate exceeds or falls short of six per cent per annum.

EXAMPLES.

1. What is the interest on $179,25, at 7 per cent per annum, for 3 years and 4 months?

Multiplier for 3 years = .06 × 3 = .18
" " 4 months = .02
Entire multiplier .20

Hence, $179,25 × .20 = $35,8500 interest at 6 per cent.
Add $\frac{1}{6}$ 5,9750
$41,8250 interest at 7 per cent.

2. What is the interest on $974,50 for 9 years, 6 months, and 18 days, at 4 per cent per annum?

Multiplier for 9 years at 6 per cent = 9 × .06 = .54
" " 6 months = .03
" " 18 days = 18 ÷ 6 = 3 = .003
Entire multiplier - - - - - .573

Hence, $974,50 × .573 = $558,3885
Subtract one-third 186,1295
Int. at 4 per cent $372,2590

3. What is the interest on $874,42, at 3 per cent, for 19 years and 6 months?

QUEST.—213. How do you find the interest when there are months and days, and the rate greater than 6 per cent? How do you find the interest when it is less?

4. What is the interest on $358,50, at 7 per cent, for 6 years and 8 months?

5. What is the interest on $1975,98, at 5 per cent, for 10 years 4 months and 18 days?

6. What is the interest on $1461,75 for 4 years and 9 months, at 8 per cent?

7. What is the interest on $45000 for 1 year and 4 months, at 7 per cent?

8. What will be the total amount of $2238,96 after 2 years and 7 months, at 7 per cent?

9. What is the interest of $1200 for 1 month and 12 days, at 5 per cent?

10. What is the interest on $1064,82 for 6 years and 6 months, at $4\frac{1}{2}$ per cent?

11. What is the interest on $1752,96, at 7 per cent, for 4 years 9 months and 14 days?

12. What is the interest on $17518,54, at $7\frac{1}{2}$ per cent, for 3 years and 9 days?

13. What is the interest on $15138,22 for 3 years, 4 months and 18 days at $6\frac{1}{2}$ per cent per annum?

14. What is the interest on $4187,635, at 5 per cent, for 5 years 5 months and 5 days?

15. What is the interest on $167,50 for 7 months and 17 days, at 7 per cent per annum?

16. What is the interest on $2934,25 for 2 years 8 months and 19 days, at $8\frac{1}{2}$ per cent?

17. What is the interest on $19345,31, at $4\frac{1}{2}$ per cent, for 5 years 6 months and 15 days?

214. NOTE.—In computing interest, it is often very convenient to find the interest for the months by considering them as aliquot parts of a year, and the interest for days by considering them as aliquot parts of a month.

QUEST.—214. Explain the second method of computing interest for months and days. What part of a year are 3 months? Four months? Six? Eight? Nine? What part of a month are 6 days? Five days? Ten days?

EXAMPLES.

1. What is the interest of $806,90 for 1 year 10 months and 10 days, at 6 per cent?

```
    $806,90
        .06
6)$48,4140  = int. for 1 year.         $8,069
  2)8,069   = int. for 2 months.            5
  3)4,034+ = int. for 1 month.        $40,345 int. for 10 mo.
    1,344+ = int. for 10 days.

           Interest for 1 year  - -  $48,4140
              "     " 10 months -     40,345
              "     " 10 days - -      1,344+
                  Total interest     $90,103+
```

2. What is the interest of $200 for 10 years 3 months and 6 days, at 7 per cent?

```
   $200
    .07
4)14,00   = int. for 1 year.           $14,00
 3)3,50   = int. for 3 months.             10
 5)1,16+ = int. for 1 month.          $140,00 for 10 years.
    ,23+ = int. for 6 days.
        $140,00  interest for 10 years.
           3,50  interest for 3 months.
            ,23+ interest for 6 days.
   Ans. $143,73+
```

3. What is the interest of $264,52 for 2 years 8 months and 20 days, at 6 per cent per annum?

4. What is the interest of $76,50 for 1 year 9 months and 12 days, at 6 per cent?

5. What is the interest of $1041,75 for 1 year 1 month and 6 days, at 4 per cent per annum? Also, at 5 per cent? At 5½ per cent? At 6 per cent? At 7 per cent? At 7½ per cent? At 8 per cent? At 8½ per cent? And at 9 per cent?

6. What is the interest, at 6 per cent per annum, on $241,60, for 3 years 3 months and 15 days?

7. What is the interest, at 8 per cent per annum, on 1351,74, for 3 years 6 months and 6 days?

8. What is the interest, at 7 per cent, on $1761,75, for 5 years 5 months and 5 days?

9. What is the interest on $135178,40 for 3 years 9 months and 12 days, at 5 per cent per annum?

CASE VI.

215. To find the interest, when the sum on which the interest is to be cast is in pounds, shillings, and pence.

I. *Reduce the shillings and pence to the decimal of a pound* (Art. **161**).

II. *Then find the interest as though the sum were dollars and cents; after which reduce the decimal part of the answer to shillings and pence* (Art. **162**).

EXAMPLES.

1. What is the interest, at 6 per cent, of £27 15*s*. 9*d*. for 2 years?

We first find the interest for one year. We then multiply by 2, which gives the interest for two years. We then reduce to pounds, shillings, and pence.

OPERATION.

£27 15*s*. 9*d*. = £27,7875
.06
1.667250
2
£3.334500
20
6.690000
12
8.280000
4
1.120000
Ans. £3 6*s*. $8\frac{1}{4}$*d*

QUEST—215. How do you determine the interest when the sum is in pounds, shillings, and pence?

2. What is the interest on £203 18*s.* 6*d.*, at 6 per cent, for 3 years 8 months 16 days?

3. What is the interest on £255 10*s.* 8*d.* at 6 per cent, for 3 years and 3 months?

4. What is the interest of £215 13*s.* 8*d.*, at 6 per cent, for 3 years 6 months and 6 days?

5. What will £559 7*s.* 4*d.* amount to in 3 years and a half, at $5\frac{1}{4}$ per cent per annum?

6. What is the interest of £1543 10*s.* 6*d.* for 3 years and a half, at 4 per cent?

7. What is the interest of £1047 3*s.* for 3 years and a half, at 6 per cent?

8. What is the interest on £511 1*s.* 4*d.*, at 6 per cent per annum, for 6*yr.* 6*mo.*?

9. What is the interest on £161 15*s.* 3*d.*, at 6 per cent, for 7*yr.* 13*da.*?

APPLICATIONS.

216. For computing the interest on notes, the time may be found by the table in Art. 38.

The day on which a note is dated and the day on which it falls due, are not both reckoned in determining the time; *but one of them is always excluded.*

Thus, a note dated on the 1st of May, and falling due on the 16th of June, will bear interest but one month and 15 days.

Calculate the interest on the following notes.

$382,50 Philadelphia, January 1st, 1846.

1. For value received I promise to pay on the 10th day of June next, to C. Hanford or order, the sum of three hundred and eighty-two dollars and fifty cents with interest from date, at 7 per cent. *John Liberal.*

QUEST.—216. How may the time be found for computing interest on notes? What days named in a note are reckoned and what excluded, in reckoning the time? If a note is dated on the first and payable on the 15th, how many days will the interest run?

$612 Baltimore, January 1st, 1833.

2. For value received I promise to pay on the 4th of July, 1835, to Wm. Johnson or order, six hundred and twelve dollars with interest at 6 per cent from the 1st of March, 1833.

John Liberal.

$3120 Charleston, July 3d, 1846.

3. Six months after date, I promise to pay to C. Jones or order, three thousand one hundred and twenty dollars with interest from the 1st of January last, at 7 per cent

Joseph Springs.

$786,50 Mobile, July 7th, 1845.

4. Twelve months after date, I promise to pay to Smith & Baker or order, seven hundred and eighty-six $\frac{50}{100}$ dollars for value received with interest from December 3d, 1845, at 8 per cent. *Silas Day.*

$4560,72 Cincinnati, March 10th, 1846.

5. Nine months after date, for value received, I promise to pay to Redfield, Wright, & Co. or order, four thousand five hundred and sixty $\frac{72}{100}$ dollars with interest after 6 months, at 7 per cent. *Frederick Stillman.*

$1854,83 Boston, July 17th, 1846.

6. Eleven months after date, for value received, we promise to pay to the order of Fondy, Burnap, & Co., one thousand eight hundred and fifty-four $\frac{83}{100}$ dollars with interest from May 13th, 1846, at 6 per cent. *Palmer & Blake.*

PARTIAL PAYMENTS.

217. We shall now give the rule established in New York (See Johnson's Chancery Reports, Vol. I. page 17) for computing the interest on a bond or note, when partial payments have been made. The same rule is also adopted in Massachusetts, and in most of the other states.

I. *Compute the interest on the principal to the time of the first payment, and if the payment exceed this interest, add the interest to the principal and from the sum subtract the payment: the remainder forms a new principal.*

II. *But if the payment is less than the interest, take no notice of it until other payments are made, which in all, shall exceed the interest computed to the time of the last payment: then add the interest, so computed, to the principal, and from the sum subtract the sum of the payments: the remainder will form a new principal on which interest is to be computed as before.*

EXAMPLES.

$349,998 Richmond, Va., May 1st, 1826.

1. For value received I promise to pay James Wilson or order, three hundred and forty-nine dollars ninety-nine cents and eight mills with interest, at 6 per cent.

James Paywell.

On this note were endorsed the following payments:

Dec. 25th, 1826	received	$49,998
July 10th, 1827	"	$ 4,998
Sept. 1st, 1828	"	$15,008
June 14th, 1829	"	$99,999

What was due April 15th, 1830?

Principal on int. from May 1st, 1826 - -	$349,998
Interest to Dec. 25th, 1826, time of first payment, 7 months 24 days - - - - - -	13,649+
Amount - -	$363,647+
Payment Dec. 25th, exceeding interest then due - - - - - - - - - - - - -	$ 49,998
Remainder for a new principal - - - - -	$313,649
Interest of $313,649 from Dec. 25th, 1826, to June 14th, 1829, 2 years 5 months 20 days	$ 46,524+
Amount - -	$360,173

QUEST.—217. What is the rule in regard to partial payments?

Payment, July 10th, 1827, less than interest then due - -	$ 4,998	
Payment, Sept. 1st, 1828 - - -	15,008	
Their sum - - - - - - less than interest then due -	$20,006	
Payment, June 14th, 1829 - -	99,999	
Their sum exceeds the interest then due -		$120,005
Remainder for a new principal, June 14th, 1829 - - - - - - - - - - -		240,168
Interest of $240,168 from June 14th, 1829, to April 15th, 1830, 10 months 1 day - -		$ 12,048
Total due, April 15th, 1830 - -		$252,216+

$6478,84 New Haven, Feb. 6th, 1825.

2. For value received I promise to pay William Jenks or order, six thousand four hundred and seventy-eight dollars and eighty-four cents with interest from date, at 6 per cent.

John Stewart.

On this note were endorsed the following payments:

May 16th, 1828,	received	$545,76
May 16th, 1830,	"	$1276
Feb. 1st, 1831,	"	$2074,72.

What remained due August 11th, 1832?

3. A's note of $7851,04 was dated Sept. 5th, 1837, on which were endorsed the following payments, viz:—Nov. 13th, 1839, $416,98; May 10th, 1840, $152: what was due March 1st, 1841, the interest being 6 per cent?

$8974,56 New York, Jan. 3d, 1842.

4. For value received I promise to pay to James Knowles or order, eight thousand nine hundred and seventy-four dollars and fifty-six cents, with interest from date at the rate of 7 per cent. *Stephen Jones.*

On this note were endorsed the following payments:

Feb. 16th, 1843, received $1875,40
Sept. 15th, 1844, " $3841,26
Nov. 11th, 1845, " $1809,10
June 9th, 1846, " $2421,04.

What was due July 1st, 1846?

QUESTIONS IN INTEREST.

218. In all the questions relating to interest four things have been considered, viz.:

1st. The principal; 2d. The rate of interest; 3d. The time; and 4th. The amount of interest. Now, these four quantities are so connected with each other, that if three of them be known the fourth can always be found.

CASE I.

219. The principal, the rate of interest, and the time being known, to find the interest.

This case has already been considered.

CASE II.

220. Having given the interest, the time, and the rate of interest, to find the principal.

When the time and rate are the same, the interest on any principal, is to the principal, as any other interest is to its principal; that is,

Interest of $1 : $1 : : given interest : principal

Hence, to find the principal,

Cast the interest on one dollar for the given time and divide the given interest by the interest so found, and the quotient will be the principal.

QUEST.—218. How many things are considered in all questions relating to interest? How many of these must be given before the remaining ones can be determined? 219. What are given in Case I.? What required? 220. What are given in Case II.? What required? How do you find the principal?

EXAMPLES.

1. The interest on a certain sum for 1 year and 4 months, at 6 per cent, is $3007,7136: what is the principal?

The interest on $1 for the same time is $0,08. Hence,

$3007,7136 ÷ 0,08 = $37596,42 = principal.

2. The interest on a certain sum for nine months, at 6 per cent, is $178,9582: what is the principal?

3. The interest for 29 days is $2,78, at 6 per cent: what is the principal?

4. The interest for 17 days, at 6 per cent, is $4,0366: what is the principal?

5. The interest on a certain sum for 1 year 1 month and 6 days, at 7 per cent, is $26,7381: what is the sum? If the interest for the same time be $22,9184 at the rate of 6 per cent, what will be the sum? For the same time, what will be the principal, when the rate is 4 per cent and interest $15,2790? When the rate is 5 per cent and interest $19,0987? When the interest is $21,0086 and rate $5\frac{1}{2}$ per cent? When the rate is $7\frac{1}{2}$ per cent and interest $28,6479? When the rate is 8 per cent and interest $30,5578?

CASE III.

221. Having given the interest, the principal, and time, to find the rate per cent of interest.

If interest be cast at different rates, on the same sum and for the same time, the amounts of interest will be proportional to the rates. Therefore, cast the interest on the principal for the given time, at 1 per cent per annum; then,

Interest at 1 per cent : given interest : : 1 per cent : rate.

Hence, to find the rate of interest,

Cast the interest on the principal for the given time at 1 per cent: then divide the given interest by the interest so found, and the quotient will be the rate of interest.

QUEST.—221. What are given in Case III? What are required? How do you find the rate of interest?

EXAMPLES.

1. The interest on \$437,21 for 9 years and 9 months is \$127,8840: what is the rate of interest?

Interest on \$437,21 for 9 years and 9 months, at 1 per cent, is \$42,6280: hence,

\$127,8840 ÷ 42,6280 = 3 per cent, the rate.

2. The interest on \$987,99, for 5 years 2 months and 9 days, is \$256,4657: what is the rate of interest?

NOTE.—In examples similar to the above, and to those of the following section, the fractions of a per cent less than a quarter, or of a day, may be omitted. Such small fractions may arise from the different methods of computation.

CASE IV.

222. Having given the principal, the interest, and the rate of interest, to find the time.

If interest be cast at the same rate and on the same principal for different times, the amounts of the interest will be proportional to the times. Hence, if the interest on the principal be cast for 1 year, we shall have,

Interest for 1 year : given Interest : : 1 year : time.

Hence, to find the time,

Cast the interest on the given principal at the given rate for one year: then divide the given interest by the interest so found, and the quotient will be the time.

EXAMPLES.

1. The interest on \$15000, at 7 per cent per annum, is \$700: what is the time?

Interest on \$15000 for 1 year at 7 per cent = \$1050: hence, $\frac{700}{1050} = \frac{70}{105} = \frac{2}{3}$ of a year = 8 months.

2. The interest on \$1119,48, at 7 per cent per annum, is \$195,909: what is the time?

QUEST.—222. What are given in Case IV.? What are required? How do you find the time?

A TABLE, showing the number of shillings in a dollar in each State, and the rate of interest: also, the value of a dollar expressed in parts of a pound, which is found by dividing the number of pence in a dollar by the number in a pound.

	STATES.	No. of shillings to the dollar.	Value of the dollar in pounds.	Legal rate of interest.
1	Maine	6 shillings	$\$1=£\frac{72}{240}=£\frac{3}{10}$	6 per cent.
2	N. Hampshire	6 shillings	$\$1=£\frac{72}{240}=£\frac{3}{10}$	6 per cent.
3	Vermont	6 shillings	$\$1=£\frac{72}{240}=£\frac{3}{10}$	6 per cent.
4	Massachusetts	6 shillings	$\$1=£\frac{72}{240}=£\frac{3}{10}$	6 per cent.
5	Rhode Island	6 shillings	$\$1=£\frac{72}{240}=£\frac{3}{10}$	6 per cent.
6	Connecticut	6 shillings	$\$1=£\frac{72}{240}=£\frac{3}{10}$	6 per cent.
7	New York	8 shillings	$\$1=£\frac{96}{240}=£\frac{2}{5}$	7 per cent.
8	Ohio	8 shillings	$\$1=£\frac{96}{240}=£\frac{2}{5}$	6 per cent.
9	New Jersey	7*s.* 6*d.*	$\$1=£\frac{90}{240}=£\frac{3}{8}$	6 per cent.
10	Pennsylvania	7*s.* 6*d.*	$\$1=£\frac{90}{240}=£\frac{3}{8}$	6 per cent.
11	Delaware	7*s.* 6*d.*	$\$1=£\frac{90}{240}=£\frac{3}{8}$	6 per cent.
12	Maryland	7*s.* 6*d.*	$\$1=£\frac{90}{240}=£\frac{3}{8}$	6 per cent.
13	Michigan	8 shillings	$\$1=£\frac{96}{240}=£\frac{2}{5}$	7 per cent.
14	Indiana	6 shillings	$\$1=£\frac{72}{240}=£\frac{3}{10}$	6 per cent.
15	Illinois	6 shillings	$\$1=£\frac{72}{240}=£\frac{3}{10}$	6 per cent.
16	Missouri	6 shillings	$\$1=£\frac{72}{240}=£\frac{3}{10}$	6 per cent.
17	Virginia	6 shillings	$\$1=£\frac{72}{240}=£\frac{3}{10}$	6 per cent.
18	Kentucky	6 shillings	$\$1=£\frac{72}{240}=£\frac{3}{10}$	6 per cent.
19	Tennessee	6 shillings	$\$1=£\frac{72}{240}=£\frac{3}{10}$	6 per cent.
20	North Carolina	10 shillings	$\$1=£\frac{120}{240}=£\frac{1}{2}$	6 per cent.
21	South Carolina	4*s.* 8*d.*	$\$1=£\frac{56}{240}=£\frac{7}{30}$	7 per cent.
22	Georgia	4*s.* 8*d.*	$\$1=£\frac{56}{240}=£\frac{7}{30}$	7 per cent.
23	Alabama	Fed. money		8 per cent.
24	Mississippi	6 shillings	$\$1=£\frac{72}{240}=£\frac{3}{10}$	6 per cent.
25	Louisiana	Fed. money		6 per cent.
26	Arkansas	Fed. money		6 per cent.
27	Florida	Fed. money		6 per cent.
28	Texas	6 shillings	$\$1=£\frac{72}{240}=£\frac{3}{10}$	8 per cent.
29	Nov. Scotia and Canada	5 shillings	$\$1=£\frac{60}{240}=£\frac{1}{4}$	6 per cent.

REDUCTION OF CURRENCIES.

It has already been shown (Art. 16), that Federal Money is the currency of the United States; the pound, however, is occasionally used.

There are two principal reductions:

1st. To change any sum expressed in Federal money into pounds shillings and pence.

2d. To change any sum expressed in pounds shillings and pence, into Federal money.

For the first,

Multiply the sum in dollars cents and mills, by the value of $1 expressed in the fraction of a pound, and the product will be the corresponding value in pounds and the decimal of a pound.

For the second,

Reduce the shillings and pence to the decimal of a pound by Art. **161**, *and annex the decimal to the entire pounds. Then multiply by the fraction with its terms inverted, which expresses the value of $1 in terms of a pound, and the product will be dollars cents and mills.*

EXAMPLES.

1. What is the value of $375,87, in pounds shillings and pence, New York Currency?

We first multiply by $\frac{2}{5}$, and then reduce the decimal of a pound to shillings and pence.

OPERATION.

$375,87\times\frac{2}{5}=£150.348$
$=£150\ 6s\ 11\frac{1}{2}d.+$

2. What is the value of £127 18*s*. 6*d*., in Federal money, if the currency be 6 shillings to the dollar?

We first reduce the shillings and pence to the fraction of a £, and then multiply by the fraction of a dollar with its terms inverted.

OPERATION.

£127 18*s*. 6*d*.$=127.925$
$127.925\times\frac{10}{3}=\$426,416+$.

3. What is the value of $2863,75 in pounds shillings and pence, Pennsylvania currency?

4. What is the value of £459 3*s*. 6*d*., Georgia currency, in dollars and cents?

5. What is the value of $9763,28, in pounds shillings and pence, North Carolina currency?

6. What is the value, in dollars and cents, of £637 18*s*. 8*d*., Nova Scotia currency?

COMPOUND INTEREST.

223. Compound Interest is when the interest on a sum of money becoming due, and not being paid, is added to the principal, and the interest then calculated on this amount, as on a new principal. For example, suppose I were to borrow of Mr. Wilson $200 for one year, at 6 per cent. If at the end of the year Mr. Wilson should add the interest, $12, to the principal, $200, making $212, and charge interest on this sum till paid, this would be Compound Interest, because it is interest upon interest. Hence,

Calculate the interest to the time at which it becomes due: then add it to the principal and calculate the interest on the amount as on a new principal: add the interest again to the principal and calculate the interest as before: do the same for all the times at which payments of interest become due: from the last result subtract the principal, and the remainder will be the compound interest.

EXAMPLES.

1. What will be the compound interest, at 7 per cent, of $3750 for 4 years, the interest being added yearly?

	$3750,000	principal for 1st year.
$3750 × .07 =	262,500	interest for 1st year.
	4012,500	principal for 2d "
$4012,50 × .07 =	280,875	interest for 2d "
	4293,375	principal for 3d "
$4293,375 × .07 =	300,536+	interest for 3d "
	4593,911+	principal for 4th "
$4593,911 × .07 =	321,573+	interest for 4th "
	4915,484+	amount at 4 years.
1st principal	3750,000	
Amount of interest	$1165,484+	

Quest.—223. What is compound Interest? How do you find the compound interest on any sum?

2. If the interest be computed annually, what will be the interest on $300 for three years, at 6 per cent?

3. What will be the compound interest on $590,74, at 6 per cent, for 2 years, the interest being added annually?

4. What will be the compound interest on $500 for 1 year, at 8 per cent, the interest being computed quarterly?

5. What will be the compound interest on $3758,56 for 3 years, at 7 per cent, the interest being added every 6 months?

6. What will be the compound interest on $95637,50 for 7 years, at 6 per cent, the interest being added annually?

7. What will be the compound interest on $75439,75 for 4 years, at 4½ per cent, the interest being added annually?

A TABLE,

224. Showing the interest of £1, or $1, compounded annually, for any number of years not exceeding 20.

Years.	3 per cent.	3½ per cent.	4 per cent.	4½ per cent.	5 per cent.	6 per cent.
1	.030000	.035000	.040000	.045000	.050000	.060000
2	.060900	.071225	.081600	.092025	.102500	.123600
3	.092727	.108718	.124864	.141166	.157625	.191016
4	.125509	.147523	.169859	.192519	.215506	.262477
5	.159274	.187686	.216653	.246182	.275282	.338226
6	.194052	.229255	.265319	.302260	.340096	.418519
7	.229874	.272279	.315932	.260862	.407100	.503630
8	.266770	.316809	.368569	.422100	.477455	.593848
9	.304773	.362897	.423312	.486095	.551328	.689479
10	.343916	.410599	.480244	.552969	.628895	.790848
11	.384234	.459970	.539454	.622853	.710339	.898299
12	.425761	.511069	.601032	.695881	.795856	1.012196
13	.468534	.563956	.665074	.772196	.885649	1.132928
14	.512590	.618695	.731676	.851945	.979932	1.260904
15	.557967	.675349	.800944	.935282	1.078928	1.396558
16	.604706	.733986	.872981	1.022370	1.182875	1.540352
17	.652848	.794676	.947900	1.113377	1.292018	1.692773
18	.702433	.857489	.025817	1.208479	1.406619	1.854339
19	.753506	.922501	.106849	1.307860	1.526950	2.025600
20	.806111	.989789	.191123	1.411714	1.653298	2.207135

We will now explain the method of finding the compound interest on any sum, for any time, by means of the above table.

Take from the table the interest of £1 or $1 for the same time, and at the same rate, and then multiply the number so found by the principal, and the product will be the compound interest.

EXAMPLES.

1. What will be the compound interest on $350 for three years, at 6 per cent per annum, the interest being computed annually?

Interest from the table on $1 = $0.191016;
then, $0.191016 × 350 = $66.8556.

2. What will be the compound interest on $856,95 for 15 years, at $3\frac{1}{2}$ per cent per annum?

3. What will be the compound interest on $9864,05 for 16 years, the interest being computed annually, at 4 per cent?

4. What will be the compound interest on $1675,20 for 20 years, at $4\frac{1}{2}$ per cent, the interest being computed annually?

5. What will be the compound interest on $5463,25 for 17 years, at 5 per cent, the interest being computed annually?

6. What will be the compound interest on $3769,75 for 18 years, at 3 per cent, the interest being computed annually?

7. What will be the compound interest on £24 17*s.* 6*d.* for 10 years, at 4 per cent, the interest being computed annually?

8. What will be the compound interest on $9854,50 for 12 years, at 6 per cent, the interest being computed annually?

QUEST.—224. How do you find the compound interest on any sum by the table?

LOSS AND GAIN.

225. Loss and Gain is a rule by which merchants discover the amount lost or gained in the purchase and sale of goods. It also instructs them how much to increase or diminish the price of their goods, so as to make or lose so much per cent.

EXAMPLES.

1. Bought a piece of cloth containing 75*yd.* at \$5,25 per yard, and sold it at \$5,75 per yard: how much was gained in the trade?

We first find the profit on a single yard, and then the profit on the 75 yards.

OPERATION.

\$5,75 price of 1 yard.
\$5,25 cost of 1 yard.
50*cts.* profit on 1 yard.

yd.		*yd.*		*cts.*		
1	:	75	: :	50	:	*Ans.*
				75		
				\$37,50		

Ans. \$37,50.

2. A merchant bought a bale of goods containing 125 yards for \$687,75, and sold it at auction for \$4,50 per yard: how much did he lose in all, and how much per yard?

\$687,75 = cost of the bale.
562,50 = price of 125 *yd.* at \$4,50 per *yd.*
\$125,25 = total loss.
125)\$125,25(1,00,2.

Ans. { Total loss \$125,25.
Loss on each *yd.* \$1,00,2.

QUEST.—225. What is the rule of loss and gain?

2. Bought a piece of calico containing 50*yd.* at 2*s.* 6*d.* per yard: what must it be sold for per yard to gain £1 0*s.* 10*d.*?

	50*yd.* at 2*s.* 6*d.*	= £6 5*s.*
	Profit	= £1 0*s.* 10*d.*
It must sell for		£7 5*s.* 10*d.*
		50)£7 5*s.* 10*d.*(2*s.* 11*d.*
		Ans. 2*s.* 11*d.*

3. Bought a hogshead of brandy at $1,25 per gallon, and sold it for $78: was there a loss or gain?

4. A merchant purchased 3275 bushels of wheat for which he paid $3517,10, but finding it damaged is willing to lose 10 per cent: what must it sell for per bushel?

226. In the sale of goods, knowing the per cent of gain, and the amount received, to find the principal or cost.

I sold a parcel of goods for $195,50, on which I made 15 per cent: what did they cost me?

It is evident that the cost added to 15 hundredths of the cost will be equal to what the goods brought, viz., $195,50. If we call the cost 1, then 1 plus $\frac{15}{100}$ of the cost will be equal to what they bring: that is,

$$1 + \frac{15}{100} = \frac{115}{100} = \$195{,}50;$$

or, cost equals $\$195{,}50 \times 100 \div 115 = \$170$.

Hence, to find the cost,

Multiply the amount by 100 *and divide the product by* 100 *plus the per cent of gain, and the quotient will be the cost.*

227. When there is a loss, we have the following method:

If I sell a parcel of goods for $170, by which I lose 15 per cent, what did they cost?

It is evident that the cost, less 15 per cent, that is, less 15 hundredths of the cost, is equal to $170. Hence, 85 hun-

---

Quest.—226. Knowing the per cent of gain and the amount received, how do you find the cost? 227. Knowing the per cent and the amount lost, how do you find the cost?

dredths of the cost is equal to $170; and consequently, the cost is equal to

$$\$170 \times 100 \div 85 = \$200 \text{ cost.}$$

Hence, to find the cost when there is a loss,

Multiply the amount received by 100 *and divide the product by the difference between* 100 *and the per cent lost, and the quotient will be the cost.*

EXAMPLES.

1. Sold cloth at $1,25 per yard and lost 15 per cent: for what should I have sold it to have gained 12 per cent?

2. Sold cloth at $1,25 per yard and lost 15 per cent: what per cent should I have gained had I sold it at $1,6470\frac{10}{17}$ per yard?

3. Sold cloth at $1,6470\frac{10}{17}$ per yard and gained 12 per cent: for what ought I to have sold it to lose 15 per cent?

4. A bought a piece of cotton containing 80 yards, at 6 cents per yard; he sold it for $7\frac{1}{2}$ cents per yard: how much did he gain, and how much per cent?

5. Bought a piece of cloth containing 150 yards for $650: what must it be sold for per yard, in order to gain $300?

6. Bought a quantity of wine at $1,25 per gallon, but it proves to be bad and I am obliged to sell it at 15 per cent less than I gave: how much must I sell it for per gallon?

7. A farmer sells 375 bushels of corn for 75*cts.* per bushel: the purchaser sells it at an advance of 20 per cent: how much did he receive for the corn?

8. A merchant buys one tun of wine for which he pays $725, and wishes to sell it by the hogshead at an advance of 20 per cent: what must he charge per hogshead?

9. A merchant buys 316 yards of calico for which he pays 20 cents per yard; one-half is so damaged that he is obliged to sell it at a loss of 6 per cent; the remainder he sells at an advance of 19 per cent: how much did he gain?

10. If I buy coffee at 16 cents and sell it at 20 cents, how much do I make per cent on the money paid?

11. If I buy tea at 4*s*. per pound and sell it at 4*s*. 9*d*. per pound, how much should I gain on a purchase of £100?

12. A merchant bought 650 pounds of cheese at 10 cents per pound, and sold it at 12 cents per pound: how much did he gain on the whole, and how much per cent on the money laid out?

13. Bought cloth at $2,50 per yard, which proving bad, I wish to sell it at a loss of 18 per cent: how much must I ask per yard?

14. Bought 150 gallons of molasses at 75 cents a gallon, 30 gallons of which leaked out. At what price per gallon must the remainder be sold that I may clear 10 per cent on the cost?

STOCKS AND CORPORATIONS.

228. Stock is a general name for the money contributed by individuals for the establishment of banks and manufacturing companies, and the individuals who contribute the money are called *Stockholders.*

229. The individuals so associated are called, in their collective capacity, a *Corporation;* and the law which defines their rights and powers, is called the Charter of the Bank or Company.

230. The amount of money paid in by the stockholders to carry on the business of the corporation, is called the *Capital.* The capital is generally divided into a certain number of equal parts called *shares*, and the written evidences of ownership of such shares, are called *certificates of stock.*

Quest.—228. What is stock? What are individuals called who own the stock? 229. What are they called in their associated capacity? What is the law called which incorporates them? 230. What is the amount of money paid in by the stockholders called? How is the capital generally divided? What is the evidence of ownership called?

231. When the General Government, or any of the states borrows money for public purposes, an evidence is given to the lender in the form of a bond, bearing a given interest. Such bonds, when given by the United States, are called United States Stock; and when given by any one of the states, are called State Stocks. These bonds or stocks are generally made transferable from one person to another.

232. The nominal or *par* value of a stock is its original cost; that is, the amount named in the certificate or bond. The *market value* is what it will bring when sold. If the market value is above the par value, the stock is said to be at a premium, or *above par;* but if the market value is below the par value, it is then said to be at a discount, or *below par*. For example, if $100 of stock will bring in the market $110, the stock is 10 per cent above par; if, on the contrary, it will bring but $90, it is 10 per cent below par: the percentage of premium or discount being always estimated on the par value.

COMMISSION AND BROKERAGE.

233. A person who buys or sells goods for another, receiving therefor a certain rate per cent, is called a factor or commission merchant; and the percentage on any purchase or sale, is called the *commission*.

234. Dealers in money or stocks are called Brokers, and the amount of their commissions on any purchase or sale, is called the brokerage. The commission for goods or moneys is generally a certain per cent or rate per hundred on the moneys paid out or received, and the amount may be determined by the rules of simple interest.

QUEST.—231. What is United States stock? What are state stocks? 232. What is the nominal or par value of a stock? What is the market value? What do you understand by a stock's being at a premium? What by its being at a discount? 233. What is the business of a commission merchant? 234. What is the business of a broker? How is the commission on goods and moneys generally estimated?

The commission for the purchase and sale of goods varies from $2\frac{1}{2}$ to $12\frac{1}{2}$ per cent, and under some circumstances even higher rates are paid. The brokerage on the purchase and sale of stocks in Wall-street, in the city of New York, is generally one-fourth per cent on the par value of the stock.

EXAMPLES.

1. What is the commission on $4396 at 6 per cent?

We here find the commission, as in simple interest, by multiplying by the decimal which expresses the rate per cent.

OPERATION.

```
 $4396
   .06
$263,76
```

Ans. $263,76.

2. A factor sells 120 bales of cotton at $425 per bale, and is to receive $2\frac{1}{2}$ per cent commission: how much must he pay over to his principal?

3. A sent to B, a broker, $3825 to be invested in stock: B is to receive 2 per cent on the amount paid for the stock: what was the value of the stock purchased?

As B is to receive 2 per cent, it follows that $102 of A's money will purchase but $100 of stock: hence, 100+ the commission, is to 100, as the given sum to the value of the stock which it will purchase. Hence, to find the value of the stock purchased,

OPERATION.

```
100
  2
102 : 100 : : 3825 : Ans
                100
       102)382500(3750
           306
            765
            714
             510
             510
```

Ans. $3750.

Multiply the amount to be invested by 100 *and divide the product by* 100 *plus the brokerage.*

QUEST.—What is the general commission on the purchase and sale of goods? How may it be determined? What is the customary brokerage on the purchase and sale of stocks?

PROOF.

Amount paid - - - - - - - -	\$3750
Brokerage on \$3750, at 2 per cent =	75
Total sum - -	\$3825

4. I have \$5000 to be laid out in stocks which are 15 per cent below par: how much will it purchase at the par or nominal value?

It is plain that every 85 dollars will purchase stock of the par value of \$100: hence,

\$85 : \$100 : : \$5000 : *Ans.*

Therefore, to find how much will be purchased at the par value, when the stock is below par,

Multiply the sum to be invested by 100 *and divide the product by* 100 *minus the discount.*

5. A person has \$7000 which he wishes to invest: what will it purchase in 5 per cent stocks, at $3\frac{1}{2}$ per cent below par, if he pays $\frac{1}{4}$ per cent brokerage?

6. How much 6 per cent stock can be purchased for \$8500, at $8\frac{1}{2}$ per cent premium, $\frac{1}{4}$ per cent being paid to the broker?

7. A factor receives \$708,75, and is directed to purchase iron at \$45 per ton: he is to receive 5 per cent on the money paid: how much iron can he purchase?

8. Messrs. P, W, & K buy 200 shares of United States stock for Mr. A. The par value is \$1000 dollars a share, the stock is at a premium of $6\frac{1}{2}$ per cent, and their brokerage is one-fourth per cent. How much must A pay them for his stock?

9. Messrs. P, W, & K receive \$28750 to be invested in stock. They charge $\frac{1}{2}$ per cent commission on the amount paid: what is the value of the stock purchased?

10. The par value or first cost of 257 shares of bank stock was \$200 per share: what is the present value, if the stock is at a premium of 15 per cent, that is, 15 per cent above par?

11. What would be the value of the stock named in the last example, if it were at a discount of 10 per cent?

12. One hundred shares of United States Bank stock was worth $18\frac{1}{2}$ per cent premium: the par value being $200 per share, what was the value of the 100 shares?

13. A bank fails, and has in circulation bills to the amount of $267581. It can pay $9\frac{1}{2}$ per cent: how much money is there on hand?

14. Sixty-nine shares of bank stock, of which the par value is $125, is at a discount of 8 per cent: what is its value?

15. My commission merchant sells goods to the amount of $1000, on which I allow him a commission of 2 per cent; and as he pays over before the money becomes due, I allow him $1\frac{1}{2}$ per cent: how much am I to receive?

16. My broker receives from me $2000 to be laid out in stocks: what will be the value of my stocks after allowing him $\frac{1}{4}$ per cent commission?

17. I sold $13921,60 worth of goods for a merchant at a commission of $2\frac{1}{2}$ per cent: how much ought I to pay over to my principal?

18. I remitted to my agent $14760 to lay out in the purchase of iron. He takes $3\frac{1}{2}$ per cent on the whole sum for his commission, and then buys iron at 95 dollars per ton: how much does he purchase?

BANKING.

235. Banks are corporations created by law for the purpose of receiving deposites, loaning money, and furnishing a paper circulation represented by specie.

The notes made by a bank circulate as money, because they are payable in specie on presentation at the bank. They are called *bank notes*, or *bank bills*.

Quest.—235. What are banks? Why do the notes of a bank circulate as money? What are they called?

236. The note of an individual, or as it is generally called, a promissory note or note of hand, is a positive engagement, in writing, to pay a given sum at a time specified, and to a person named in the note, or to his order, or sometimes to the bearer at large.

FORMS OF NOTES.

Negotiable Note.

No. 1.
$25,50. Providence, May 1, 1846.

For value received I promise to pay on demand, to Abel Bond, or order, twenty-five dollars and fifty cents.

REUBEN HOLMES.

Note Payable to Bearer.

No. 2.
$875,39. St. Louis, May 1, 1845.

For value received I promise to pay, six months after date, to John Johns, or bearer, eight hundred and seventy-five dollars and thirty-nine cents.

PIERCE PENNY.

Note by two Persons.

No. 3.
$659,27. Buffalo, June 2, 1846.

For value received we, jointly and severally, promise to pay to Richard Ricks, or order, on demand, six hundred and fifty nine dollars and twenty-seven cents.

ENOS ALLAN.
JOHN ALLAN.

Note Payable at a Bank.

No. 4.
$20,25. Chicago, May 7, 1846.

Sixty days after date, I promise to pay John Anderson, or order, at the Bank of Commerce in the city of New York, twenty dollars and twenty-five cents, for value received.

JESSE STOKES.

QUEST.—236. What is a promissory note?

Remarks relating to Notes.

1. The person who signs a note, is called the *drawer* or *maker* of the note; thus Reuben Holmes is the drawer of note No. 1.

2. The person who has the rightful possession of a note, is called the *holder* of the note.

3. A note is said to be *negotiable* when it is made payable to A B, or order, who is called the payee, (see No. I.) Now, if Abel Bond, to whom this note is made payable, writes his name on the back of it, he is said to *endorse* the note, and he is called the endorser; and when the note becomes due, the holder must first demand payment of the maker, Reuben Holmes, and if he declines paying it, the holder may then require payment of Abel Bond, the endorser.

4. If the note is made payable to A B, or bearer, then the drawer alone is responsible, and he must pay to any person who holds the note.

5. The time at which a note is to be paid should always be named, but if no time is specified, the drawer must pay when required to do so, and the note will draw interest after the payment is demanded.

6. When a note, payable at a future day, becomes due, it will draw interest, though no mention is made of interest.

7. In each of the States there is a *rate* of interest established by law, which is called the legal interest, and when no rate is specified, the note will always draw legal interest. If a rate *higher* than legal interest be taken, the drawer, in most of the States, is not bound to pay the note.

8. If two persons jointly and severally give their note, (see No. 3,) it may be collected of either of them.

9. The words "For value received," should be expressed in every note.

QUEST.—1. What is the person called who signs a note? 2. What is the person called who owns it? 3. When is a note said to be negotiable? What is the person called to whom a note is made payable? When the payee writes his name on the back, what is he said to do? What is he then called? 4. If a note is made payable to A B, who is responsible for its payment? 5. If no time is specified, when is a note to be paid? 6. Will a note draw interest after it falls due, if not stated in the note? 7. If the rate of interest named in a note is higher than the legal rate, can the amount of the note be collected? 8. If two persons jointly and severally give a note, of whom may it be collected? 9. What words should be put in every note?

10. When a note is given, payable on a fixed day, and in a specific article, as in wheat or rye, payment must be offered at the specified time, and if it is not, the holder can demand the value in money.

237. By mercantile usage and the custom of banks, a note does not really fall due until the expiration of 3 days after the time mentioned on its face. For example, Note No. 1 would be due on the 4th of November, and the three additional days are called *days of grace.*

When the last day of grace happens to be a Sunday, or a holiday, such as New Year's or the 4th of July, the note must be paid the day before; that is, on the second day of grace.

BANK DISCOUNT.

238. BANK DISCOUNT is the charge made by a bank for the payment of money on a note before it becomes due. By the custom of banks, this discount is the interest on the amount named in the note, from the time the note is discounted to the time when it falls due, in which time the three days of grace are always included. The amount named in a note is called the face of it.

The PRESENT VALUE of a note is the difference between the face of the note and the discount.

239. There are two kinds of notes discounted at banks: 1st. Notes given by one individual to another for property actually sold—these are called *business notes,* or *business paper.* 2d. Notes made for the purpose of borrowing money,

---

QUEST.—10. If a note is made payable on a fixed day and in a specified article, and is not paid, what may be done? 237. How long is the time for the payment of a note extended by mercantile usage? What are these days called? When the last day of grace falls on a Sunday, or holiday, when must the note be paid? 238. What is bank discount? How is it estimated? How is it estimated by the custom of banks? What is the face of a note? What is the present value of a note? 239. How many kinds of notes are discounted at banks? What distinguishes one kind from the other, and what are they called?

which are called *accommodation notes*, or *accommodation paper*. The first class of paper is much preferred by the banks, as more likely to be paid when it falls due, or in mercantile phrase, " when it comes to maturity."

Hence, to find the bank discount on a note,

Add 3 days to the time which the note has to run before it becomes due, and calculate the interest for this time at the given rate per cent.

EXAMPLES.

1. What is the bank discount of a note of $1000 payable in 60 days, at 6 per cent interest? This note will have 63 days to run.

2. A merchant sold a cargo of cotton for $15720, for which he receives a note at 6 months: how much money will he receive at a bank for this note, discounting it at 6 per cent interest?

3. What is the bank discount on a note of $556,27 payable in 60 days, discounted at 6 per cent per annum?

4. A has a note against B for $3456, payable in three months; he gets it discounted at 7 per cent interest: how much does he receive?

5. What is the bank discount on a note of $367,47, having 1 year, 1 month, and 13 days to run; as shown by the face of the note, discounted at 7 per cent?

6. For value received I promise to pay to John Jones, four months from the 17th of July next, six thousand five hundred and seventy-nine dollars and 15 cents. What will be the discount on this, if discounted on the 1st of August, at 6 per cent per annum?

240. It is often necessary to make a note, of which the present value shall be a given amount. For example, if I wish to receive at bank the sum of two hundred dollars, for what amount must I give my note payable in three months?

QUEST.—Which kind is preferred? How do you find the bank discount on a note? 240. What is often necessary in bank business?

If we calculate the interest on one dollar for the time, which will be 3 months added to the 3 days of grace, and at the same rate per cent, this will be the bank discount on $1 payable in 3 months ; and if this discount be subtracted from one dollar, the remainder will be the present value of one dollar, to be paid at the end of 3 months. Therefore,

Pres. val. of $1 : pres. val. of note : : $1 : amount of note.

Hence, to find the face of a note, due at a future time and bearing a given interest, that shall have a known present value,

*Find the present value of $1 for the same time and at the same rate of interest, by which divide the present value of the note, and the quotient will be the face of the note.**

EXAMPLES.

1. For what sum must a note be drawn at 3 months, so that when discounted at a bank, at 6 per cent, the amount received shall be $500?

Interest on $1 for the time, 3*mo.* and 3*da.*=$0,0155, which taken from $1, gives present value of $1 = 0,9845 ; then

$500 ÷ 0,9845 = 507,872 + = face of note.

PROOF.

Bank interest on $507,872 for 3 months, including 3 days of grace, at 6 per cent = 7,872, which being taken from the face of the note, leaves $500 for its present value.

2. For what sum must a note be drawn, at seven per cent, payable on its face in 1 year 6 months and 14 days, so that when discounted at bank it shall produce $307,27 ?

3. A note is to be drawn having on its face 8 months and 12 days to run, and to bear an interest of 7 per cent, so that it will pay a debt of $5450 : what is the amount ?

* The first rule founded on the above well-known principle, was, it is believed, first published by Roswell C. Smith, in his New Arithmetic.

QUEST.—What will be the present value of one dollar due in 3 months? How will you find the face of a note, of a given present value, that shall be payable at a future time?

4. What sum, 6 months and 9 days from July 18th, 1846, drawing an interest of 6 per cent, will pay a debt of $674,89 at bank, on the 1st of August, 1846?

5. Mr. Johnson has Mr. Squires' note for $874,57, having 4 months to run, from July 13th, and bearing an interest of 5 per cent. On the 1st of October he wishes to pay a debt at bank of $750,25, and gives the note in payment: how much must he receive back from the bank?

6. What must be the amount of a note discounted at 6 pr ct. having 4 months and 7 days to run, to pay a debt of $1475,50?

7. Mr. Jones, on the 1st of June, desires to pay a debt at bank by a note dated May 16th, having 6 months to run and drawing 7 per cent interest: for what amount must the note be drawn the debt being $1683,75?

8. What amount at the end of one year, with grace, interest at 5 per cent, will pay $1004,20 at bank?

DISCOUNT.

241. If I give my note to Mr. Wilson for $106, payable in one year, the true present value of the note will be less than $106 by the interest on its *present value* for one year; that is, its true present value will be $100.

The true present value of a note is that sum which being put at interest until the note becomes due, would increase to an amount equal to the face of the note. Thus, $100 is the true present value of the note to Mr. Wilson.

The discount is the difference between the face of a note and its true present value. Thus, $6 is the discount on the note to Mr. Wilson.

To find the true present value of a note due at a future time, find the interest of $1 for the same time; then,

$1 + its interest : $1 : given sum : its present value.

Quest.—241. What is the true present value of a note? What is the true discount? How do you find the true present value of a note due at a future time?

Hence, to find the present value of any sum,

Add one dollar to its interest for the given time and divide the given amount by this number, and the quotient will be the present value.

EXAMPLES.

1. What is the present value of a note for $1828,75, due in one year, without grace, and bearing an interest of 4½ per cent per annnm?

$1 + its interest for the given time = $1,045:

Hence, $1828,75 ÷ $1,045 = $1750 the present value.

PROOF.

Int. on $1750 for 1 year, at 4½ per cent =	$78,75
Add principal - - - - - - - - - -	1750
Amount - -	$1828,75

2. A note of $1651,50 is due in 11 months, without grace, but the person to whom it is payable sells it with the discount off at 7 per cent: how much shall he receive?

3. How much ought Mr. Ready to pay in cash for his note of £36, due 15 months hence, without grace, it being discounted at 5 per cent?

242. Note.—When payments are to be made at different times, *find the present value of the sums separately, and their sum will be the present value of the note.*

4. What is the present value of a note for $10500, on which $900 are to be paid in six months; $2700 in one year; $3900 in eighteen months; and the residue at the expiration of two years, all without grace, the rate of interest being 6 per cent per annum?

5. What is the discount of £4500, one-half payable in 6 months and the other half at the expiration of a year, without grace, at 7 per cent per annum?

Quest.—242. When payments are made at different times, how do you find the true present value?

6. What is the present value of $5760, one-half payable in 3 months, one-third in 6 months, and the rest in 9 months, without grace, at 6 per cent per annum?

7. Mr. A gives his note to B for $720, one-half payable in 4 months and the other half in 8 months, without grace: what is the present value of said note, discount at 5 per cent per annum?

8. What is the present value of £825 payable as follows: one-half in 3 months, one-third in 6 months, and the rest in 9 months, without grace, the discount being 6 per cent per annum?

9. Bought goods for £750 ready money, and sold them for £900 payable by a note at 6 months, without grace: now, if I discount the note at 6 per cent per annum, will I make or lose?

10. What is the present value of $4000 payable in 9 months, without grace, discount $4\frac{1}{2}$ per cent per annum?

11. How much corn must I carry to a miller that I may receive a bushel of meal, $\frac{1}{16}$ being allowed for toll and waste?

12. Mr. Johnson has a note against Mr. Williams for $2146,50, dated August 17th, 1838, which becomes due Jan. 11th, 1839: if the note is discounted at 6 per cent, what ready money must be paid for it September 25th, 1838?

13. C owes D $3456, to be paid October 27th, 1842: C wishes to pay on the 24th of August, 1838, to which D consents: how much ought D to receive, interest at 6 per cent?

14. What is the present value of a note of $4800, due 4 years hence, without grace, the interest being computed at 5 per cent per annum?

15. A man having a horse for sale, offered it for $225 cash in hand, or 230 at 9 months, without grace; the buyer chose the latter: did the seller lose or make by his offer, supposing money to be worth 7 per cent?

INSURANCE.

243. Insurance is an agreement, generally in writing, by which an individual or company bind themselves to exempt the owners of certain property, such as ships, goods, houses, &c., from loss or hazard.

The written agreement made by the parties, is called the *policy*.

The amount paid by him who owns the property to those who insure it, as a compensation for their risk, is called the *premium*. The premium is generally so much per cent on the property insured, and is found by the rules for simple interest.

EXAMPLES.

1. What would be the premium for the insurance of a house valued at $8754 against loss by fire for 1 year, at $\frac{1}{2}$ per cent?

By multiplying by .01, we have the insurance at 1 per cent - - - - - - - 87,54

The half, is the insurance at half per cent - $43,77.

2. What would be the premium for insuring a ship and cargo, valued at $147674, from New York to Liverpool, at $3\frac{1}{2}$ per cent?

3. What would be the insurance on a ship valued at $47520, at $\frac{1}{2}$ per cent? Also at $\frac{1}{3}$ per cent?

4. What would be the insurance on a house valued at $16800, at $1\frac{1}{2}$ per cent? Also at $\frac{3}{4}$ per cent? At $\frac{1}{2}$ per cent? At $\frac{1}{3}$ per cent? At $\frac{1}{4}$ per cent?

5. What is the insurance on a store and goods valued at $47000, at $2\frac{1}{4}$ per cent? At 2 per cent? At $1\frac{1}{2}$ per cent? At $\frac{3}{4}$ per cent? At $\frac{1}{2}$ per cent? At $\frac{1}{4}$ per cent? At $\frac{1}{5}$ per cent? At $\frac{1}{6}$ per cent?

Quest.—243. What is insurance? What is the written agreement called? What is the amount paid for the insurance called? How are the premiums generally estimated? How are they found?

6. A merchant wishes to insure on a vessel and cargo at sea, valued at $28800: what will be the premium at $1\frac{3}{4}$ per cent?

7. What is the premium on $2250 at $1\frac{3}{4}$ per cent?

8. What is the premium on $8750 at $3\frac{1}{2}$ per cent?

9. A merchant owns three-fourths of a ship valued at $24000, and insures his interest at $2\frac{1}{2}$ per cent: what does he pay for his policy?

10. A merchant learns that his vessel and cargo, valued at $36000, have been injured to the amount of $12000; he effects an insurance on the remainder at $5\frac{1}{2}$ per cent: what premium does he pay?

11. What is the insurance on my house, valued at $7500, at $\frac{1}{4}$ per cent?

ASSESSING TAXES.

244. A TAX is a certain sum required to be paid by the inhabitants of a town, county, or state, for the support of government. It is generally collected from each individual, in proportion to the amount of his property.

In some states, however, every white male citizen over the age of twenty-one years is required to pay a certain tax. This tax is called a poll-tax; and each person so taxed is called a *poll.*

245. In assessing taxes, the first thing to be done is to make a complete inventory of all the property in the town on which the tax is to be laid. If there is a poll tax, make a full list of the polls and multiply the number by the tax on each poll, and subtract the product from the whole tax to be raised by the town; the remainder will be the amount to be raised on the property. Having done this, *divide the whole*

QUEST.—244. What is tax? How is it generally collected? What is a poll-tax? 245. What is the first thing to be done in assessing a tax? If there is a poll-tax, how do you find the amount? How then do you find the per cent of tax to be levied on a dollar?

tax to be raised by the amount of taxable property, and the quotient will be the tax on $1. Then multiply this quotient by the inventory of each individual, and the product will be the tax on his property.

EXAMPLES.

1. A certain town is to be taxed $4280; the property on which the tax is to be levied is valued at $1000000. Now there are 200 polls, each taxed $1,40. The property of A is valued at $2800, and he pays 4 polls,

B's at $2400, pays 4 polls,	E's at $7242, pays 4 polls,
C's at $2530, pays 2 "	F's at $1651, pays 6 "
D's at $2250, pays 6 "	G's at $1600,80 pays 4 "

What will be the tax on one dollar, and what will be A's tax, and also that of each on the list?

First, $1,40 × 200 = $280 amount of poll-tax.

$4280 − $280 = $4000 amount to be levied on property.

Then, $4000 ÷ $1000000 = 4 mills on $1.

Now, to find the tax of each, as A's, for example,

A's inventory - - -	$2800
	,004
	11,20
4 polls at $1,40 each -	5,60
A's whole tax - - -	$16,80

In the same manner the tax of each person in the township may be found.

246. Having found the per cent, or the amount to be raised on each dollar, form a table showing the amount which certain sums would produce at the same rate per cent. Thus, after having found, as in the last example, that four mills are to be raised on every dollar, we can, by multiplying in succession by the numbers 1, 2, 3, 4, 5, 6, 7, 8, &c., form the following

QUEST.—How do you then find the amount to be levied on each individual? 246. How do you form an assessment table?

TABLE.

$		$	$		$	$		$
1	gives	0,004	20	gives	0,080	300	gives	1,200
2	"	0,008	30	"	0,120	400	"	1,600
3	"	0,012	40	"	0,160	500	"	2,000
4	"	0,016	50	"	0,200	600	"	2,400
5	"	0,020	60	"	0,240	700	"	2,800
6	"	0,024	70	"	0,280	800	"	3,200
7	"	0,028	80	"	0,320	900	"	3,600
8	"	0,032	90	"	0,360	1000	"	4,000
9	"	0,036	100	"	0,400	2000	"	8,000
10	"	0,040	200	"	0,800	3000	"	12.000

This table shows the amount to be raised on each sum in the columns under $'s.

1. To find the amount of B's tax from this table.

B's tax on $2000	- -	is -	$8,000
B's tax on 400	- -	is -	$1,600
B's tax on 4 polls, at $1,40		-	$5,600
B's total tax	- -	is -	$15,200

2. To find the amount of C's tax from the table.

C's tax on $2000	- -	is -	$8,000
C's tax on 500	- -	is -	$2,000
C's tax on 30	- -	is -	$0,120
C's tax on 2 polls	- -	is -	$2,800
C's total tax	- -	is -	$12,920

In a similar manner, we might find the taxes to be paid by D, E, &c.

2. In a county embracing 350 polls, the amount of property on the tax list is $318200; the amount to be raised is as follows: for state purposes $1465,50; for county purposes $350,25; and for town purposes $200,25. By a vote of the county, a tax is levied on each poll of $1,50: how much per cent will be laid upon the property?

3. In a county embracing a population of 98415 persons, a tax is levied for town, county, and state purposes, amount-

ing to $100406. Of this sum, a part is to be raised by a tax of 25 cents on each poll, and the remainder by a tax of two mills on the dollar: what was the amount of property on the tax list?

EQUATION OF PAYMENTS.

247. I owe Mr. Wilson $2 to be paid in 6 months; $3 to be paid in 8 months; and $1 to be paid in 12 months. I wish to pay his entire dues at a single payment, to be made at such a time, that neither he nor I shall lose interest: at what time must the payment be made?

The method of finding the mean time of payment of several sums due at different times, is called *Equation of Payments.*

Taking the example above,

Int. of $2 for 6*mo.*	= int. of $1 for 12*mo.*	$2 \times 6 = 12$
" of $3 for 8*mo.*	= int. of $1 for 24*mo.*	$3 \times 8 = 24$
" of $1 for 12*mo.*	= int. of $1 for 12*mo.*	$1 \times 12 = 12$
$6	48	48

The interest on all the sums, to the times of payment, is equal to the interest of $1 for 48 months. But 48 is equal to the sum of all the products which arise from multiplying each sum by the time at which it becomes due: hence, the sum of the products is equal to the time which would be necessary for $1 to produce the same interest as would be produced by all the sums.

Now, if $1 will produce a certain interest in 48 months, in what time will $6 (or the sum of the payments) produce the same interest? The time is obviously found by dividing 48 (the sum of the products) by $6, (the sum of the payments.)

Quest.—247. What is Equation of Payments? What is the sum of the products, which arise from multiplying each payment by the time to which it becomes due, equal to?

Hence, to find the mean time,

Multiply each payment by the time before it becomes due, and divide the sum of the products by the sum of the payments: the quotient will be the mean time.

EXAMPLES.

1. B owes A $600: $200 is to be paid in two months, $200 in four months, and $200 in six months: what is the mean time for the payment of the whole?

We here multiply each sum by the time at which it becomes due, and divide the sum of the products by the sum of the payments.

OPERATION.

$$\begin{array}{lr} 200 \times 2 = & 400 \\ 200 \times 4 = & 800 \\ 200 \times 6 = & 1200 \\ \hline 6|00 &)24|00 \\ \hline & 4 \end{array}$$

Ans. 4 *months*

2. A merchant owes $1200, of which $200 is to be paid in 4 months, $400 in 10 months, and the remainder in 16 months: if he pays the whole at once, at what time must he make the payment?

3. A merchant owes $1800 to be paid in 12 months, $2400 to be paid in 6 months, and $2700 to be paid in 9 months: what is the equated time of payment?

4. A owes B $2400; one-third is to be paid in 6 months, one fourth in 8 months, and the remainder in 12 months: what is the mean time of payment?

5. A merchant has due him $600 to be paid in 30 days, $1000 to be paid in 60 days, and 1500 to be paid in 90 days: what is the equated time for the payment of the whole?

6. A merchant has due him $4500; one-sixth is to be paid in 4 months, one-third in 6 months, and the rest in 12 months: what is the equated time for the payment of the whole?

QUEST.—How do you find the mean time of payment? When you reckon the time from the date at which the first payment becomes due, do you include the first payment?

NOTE 1.—If one of the payments is due on the day from which the equated time is reckoned, its corresponding product will be nothing, but the payment must still be added in finding the sum of the payments.

7. I owe $1000 to be paid on the 1st of January, $1500 on the 1st of February, $3000 on the 1st of March, and $4000 on the 15th of April: reckoning from the 1st of January, and calling February 28 days, on what day must the money be paid?

NOTE 2.—In finding the equated time of payments for several sums, due at different times, any day may be assumed as the one from which we reckon. Thus, if I owe Mr. Wilson $100 to be paid on the 15th of July, $208 on the 15th of August, and $300 on the 9th of September, and we require the mean time of a single payment, it would be most convenient to estimate from the 1st of July.

From 1st of July to 1st payment 14 days
" " " to 2d payment 45 days
" " " to 3d payment 70 days.

Then, by rule given above, we have,

$$\begin{array}{lr} 100 \times 14 = & 1400 \\ 200 \times 45 = & 9000 \\ 300 \times 70 = & 21000 \\ \hline 600 & 6|00)314|00 \\ \hline & 52\frac{1}{3} \end{array}$$

Hence, the amount will fall due in $52\frac{1}{3}$ days from the 1st of July; that is, on the 22d day of August.

But we may, if we please, demand at what time the payment would be due from the 1st of June.

From June 1st to 1st payment 44 days
" " " to 2d payment 75 days
" " " to 3d payment 100 days.

Thus,

$$\begin{array}{lr} 100 \times \ \ 44 = & 4400 \\ 200 \times \ \ 75 = & 15000 \\ 300 \times 100 = & 30000 \\ \hline 600 & 6|00)494|00 \\ \hline & 82\frac{1}{3} \end{array}$$

Hence, the payment becomes due in $82\frac{1}{3}$ days from June 1st, or on the 22d of August—the same as before.

Any day may, therefore, be taken as the one from which the mean time is estimated.

9. Mr. Jones purchased of Mr. Wilson, on a credit of six months, goods to the following amounts:

15th of January, a bill of $3750,
10th of February, a bill of 3000,
6th of March, a bill of 2400,
8th of June, a bill of 2250.

He wishes, on the 1st of July, to give his note for the amount; at what time must it be made payable?

10. Mr. Gilbert bought $4000 worth of goods: he was to pay $1600 in five months, $1200 in six months, and the remainder in eight months: what will be the time of credit, if he pays the whole amount at a single payment?

11. A owes B $1200, of which $240 is to be paid in three months, $360 in five months, and the remainder in ten months what is the mean time of payment?

12. A merchant bought several lots of goods, as follows:

A bill of $650, June 6th,
Do. of 890, July 8th,
Do. of 7940, August 1st.

Now, if the credit is 6 months, at what time will the whole become due?

13. Mr. Swain bought goods to the amount of $3840, to be paid for as follows, viz.: one-fourth in cash, one-fourth in 6 months, one-fourth in 7 months, and the remainder in one year: what is the average time of payment?

14. Mr. Johnson sold, on a credit of 8 months, the following bills of goods:

April 1st, a bill of $4350,
May 7th, a bill of 3750,
June 5th, a bill of 2550.

At what time will the whole become due?

PARTNERSHIP OR FELLOWSHIP.

248. Partnership or Fellowship is the joining together of several persons in trade, with an agreement to share the losses and profits according to the amount which each one puts into the partnership. The money employed is called the *Capital Stock.*

The gain or loss to be shared is called the *Dividend.*

It is plain that the whole stock which suffers the gain or loss, must be to the gain or loss, as the stock of any individual to his part of the gain or loss. Hence,

As the whole stock is to each man's share, so is the whole gain or loss to each man's share of the gain or loss.

PROOF.

Add all the separate profits or shares together; their sum should be equal to the gross profit or stock.

EXAMPLES.

1. A and B buy certain merchandise amounting to £160, of which A pays £90, and B £70: they gain by the purchase £32: what is each one's share of the profits?

A - - £90
B - - £70

£160 : { 90 / 70 } : : £32 : { £18 A's share. / £14 B's share. }

2. A and B have a joint stock of $4200, of which A owns $3600, and B $600: they gain in a year $2000: what is each one's share of the profits?

3. A, B, C, and D have £40,000 in trade: at the end of six months their profits amount to £16,000: what is each one's share, supposing A to receive £50 and D £30 out of the profits, for extra services?

Quest.—248. What is Partnership, or Fellowship? What is the gain or loss called? What is the rule for finding each one's share?

4. Five persons, A, B, C, D, and E, have to share between them an estate of $20,000: A is to have one-fourth, B one-eighth, C one-sixth, D one-eighth, and E what is left: what will be the share of each?

DOUBLE FELLOWSHIP.

249. WHEN several persons who are joined together in trade, employ their capital for different periods of time, the partnership is called *Double Fellowship.*

For example, suppose A puts $100 in trade for 5 years, B $200 for 2 years, and C $300 for 1 year: this would make a case of double fellowship.

Now it is plain that there are two circumstances which should determine each one's share of the profits: 1*st*, *The amount of capital he puts in; and* 2*dly*, *The time which it is continued in the business.*

Hence, each man's share should be proportional to the capital he puts in, multiplied by the time it is continued in trade. Therefore, to find each share,

Multiply each man's stock by the time he continues it in trade; then say, as the sum of the products is to each particular product, so is the whole gain or loss to each man's share of the gain or loss.

EXAMPLES.

1. A and B enter into partnership: A puts in £840 for 4 months, and B puts in £650 for 6 months: they gain £300: what is each one's share of the profits?

A's stock £840 × 4 = 3360
B's stock £650 × 6 = 3900

£7260 : { 3360 / 3900 } :: £300 : { £138 16s. 10d. / £161 3s. 1d. }

	£	s.	d.
A	138	16	10
B	161	3	1

QUEST.—249. What is Double Fellowship? What two circumstances determine each one's share of the profits? Give the rule for finding each one's share.

2. A put in trade £50 for 4 months, and B £60 for 5 months: they gained £24: how is it to be divided between them?

3. C and D hold a pasture together, for which they pay £54: C pastures 23 horses for 27 days, and D 21 horses for 39 days: how much of the rent ought each one to pay?

GENERAL EXAMPLES IN FELLOWSHIP.

1. A bankrupt is indebted $2729, viz.: to A $509,37; to B $228; to C $1291,23; and to D $709,40; but his estate is only worth $2046,75. How much can he pay on the dollar, and how much will each creditor receive?

2. A, B, and C send a ship to sea, which together with her cargo was worth $15000. A and B owned each one-fifth, and C the rest. They gained $1250: how much did each pay towards the ship and cargo, and what did each receive of the profits?

3. A man bequeathed his estate to his four sons in the following manner, viz.: to his first $5000; to his second $4500; to his third $4500; and to his fourth $4000. But on settling his estate, it was found that after paying debts, charges, &c., only $12000 remained to be divided: how much must each receive?

4. A widow and her two sons have a legacy of $4500, of which the widow is to have one-half and the sons each one-fourth. Now suppose the eldest son to relinquish his share, and the whole to be divided in the above proportions between the mother and youngest son, what will each receive?

5. Suppose premiums to the value of $12 are to be distributed in a school in the following manner. The premiums are divided into three grades. The value of a premium of the first grade is twice the value of one of the second; and the value of one of the second grade twice that of the third. Now there are 6 to receive premiums of the first grade, 12 of the second, and 6 of the third: what will be the value of a single premium of each grade?

6. Four traders form a company: A puts in $300 for 5 months, B $600 for 7 months, C $960 for 8 months, D $1200 for 9 months. In the course of trade they lost $750: how much falls to the share of each?

7. A and B lay out certain sums in merchandise amounting to $320, of which A pays $180 and B $140; they gain by the purchase $64: what is each one's share?

ALLIGATION MEDIAL.

250. A MERCHANT mixes 8*lb.* of tea, worth 75*cts.* per pound, with 16*lb.* worth $1,02 per pound: what is the value of the mixture per pound?

The manner of finding the price of this mixture is called *Alligation Medial.* Hence,

ALLIGATION MEDIAL *teaches the method of finding the price of a mixture when the simples of which it is composed, and their prices, are known.*

In the example above, the simples 8*lb.* and 16*lb.*, and also their prices per pound, 75*cts.* and $1,02, are known.

8*lb.* of tea at 75*cts.* per *lb.* - - - - - -	6,00
16*lb.* " " $1,02 per *lb.* - - - - - -	16,32
24 sum of simples. Total cost	$22,32

Now, if the entire cost of the mixture, which is $22,32, be divided by 24, the number of pounds or sum of the simples, the quotient 93*cts.* will be the price per pound. Hence, to find the price of the mixture,

OPERATION.

```
24)22,32(93cts.
   216
   ---
    72
    72
    --
```

Divide the entire cost of the whole mixture by the sum of the simples, and the quotient will be the price of the mixture.

QUEST.—250. What is Alligation Medial? How do you find the price of the mixture?

EXAMPLES.

1. A farmer mixes 30 bushels of wheat worth 5*s.* per bushel, with 72 bushels of rye at 3*s.* per bushel, and with 60 bushels of barley worth 2*s.* per bushel: what is the value of a bushel of the mixture?

30 bushels of wheat at 5*s.*	- -	150*s.*
72 " " rye at 3*s.*	- -	216*s.*
60 " " barley at 2*s.*	- -	120*s.*
162		162)486(3*s.*
		486
		Ans. 3*s.*

2. A wine merchant mixes 15 gallons of wine at $1 per gallon with 25 gallons of brandy worth 75 cents per gallon: what is the value of a gallon of the compound?

3. A grocer mixes 80 gallons of whiskey worth 31*cts.* per gallon with 6 gallons of water, which costs nothing: what is the value of a gallon of the mixture?

4. A goldsmith melts together 2*lb.* of gold of 22 carats fine, 6*oz.* of 20 carats fine, and 6*oz.* of 16 carats fine: what is the fineness of the mixture?

5. On a certain day the mercury in the thermometer was observed to average the following heights: from 6 in the morning to 9, 64°; from 9 to 12, 74°; from 12 to 3, 84°; and from 3 to 6, 70°: what was the mean temperature of the day?

ALLIGATION ALTERNATE.

251. A FARMER would mix oats worth 3*s.* per bushel with wheat worth 9*s.* per bushel, so that the mixture shall be worth 5*s.* per bushel: what proportion must be taken of each sort?

The method of finding how much of each sort must be taken is called *Alligation Alternate.* Hence,

Alligation Alternate *teaches the method of finding what proportion must be taken of several simples, whose prices are known, to form a compound of a given price.*

Alligation Alternate is the reverse of Alligation Medial, and may be proved by it.

For a first example, let us take the one above stated. If oats worth 3*s.* per bushel be mixed with wheat worth 9*s.*, how much must be taken of each sort that the compound may be worth 5*s.* per bushel?

5	3	4 Oats.
	9	2 Wheat.

If the price of the mixture were 6*s.*, half the sum of the prices of the simples, it is plain that it would be necessary to take just as much oats as wheat.

But since the price of the mixture is *nearer* to the price of the oats than to that of the wheat, less wheat will be required in the mixture than oats.

Having set down the prices of the simples under each other, and linked them together, we next set 5*s.*, the price of the mixture, on the left. We then take the difference between 9 and 5 and place it opposite 3, the price of the oats, and also the difference between 5 and 3, and place it opposite 9, the price of the wheat. The difference standing opposite each kind shows how much of that kind is to be taken. In the present example, the mixture will consist of 4 bushels of oats and 2 of wheat; and any other quantities, bearing the same proportion to each other, such as 8 and 4, 20 and 10, &c., will give a mixture of the same value.

PROOF BY ALLIGATION MEDIAL.

4 bushels of oats at 3*s.* - - -	12*s.*
2 bushels of wheat at 9*s.* - -	18*s.*
6	6)30
	Ans. 5*s.*

Quest.—251. What is Alligation Alternate? How do you prove Alligation Alternate?

CASE I.

252. To find the proportion in which several simples of given prices must be mixed together, that the compound may be worth a given price.

I. *Set down the prices of the simples one under the other, in the order of their values, beginning with the lowest.*

II. *Link the least price with the greatest, and the one next to the least with the one next to the greatest, and so on, until the price of each simple which is less than the price of the mixture is linked with one or more that is greater; and every one that is greater with one or more that is less.*

III. *Write the difference between the price of the mixture and that of each of the simples opposite that price with which the particular simple is linked; then the difference standing opposite any one price, or the sum of the differences when there is more than one, will express the quantity to be taken of that price.*

EXAMPLES.

1. A merchant would mix wines worth 16*s*., 18*s*., and 22*s*. per gallon in such a way, that the mixture may be worth 20*s* per gallon: how much must be taken of each sort?

20	16	2 at 16*s*.
	18	2 at 18*s*.
	22	4+2=6 at 22*s*.

Ans. 2*gal.* at 16*s*., 2 at 18*s*., and 6 at 22*s*.: or any other quantities bearing the proportion of 2, 2, and 6.

2. What proportions of coffee at 8*cts*., 10*cts*., and 14*cts*. per *lb*. must be mixed together so that the compound shall be worth 12*cts*. per *lb*.?

3. A goldsmith has gold of 16, of 18, of 23, and of 24 carats fine: what part must be taken of each so that the mixture shall be 21 carats fine?

QUEST.—252. How do you find the proportions so that the compound may be of a given price?

4. What portion of brandy at 14*s.* per gallon, of old Madeira at 24*s.* per gallon, of new Madeira at 21*s.* per gallon, and of brandy at 10*s.* per gallon, must be mixed together so that the mixture shall be worth 18*s.* per gallon?

CASE II.

253. When a given quantity of one of the simples is to be taken.

I. *Find the proportional quantities of the simples as in Case I.*

II. *Then say, as the number opposite the simple whose quantity is given, is to either proportional quantity, so is the given quantity, to the proportional part of the corresponding simple.*

EXAMPLES.

1. How much wine at 5*s.*, at 5*s.* 6*d.*, and 6*s.* per gallon must be mixed with 4 gallons at 4*s.* per gallon, so that the mixture shall be worth 5*s.* 4*d.* per gallon?

64	48	8	- - simple whose quantity is known.
	60	2	proportional quantities.
	66	4	
	72	16	

Then 8 : 2 : : 4 : 1
8 : 4 : : 4 : 2
8 : 16 : : 4 : 8

Ans. 1*gal.* at 5*s.*, 2 at 5*s.* 6., and 8 at 6*s*

PROOF BY ALLIGATION MEDIAL.

4*gal.*	at 4*s.* per gallon	- -	192*d.*
1 "	5*s.* "	- - -	60
2 "	5*s.* 6*d.* "	- - -	132
8 "	6*s.* "	- - -	576
15			15)960(64*d.* price of mixture.

QUEST.—253. How do you find the proportion when the quantity of one of the simples is given?

2. A farmer would mix 14 bushels of wheat at $1,20 per bushel, with rye at 72*cts.*, barley at 48*cts.*, and oats at 36*cts.*: how much must be taken of each sort to make the mixture worth 64 cents per bushel?

3. There is a mixture made of wheat at 4*s.* per bushel, rye at 3*s.*, barley at 2*s.*, with 12 bushels of oats at 18*d.* per bushel: how much has been taken of each sort when the mixture is worth 3*s.* 6*d.*?

4. A distiller would mix 40*gal.* of French brandy at 12*s.* per gallon, with English at 7*s.* and spirits at 4*s.* per gallon: what quantity must be taken of each sort, that the mixture may be afforded at 8*s.* per gallon?

CASE III.

254. When the quantity of the compound is given as well as the price.

I. *Find the proportional quantities as in Case I.*

II. *Then say, as the sum of the proportional quantities, is to each proportional quantity, so is the given quantity, to the corresponding part of each.*

EXAMPLES.

1. A grocer has four sorts of sugar worth 12*d.*, 10*d.*, 6*d.*, and 4*d.* per pound; he would make a mixture of 144*lb.* worth 8*d.* per pound: what quantity must be taken of each sort?

8	4	4	12	:	4	: :	144	:	48
	6	2	12	:	2	: :	144	:	24
	10	2	12	:	2	: :	144	:	24
	12	4	12	:	4	: :	144	:	48
Sum of the proportional parts		12							

Ans. { 48*lb.* at 4*d.*; 24*lb.* at 6*d.*; 24*lb.* at 10*d.*; and 48*lb.* at 12*d.*

QUEST.—254. How do you determine the proportion when the quantity of the compound is given as well as the price?

PROOF BY ALLIGATION MEDIAL.

48*lb.* at 4*d.*	- - - - -	192*d.*
24*lb.* " 6*d.*	- - - - -	144*d.*
24*lb.* " 10*d.*	- - - - -	240*d.*
48*lb.* " 12*d.*	- - - - -	576*d.*
144		144)1152(8*d.*

Hence, the average cost is 8*d.*

2. A grocer having four sorts of tea worth 5*s.*, 6*s.*, 8*s.*, and 9*s.* per *lb.*, wishes a mixture of 87*lb.* worth 7*s.* per *lb.*: how much must be taken of each sort?

3. A vintner has four sorts of wine, viz., white wine at 4*s.* per gallon, Flemish at 6*s.* per gallon, Malaga at 8*s.* per gallon, and Canary at 10*s.* per gallon: he would make a mixture of 60 gallons to be worth 5*s.* per gallon: what quantity must be taken of each?

4. A silversmith has four sorts of gold, viz., of 24 carats fine, of 22 carats fine, of 20 carats fine, and of 15 carats fine: he would make a mixture of 42*oz.* of 17 carats fine: how much must be taken of each sort?

CUSTOM HOUSE BUSINESS.

255. PERSONS who bring goods, or merchandise, into the United States, from foreign countries, are required to land them at particular places or ports, called Ports of Entry, and to pay a certain amount on their value, called a *Duty*. This duty is imposed by the General Government, and must be the same on the same articles of merchandise, in every part of the United States.

Besides the duties on merchandise, vessels employed in commerce are required, by law, to pay certain sums for the privilege of entering the ports. These sums are large or

---

QUEST.—255. What is a port of entry? What is a duty? By whom are duties imposed? What charges are vessels required to pay? What are the moneys arising from duties and tonnage called?

small, in proportion to the size or tonnage of vessels. The moneys arising from duties and tonnage, are called *revenues.*

256. The revenues of the country are under the general direction of the Secretary of the Treasury, and to secure their faithful collection, the government has appointed various officers at each port of entry or place where goods may be landed.

257. The office established by the government at any port of entry, is called a *Custom House*, and the officers attached to it are called Custôm House Officers.

258. All duties levied by law on goods imported into the United States, are collected at the various custom houses, and are of two kinds, *Specific* and *Ad valorem.*

A *specific* duty is a certain sum on a particular kind of goods named; as so much per square yard on cotton or woollen cloths, so much per ton weight on iron, or so much per gallon on molasses.

An *ad valorem* duty is such a per cent on the actual cost of the goods in the country from which they are imported. Thus, an ad valorem duty of 15 per cent on English cloths, is a duty of 15 per cent on the cost of cloths imported from England.

259. The laws of Congress provide, that the cargoes of all vessels freighted with foreign goods or merchandise, shall be weighed or gauged by the custom house officers at the port to which they are consigned. As duties are only to be paid on the articles, and not on the boxes, casks, and bags which contain them, certain deductions are made from the weights and measures, called *Allowances.*

Gross Weight is the whole weight of the goods, together

QUEST.—256. Under whose direction are the revenues of the country? 257. What is a custom house? What are the officers attached to it called? 258. Where are the duties collected? How many kinds are there, and what are they called? What is a specific duty? An ad valorem duty? 259. What do the laws of Congress direct in relation to foreign goods? Why are deductions made from their weight? What are these deductions called? What is gross weight?

with that of the hogshead, barrel, box, bag, &c., which contains them.

Draft is an allowance from the gross weight on account of waste, where there is not actual tare.

	lb.				*lb.*
On	112			it is	1,
From	112	to	224	"	2,
"	224	to	336	"	3,
"	336	to	1120	"	4,
"	1120	to	2016	"	7,
Above	2016	any weight		"	9;

consequently, 9*lb.* is the greatest draft allowed.

Tare is an allowance made for the weight of the boxes, barrels, or bags containing the commodity, and is of three kinds. 1st. Legal tare, or such as is established by law; 2d. Customary tare, or such as is established by the custom among merchants; and 3d. Actual tare, or such as is found by removing the goods and actually weighing the boxes or casks, in which they are contained.

On liquors in casks, *customary tare* is sometimes allowed on the supposition that the cask is not full, or what is called its *actual wants;* and then an allowance of 5 per cent for leakage.

A tare of 10 per cent is allowed on porter, ale, and beer, in bottles, on account of breakage, and 5 per cent on all other liquors in bottles. At the custom house, bottles of the common size are estimated to contain $2\frac{3}{4}$ gallons the dozen. For tables of Tare and Duty, see Ogden on the Tariff of 1842.

EXAMPLES.

1. What will be the duty on 125 cartons of ribbons, each containing 48 pieces, and each piece weighing 3*oz.* net, and paying a duty of $2,50 per *lb.*?

QUEST.—What is draft? What is the greatest draft allowed? What is tare? What are the different kinds of tare? What allowances are made on liquors?

2. What will be the duty on 225 bags of coffee, each weighing gross 160*lb.*, invoiced at 6 cents per *lb.*; 2 per cent being the legal rate of tare, and 20 per cent the duty?

3. What duty must be paid on 275 dozen bottles of claret, estimated to contain $2\frac{3}{4}$ gallons per dozen, 5 per cent being allowed for breakage, and the duty being 35 cents per gallon?

4. A merchant imports 175 cases of indigo, each case weighing 196*lb.* gross: 15 per cent is the customary rate of tare, and the duty 5 cents per *lb.* What duty must he pay on the whole?

5. What is the tare and duty on 75 casks of Epsom salts, each weighing gross 2*cwt.* 2*qr.* 27*lb.*, and invoiced at $1\frac{7}{8}$ cents per lb., the customary tare being 11 per cent, and the rate of duty 20 per cent?

FORMS RELATING TO BUSINESS IN GENERAL.

FORMS OF ORDERS

MESSRS. M. JAMES & Co.

Please pay John Thompson, or order, five hundred dollars, and place the same to my account, for value received.

PETER WORTHY.

Wilmington, N. C., June 1, 1846.

MR. JOSEPH RICH,

Please pay, for value received, the bearer, sixty-one dollars and twenty cents, in goods from your store, and charge the same to the account of your Obedient Servant,

JOHN PARSONS.

Savannah, Ga., July 1, 1846.

FORMS OF RECEIPTS.

Receipt for Money on Account.

Received, Natchez, June 2d, 1845, of John Ward, sixty dollars on account.

$60,00 JOHN P. FAY.

Receipt for Money on a Note.

Received, Nashville, June 5, 1846, of Leonard Walsh, six hundred and forty dollars, on his note for one thousand dollars, dated New York, January 1, 1845.

$640,00 — J. N. Weeks.

A BOND FOR ONE PERSON, WITH A CONDITION.

KNOW ALL MEN BY THESE PRESENTS, That, *I James Wilson of the City of Hartford and State of Connecticut, am* held and firmly bound unto *John Pickens of the Town of Waterbury, County of New Haven and State of Connecticut,* in the sum *of Eighty dollars* lawful money of the United States of America, to be paid to the said *John Pickens, his* executors, administrators, or assigns: for which payment well and truly to be made *I* bind *myself, my* heirs, executors, and administrators, firmly by these presents. Sealed with *my* Seal. Dated the *Ninth day* of *March* one thousand eight hundred and *thirty-eight.*

THE CONDITION of the above obligation is such, that if the above bounden *James Wilson, his* heirs, executors, or administrators, shall well and truly pay or cause to be paid, unto the above named *John Pickens, his* executors, administrators, or assigns, the just and full sum of

Here insert the condition.

then the above obligation to be void, otherwise to remain in full force and virtue.

Sealed and delivered in the presence of
John Frost,
Joseph Wiggins,

James Wilson, 

Note.—The part in Italic to be filled up according to circumstance.

If there is no condition to the bond, then all to be omitted after and including the words "THE CONDITION, &c."

A BOND FOR TWO PERSONS, WITH A CONDITION.

KNOW ALL MEN BY THESE PRESENTS, That, We *James Wilson and Thomas Ash of the City of Hartford and State of Connecticut*, are held and firmly bound unto *John Pickens of the Town of Waterbury County of New Haven and State of Connecticut*, in the sum *of Eighty dollars* lawful money of the United States of America, to be paid to the said *John Pickens*, *his* executors or assigns: for which payment well and truly to be made *We* bind *ourselves*, *our* heirs, executors, and administrators, firmly by these presents. Sealed with *our* Seal. Dated the *Ninth day* of *March* one thousand eight hundred and *thirty-eight.*

THE CONDITION of the above obligation is such, that if the above bounden *James Wilson and Thomas Ash*, *their* heirs, executors, or administrators, shall well and truly pay or cause to be paid, unto the above named *John Pickens*, *his* executors, administrators, or assigns, the just and full sum of

Here insert the condition.

then the above obligation to be void, otherwise to remain in full force and virtue.

Sealed and delivered in
the presence of

John Frost, } *James Wilson*,
Joseph Wiggins, } *Thomas Ash.*

Note.—The part in Italic to be filled up according to circumstance.

If there is no condition to the bond, then all to be omitted after and including the words "THE CONDITION, &c."

GENERAL AVERAGE.

260. AVERAGE is a term of commerce and navigation, to signify a contribution by individuals, where the goods of a particular merchant are thrown overboard in a storm, to save the ship from sinking, or where the masts, cables, anchors, or other furniture of the ship are cut away or destroyed for the preservation of the whole. In these and like cases, where any sacrifices are deliberately made, or any expenses voluntarily incurred, to prevent a total loss, such sacrifice or expense is the proper subject of a general contribution, and ought to be rateably borne by the owners of the ship, the freight, and the cargo, so that the loss may fall proportionably on all. The amount sacrificed is called the *jettison.*

261. Average is either *general* or *particular;* that is, it is either chargeable to all the interests, viz., the ship, the freight, and the cargo, or only to some of them. As when losses occur from ordinary wear and tear, or from the perils incident to the voyage, without being *voluntarily* incurred; or when any particular sacrifice is made for the sake of the *ship only* or the *cargo only*, these losses must be borne by the parties immediately interested, and are consequently defrayed by *a particular* average. There are also some small charges called *petty* or *accustomed* averages, one-third of which is usually charged to the ship and two-thirds to the cargo.

No general average ever takes place, except it can be shown that *the danger was imminent, and that the sacrifice was made indispensable, or supposed to be so by the captain and officers, for the safety of the ship.*

262. In different countries different modes are adopted of valuing the articles which are to constitute a general average. In general, however, the value of the freight is held to be the clear sum which the ship has earned after seamen's

---

QUEST.—260. What does the term average signify? 261. How many kinds of average are there? What are the small charges called? Under what circumstances will a general average take place? 262. How is the freight valued? How much is charged on account of the seamen's wages?

wages, pilotage, and all such other charges as came under the name of petty charges, are deducted; one-third, and in some cases one-half, being deducted for the wages of the crew.

The goods lost, as well as those saved, are valued at the price they would have brought in ready money at *the place of delivery*, on the ship's arriving there, freight, duties, and all other charges being deducted: indeed, they bear their proportions, the same as the goods saved. The ship is valued at the price she would bring on her arrival at the port of delivery. But when the loss of masts, cables, and other furniture of the ship is compensated by general average, it is usual, as the new articles will be of greater value than the old, to deduct one-third, leaving two thirds only to be charged to the amount to be contributed.

EXAMPLES.

1. The vessel Good Intent, bound from New York to New Orleans, was lost on the Jersey beach the day after sailing. She cut away her cables and masts, and cast overboard a part of her cargo, by which another part was injured. The ship was finally got off, and brought back to New York.

AMOUNT OF LOSS.

Goods of A cast overboard - - - -		$500
Damage of the goods of B by the jettison -		200
Freight of the goods cast overboard - -		100
Cable, anchors, mast, &c., worth	$300	200
Deduct one-third - - - - -	100	
Expenses of getting the ship off the sands		56
Pilotage and port duties going in and out of the harbor, commissions, &c. -		100
Expenses in port - - - - - - -		25
Adjusting the average - - - - -		4
Postage - - - - - - - - -		1
	Total loss	$1186

QUEST.—How is the cargo valued? Does the part lost bear its part of the loss? How is the ship valued? When parts of the ship are lost, how are they compensated for? How do you explain the example?

ARTICLES TO CONTRIBUTE.

Goods of A cast overboard - - - - - - -	$500
Value of B's goods at N. O., deducting freight, &c.	1000
" of C's " " " "	500
" of D's " " " "	2000
" of E's " " " "	5000
Value of the ship - - - - - - - - - -	2000
Freight after deducting one-third - - - - - -	800
	$11,800

Then, total value : total loss : : 100 : per cent of loss.
$11800 : 1180 : : 100 : 10;

hence, each loses 10 per cent on the value of his interest in the cargo, ship, or freight. Therefore, A loses $50, B $100, C $50, D $200, E $500, the owners of the ship $280—in all $1180. Upon this calculation the owners are to lose $280; but they are to receive their disbursements from the contribution, viz., freight on goods thrown overboard $100, damages to ship $200, various disbursements in expenses $180, total $480; and deducting the amount of contribution, they will actually receive $200. Hence, the account will stand:

The owners are to receive - - - - - - - -	$200
A loses $500, and is to contribute $50; hence, he receives - - - - - - - - - -	450
B loses $200, and is to contribute $100; hence, he receives - - - - - - - - -	100
Total to be received - - - -	$750

C, D, and E have lost nothing, and are to pay	C	$ 50
	D	200
	E	500
Total actually paid - - - -		$750;

so that the total to be paid is just equal to the total loss, as it should be, and A and B get their remaining and injured goods, and the three others get theirs in a perfect state, after paying their rateable proportion of the loss.

TONNAGE OF VESSELS.

263. THERE are certain custom house charges on vessels, which are made according to their tonnage. The tonnage of a vessel is the number of tons weight she will carry, and this is determined by measurement.

[From the "Digest," by Andrew A. Jones, Esq., of the N. Y. Custom House.]

Custom house charges on all ships or vessels entering from any foreign port or place.

Ships or vessels of the United States, having three-fourths of the crew and all the officers American citizens, *per ton*	$0,06
Ships or vessels of nations entitled by treaty to enter at the same rate as American vessels - - - - - -	,06
Ships or vessels of the United States not having three-fourths the crew as above - - - - - - - -	,50
On foreign ships or vessels other than those entitled by treaty - - - - - - - - - - - -	,50
Additional tonnage on foreign vessels, denominated light money - - - - - - - - - - -	,50

Licensed coasters are also liable once in each year to a duty of 50 cents per ton, being engaged in a trade from a port in one state to a port in another state, other than an adjoining state, unless the officers and three-fourths of the crew are American citizens; to ascertain which, the crews are always liable to an examination by an officer.

A foreign vessel is not permitted to carry on the coasting trade; but having arrived from a foreign port with a cargo consigned to more than one port of the United States, she may proceed coastwise with a certified manifest until her voyage is completed.

264. The government estimate the tonnage according to one rule, while the ship carpenter who builds the vessel uses another.

QUEST.—263. What is the tonnage of a vessel? What are the custom house charges on the different classes of vessels trading with foreign countries? To what charges are coasters subject?

GOVERNMENT RULE. I. *Measure, in feet, above the upper deck the length of the vessel, from the fore part of the main stem to the after part of the stern post. Then measure the breadth taken at the widest part above the main wale on the outside, and the depth from the under side of the deck plank to the ceiling in the hold.*

II. *From the length take three-fifths of the breadth and multiply the remainder by the breadth and depth, and the product divided by 95 will give the tonnage of a single decker; and the same for a double decker, by merely making the depth equal to half the breadth.*

CARPENTERS' RULE. *Multiply together the length of the keel, the breadth of the main beam, and the depth of the hold, and the product divided by 95 will be the carpenters' tonnage for a single decker; and for a double decker, deduct from the depth of the hold half the distance between decks.*

EXAMPLES.

1. What is the government tonnage of a single decker, whose length is 75 feet, breadth 20 feet, and depth 17 feet?

2. What is the carpenters' tonnage of a single decker, the length of whose keel is 90 feet, breadth 22 feet 7 inches, and depth 20 feet 6 inches?

3. What is the carpenters' tonnage of a steamship, double decker, length 154 feet, breadth 30 feet 8 inches, and depth after deducting half between decks, 14 feet 8 inches?

4. What is the government tonnage of a double decker, the length being 103 feet, breadth 25 feet 6 inches?

5. What is the carpenters' tonnage of a double decker, its length 125 feet, breadth 25 feet 6 inches, entire depth 34 feet, and distance between decks 8 feet?

QUEST.—264. What is the government rule for finding the tonnage? What the ship-builders' rule.

GAUGING.

265. CASK-GAUGING is the method of finding the number of gallons which a cask contains, by measuring the external dimensions of the cask.

266. Casks are divided into four varieties, according to the curvature of their sides. To which of the varieties any cask belongs, must be judged of by inspection.

1. Of the least curvature.

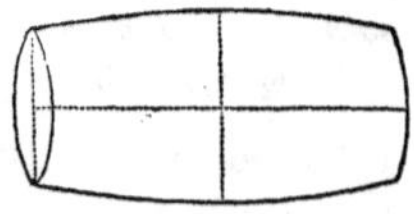

2d Variety.

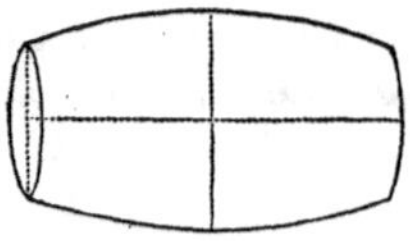

3d Variety.

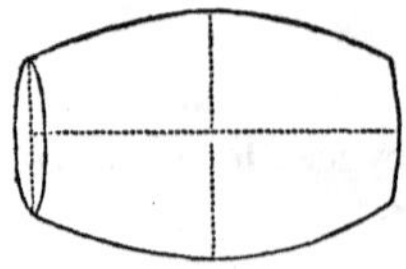

4th Variety.

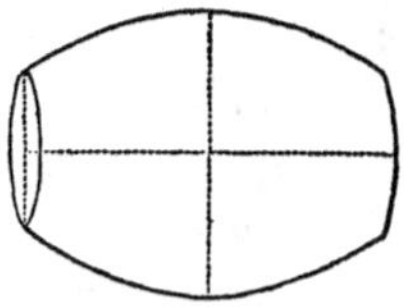

267. The first thing to be done is to find the mean diameter. To do this,

Divide the head diameter by the bung diameter, and find the quotient in the first column of the following table, marked Qu. *Then if the bung diameter be multiplied by the number on the same line with it, and in the column answering to the proper*

QUEST.—265. What is cask-gauging? 266. Into how many varieties are casks divided? 267 How do you find the mean diameter?

variety, the product will be the true mean diameter, or the diameter of a cylinder of the same content with the cask proposed, cutting off four figures for decimals.

Qu.	1st Var.	2d Var.	3d Var.	4th Var.	Qu.	1st Var.	2d Var.	3d Var.	4th Var.
50	8660	8465	7905	7637	76	9270	9227	8881	8827
51	8680	8493	7937	7681	77	9296	9258	8944	8874
52	8700	8520	7970	7725	78	9324	9290	8967	8922
53	8720	8548	8002	7769	79	9352	9320	9011	8970
54	8740	8576	8036	7813	80	9380	9352	9055	9018
55	8760	8605	8070	7858	81	9409	9383	9100	9066
56	8781	8633	8104	7902	82	9438	9415	9144	9114
57	8802	8662	8140	7947	83	9467	9446	9189	9163
58	8824	8690	8174	7992	84	9496	9478	9234	9211
59	8846	8720	8210	8037	85	9526	9510	9280	9260
60	8869	8748	8246	8082	86	9556	9542	9326	9308
61	8892	8777	8282	8128	87	9586	9574	9372	9357
62	8915	8806	8320	8173	88	9616	9606	9419	9406
63	8938	8835	8357	8220	89	9647	9638	9466	9455
64	8962	8865	8395	8265	90	9678	9671	9513	9504
65	8986	8894	8433	8311	91	9710	9703	9560	9553
66	9010	8924	8472	8357	92	9740	9736	9608	9602
67	9034	8954	8511	8404	93	9772	9768	9656	9652
68	9060	8983	8551	8450	94	9804	9801	9704	9701
69	9084	9013	8590	8497	95	9836	9834	9753	9751
70	9110	9044	8631	8544	96	9868	9867	9802	9800
71	9136	9074	8672	8590	97	9901	9900	9851	9850
72	9162	9104	8713	8637	98	9933	9933	9900	9900
73	9188	9135	8754	8685	99	9966	9966	9950	9950
74	9215	9166	8796	8732	100	10000	10000	10000	10000
75	9242	9196	8838	8780					

EXAMPLES.

1. Supposing the diameters to be 32 and 24, it is required to find the mean diameter for each variety.

Dividing 24 by 32, we obtain .75; which being found in the column of quotients, opposite thereto stand the numbers,

$$\left\{\begin{array}{l}.9242\\.9196\\.8838\\.8780\end{array}\right\}$$ which being each multiplied by 32, produce respectively, $$\left\{\begin{array}{l}29.5744\\29.4272\\28.2816\\28.0960\end{array}\right\}$$ for the corresponding mean diameters required.

2. The head diameter of a cask is 26 inches, and the bung diameter 3 feet 2 inches: what is the mean diameter, the cask being of the third variety?

3. The head diameter is 22 inches, the bung diameter 34 inches: what is the mean diameter of a cask of the fourth variety?

268. Having found the mean diameter, we multiply the square of the mean diameter by the decimal .7854, and the product by the length; this will give the solid content in cubic inches. Then if we divide by 231, we have the content in wine gallons (see Art. **31**), or if we divide by 282, we have the content in beer gallons.

For wine measure we multiply the length by the square of the mean diameter, then by the decimal .7854, and divide by 231.

OPERATION.

$$l \times d^2 \times \frac{.7854}{231} = l \times d^2 \times .0034.$$

If, then, we divide the decimal .7854 by 231, the quotient carried to four places of decimals is .0034, and this decimal multiplied by the square of the mean diameter and by the length of the cask, will give the content in wine gallons.

For similar reasons, the content is found in beer gallons by multiplying together the length, the square of the mean diameter, and the decimal .0028.

OPERATION.

$$l \times d^2 \times \frac{.7854}{282} = l \times d^2 \times .0028.$$

Hence, for gauging or measuring casks,

Multiply the length by the square of the mean diameter; then multiply by 34 *for wine, and by* 28 *for beer measure, and point off in the product four decimal places. The product will then express gallons and the decimals of a gallon.*

1. How many wine gallons in a cask, whose bung diameter is 36 inches, head diameter 30 inches, and length 50 inches; the cask being of the first variety?

QUEST.—268. How do you find the solidity? How do you find the content in wine gallons? In beer gallons?

2. What is the number of beer gallons in the last example?

3. How many wine, and how many beer gallons in a cask whose length is 36 inches, bung diameter 35 inches, and head diameter 30 inches, it being of the first variety?

4. How many wine gallons in a cask of which the head diameter is 24 inches, bung diameter 36 inches, and length 3 feet 6 inches, the cask being of the second variety?

LIFE INSURANCE.

269. Insurance for a term of years, or for the entire continuance of life, is a contract on the part of an authorized association to pay a certain sum, specified in the policy of insurance, on the happening of an event named therein, and for which the association receives a certain premium, generally in the form of an annual payment.

270. To enable the company to fix their premiums at such rates as shall be both fair to the insured and safe to the association, they must know the *average* duration of life from its commencement to its extreme limit. This average is called the "*Expectation of Life*," and this is determined by collecting from many sources the most authentic information in regard to births and deaths. The "Carlisle Table," which is subjoined, and which shows the expectation of life from birth to 103 years, is considered the most accurate. It is much used in England, and is in general use in this country.

By the "Expectation of Life," must be understood the average age of any number of individuals. Thus, if 100 infants be taken, some dying in infancy, some in childhood, some in youth, some in middle life, and some in old age, the average ages of all will be 38.72 years. So from 10 years old, the average age is 48 82 years.

Quest.—269. What is an insurance? 270. What is necessary to enable a company to fix their premiums? How is the expectation determined? What Table is generally used in this country? What do you understand by the expectation of life?

TABLE SHOWING THE EXPECTATION OF LIFE.

Age.	Expectation.	Age.	Expectation.	Age.	Expectation.	Age.	Expectation.
0	38.72	26	37.14	52	19.68	78	6.12
1	44.68	27	36.41	53	18.97	79	5.80
2	47.55	28	35.69	54	18.28	80	5.51
3	49.82	29	35.00	55	17.58	81	5.21
4	50.76	30	34.34	56	16.89	82	4.93
5	51.25	31	33.68	57	16.21	83	4.65
6	51.17	32	33.03	58	15.55	84	4.39
7	50.80	33	32.36	59	14.92	85	4.12
8	50.24	34	31.68	60	14.34	86	3.90
9	45.57	35	31.00	61	13.82	87	3.71
10	48.82	36	30.32	62	13.31	88	3.59
11	48.04	37	29.64	63	12.81	89	3.47
12	47.27	38	28.96	64	12.30	90	3.28
13	46.51	39	28.28	65	11.79	91	3.26
14	45.75	40	27.61	66	11.27	92	3.37
15	45.00	41	26.97	67	10.75	93	3.48
16	44.27	42	26.34	68	10.23	94	3.53
17	43.57	43	25.71	69	9.70	95	3.53
18	42.87	44	25.09	70	9.19	96	3.46
19	42.17	45	24.46	71	8.65	97	3.28
20	41.46	46	23.82	72	8.16	98	3.07
21	40.75	47	23.17	73	7.72	99	2.77
22	40.04	48	22.50	74	7.33	100	2.28
23	39.31	49	21.81	75	7.01	101	1.79
24	38.59	50	21.11	76	6.69	102	1.30
25	37.86	51	20.39	77	6.40	103	0.83

271. From the above table, and the value of money, which is shown by the rate of interest, a company can calculate with great exactness the amount which they should receive annually, for an insurance on a life for any number of years, or during its entire continuance.

Among the prinoipal life insurance companies in the United States, are the New York Life Insurance and Trust Company, the Girard Life Insurance, Annuity, and Trust Company of Philadelphia, and the Massachusetts Hospital Life Insurance and Trust Company of Boston.

QUEST.—Explain the table showing the expectation of life. 271. What must be known besides the expectation of life in order to find the premium? What are the principal life insurance companies in the United States? How do you find the amount which must be paid for the insurance of $100

	NEW YORK AND PHILA. COMPANIES.			MASSACHUSETTS.		
Age.	1 year.	7 years.	For life.	1 year.	7 years.	For life.
14	.72	.86	1.53	.89	1.08	1.88
15	.77	.88	1.56	.90	1.15	1.93
16	.84	.90	1.62	.96	1.23	1.99
17	.86	.91	1.65	1.06	1.30	2.04
18	.89	.92	1.69	1.16	1.38	2.09
19	.90	.94	1.73	1.25	1.43	2.14
20	.91	.95	1.77	1.36	1.48	2.18
21	.92	.97	1.82	1.44	1.50	2.23
22	.94	.99	1.88	1.46	1.53	2.26
23	.97	1.03	1.93	1.49	1.55	2.31
24	.99	1.07	1.98	1.51	1.58	2.35
25	1.00	1.12	2.04	1.53	1.60	2.40
26	1.07	1.17	2.11	1.55	1.63	2.45
27	1.12	1.23	2.17	1.58	1.66	2.50
28	1.20	1.28	2.24	1.60	1.69	2.55
29	1.28	1.35	2.31	1.64	1.71	2.61
30	1.31	1.36	2.36	1.66	1.75	2.66
31	1.32	1.42	2.43	1.69	1.78	2.73
32	1.33	1.46	2.50	1.71	1.81	2.79
33	1.34	1.48	2.57	1.75	1.84	2.85
34	1.35	1.50	2.64	1.79	1.89	2.93
35	1.36	1.53	2.75	1.81	1.94	2.99
36	1.39	1.57	2.81	1.85	1.98	3.06
37	1.43	1.63	2.90	1.89	2.05	3.14
38	1.48	1.70	3.05	1.93	2.09	3.23
39	1.57	1.76	3.11	1.96	2.15	3.31
40	1.69	1.83	3.20	2.04	2.20	3.40
41	1.78	1.88	3.31	2.10	2.26	3.49
42	1.85	1.89	3.40	2.18	2.33	3.59
43	1.89	1.92	3.51	2.23	2.39	3.69
44	1.90	1.94	3.63	2.28	2.46	3.79
45	1.91	1.96	3.73	2.34	2.54	3.90
46	1.92	1.98	3.87	2.39	2.63	4.01
47	1.93	1.99	4.01	2.45	2.71	4.13
48	1.94	2.02	4.17	2.51	2.81	4.25
49	1.95	2.04	4.49	2.61	2.93	4.39
50	1.96	2.09	4.60	2.75	3.04	4.54
51	1.97	2.20	4.75	2.86	3.14	4.68
52	2.02	2.37	4.90	2.95	3.24	4.83
53	2.10	2.59	5.24	3.05	3.35	4.98
54	2.18	2.89	5.49	3.15	3.48	5.14
55	2.32	3.21	5.78	3.25	3.60	5.31
56	2.47	3.56	6.05	3.36	3.74	5.50
57	2.70	4.20	6.27	3.49	3.88	5.70
58	3.14	4.31	6.50	3.61	4.03	5.91
59	3.67	4.63	6.75	3.75	4.19	6.14
60	4.35	4.91	7.00	3.90	4.35	6.36

The above table shows the rates at which they insure the amount of $100 for 1 year, for 7 years, or for life. It should be observed, that when a person insures for 7 years or for life, he pays *annually* the premium set opposite the age. Having found the premium for $100, it is easily found for any other amount, by simply multiplying by the amount and dividing by 100.

EXAMPLES.

1. What will be the premium per annum on the insurance of a life for 7 years, for $4500, the person being at the age of 40 years, in the New York or Philadelphia companies?

Premium per annum for 7 years on $100 = 1,83;
then, $1,83 \times 4500 \div 100 = 82,35$:
hence, $82,35 is the premium per annum.

2. What would be the premium per year if insured for life?

3. A person at 21 wishes to insure at his death $8500 to his friends: how much must he pay per annum to insure that amount at his death, in the Boston Company?

ENDOWMENTS AND ANNUITIES.

272. AN ENDOWMENT is a certain sum to be paid at the expiration of a given time, in case the person on whose life it is taken shall live till the expiration of the time named.

273. ANNUITIES are certain annual or periodical payments made to individuals by incorporated companies or associations, for a given sum paid in hand.

274. The following table shows the value of an endowment purchased for $100, at the several periods mentioned on the column of ages, the endowment to be paid if the person attains the age of 21 years.

QUEST.—272. What is an endowment? 273. What is an annuity? 274. What does the table of endowments show?

TABLE OF ENDOWMENTS.

Age.	Sum to be paid at 21, if alive.	Age.	Sum to be paid at 21, if alive,	Age.	Sum to be paid at 21, if alive.
Birth	$376,84	5	$210,53	13	$144,12
3 months..	344,28	6	198,83	14	137,86
6 " ..	331,46	7	188,83	15	131,83
9 " ..	318,90	8	179,97	16	125,97
1 year....	306,58	9	171,91	17	120,31
2 "	271,03	10	164,46	18	114,89
3 "	243,69	11	157,43	19	109,70
4 "	225,42	12	150,64	20	104,74

275. The following table exhibits the sums which must be paid, at the several ages named, to purchase an annuity of $100 a year in the Massachusetts Life Insurance Co., and in the Girard Life Insurance, Annuity, and Trust Company, Philadelphia.

Age.		Age.		Age.	
20.......	$1836,30	39.......	$1527,20	58.......	$1125,00
21.......	1823,30	40.......	1507,40	59.......	1100,00
22.......	1809,50	41.......	1488,30	60.......	1070,00
23.......	1795,10	42.......	1469,40	61.......	1045,00
24.......	1780,10	43.......	1450,50	62.......	1020,00
25.......	1764,50	44.......	1430,80	63.......	995,00
26.......	1748,60	45.......	1410,40	64.......	970,00
27.......	1732,00	46.......	1388,90	65.......	940,00
28.......	1715,40	47.......	1366,20	66.......	910,00
29.......	1699,70	48.......	1341,90	67.......	880,00
30.......	1685,20	49.......	1315,30	68.......	850,00
31.......	1670,50	50.......	1300,00	69.......	820,00
32.......	1655,20	51.......	1280,00	70.......	790,00
33.......	1639,00	52.......	1260,00	71.......	780,00
34.......	1621,90	53.......	1240,00	72.......	770,00
35.......	1604,10	54.......	1220,00	73.......	760,00
36.......	1585,60	55.......	1200,00	74.......	750,00
37.......	1566,60	56.......	1175,00	75.......	740,00
38.......	1547,10	57.......	1150,00		

EXAMPLES.

1. What sum at birth will purchase an endowment at 21 of $859,61?

2. What sum at the age of 30 years will purchase an annuity of $3150?

QUEST.—275. What does the table of annuities show?

COINS AND CURRENCIES.

276. COINS are pieces of metal, of gold, silver, or copper, of fixed values, and impressed with a public stamp prescribed by the country where they are made. These are called specie, and are generally declared to be a legal tender in payment of debts. The Constitution of the United States provides, that gold and silver only shall be a legal tender.

The coins of a country and those of foreign countries having a fixed value established by law, together with bank notes redeemable in specie, make up what is called the *Currency*.

277. A foreign coin may be said to have four values:

1st. The intrinsic value, which is determined by the amount of pure metal which it contains.

2d. The custom house or legal value, which is fixed by law.

3d. The mercantile value, which is the amount it will sell for in open market.

4th. The exchange value, which is the value assigned to it in buying and selling bills of exchange between one country and another.

Let us take, as an example, the English pound sterling, which is represented by the gold sovereign. Its intrinsic value, as determined at the Mint in Philadelphia, compared with our gold eagle, is $4,861. Its legal or custom house value is $4,84. Its commercial value, that is, what it will bring in Wall street, New York, varies from $4,83 to $4,86, seldom reaching either the lowest or highest limit. The

---

QUEST.—276. What are coins? What are they called? What is declared in regard to them? What is provided by the Constitution of the United States? What do you understand by Currency? 277. How many values may a coin be said to have? What is the intrinsic value? What is the mercantile value? What is the exchange value?

exchange value of the English pound, is \$4,44$\frac{4}{9}$, and was the legal value before the change in our standard. This change raised the legal value of the pound to \$4,84, but merchants and dealers in exchange preferred to retain the old value, which became nominal, and to add the difference in the form of a premium on exchange, which is explained in Art. 292.

TABLE OF FOREIGN COINS WHOSE VALUES ARE FIXED BY LAW.

	$	cts.
Franc of France and Belgium	0	18$\frac{6}{10}$
Florin of the Netherlands		40
Guilder of do.		40
Livre Tournois of France		18$\frac{1}{2}$
Milrea of Portugal	1	12
Milrea of Madeira	1	00
Milrea of the Azores		83$\frac{1}{3}$
Marc Banco of Hamburg		35
Pound Sterling of Great Britain	4	84
Pagoda of India	1	84
Real Vellon of Spain		05
Real Plate of do.		10
Rupee Company		44$\frac{1}{2}$
Rupee of British India		44$\frac{1}{2}$
Rix Dollar of Denmark	1	00
Rix Dollar of Prussia		68$\frac{1}{2}$
Rix Dollar of Bremen		78$\frac{3}{4}$
Rouble, silver, of Russia		75
Tale of China	1	48
Dollar of Sweden and Norway	1	06
Specie Dollar of Denmark	1	05
Dollar of Prussia and Northern States of Germany		69
Florin of Southern States of Germany		40
Florin of Austria and city of Augsburg		48$\frac{1}{2}$
Lira of the Lombardo Venetian Kingdom		16
Lira of Tuscany		16
Lira of Sardinia		18$\frac{6}{10}$
Ducat of Naples		80
Ounce of Sicily	2	40
Pound of Nova Scotia, New Brunswick, Newfoundland, and Canada	4	04

QUEST.—Give the different values of the English sovereign. How came the value of the sovereign to be altered? How is the difference now made up?

TABLE OF FOREIGN COINS WHOSE VALUES ARE FIXED BY USAGE,

When a Consular's certificate of the real value or rate of exchange is not attached to the invoice.

	$	*cts.*
Berlin Rix Dollar		$69\frac{1}{2}$
Current Marc		28
Crown of Tuscany	1	05
Elberfeldt Rix Dollar		$69\frac{3}{4}$
Florin of Saxony		48
" Bohemia		48
" Elberfeldt		40
" Prussia		$22\frac{3}{4}$
" Trieste		48
" Nuremburg		40
" Frankfort		40
" Basil		41
" St. Gaul		$40\frac{36}{100}$
" Creveld		40
Florence Livre		15
Genoa do.		$18\frac{3}{4}$
Geneva do.		21
Jamaica Pound	5	00
Leghorn Dollar		90
Leghorn Livre ($6\frac{1}{3}$ to the dollar)		$15\frac{15}{19}$
Livre of Catalonia		$53\frac{1}{3}$
Neufchatel Livre		$26\frac{1}{2}$
Pezza of Leghorn		90
Rhenish Rix Dollar		$60\frac{3}{4}$
Swiss Livre		27
Scuda of Malta		40
Turkish Piastre		05

[The above Tables are taken from a work on the Tariff, by E. D Ogden, Esq., of the New York Custom House.]

EXCHANGE.

278. EXCHANGE is a term which denotes the payment of money by a person residing in one place to a person residing in another. The payment is generally made by means of a bill of exchange.

QUEST.—278 What is exchange? How is the payment generally made?

279. A BILL OF EXCHANGE is an open letter of request from one person to another, desiring the payment to a third party named therein, of a certain sum of money to be paid at a specified time and place. There are always three parties to a bill of exchange, and generally four.

1. He who writes the open letter of request, is called the *drawer* or *maker* of the bill.
2. The person to whom it is directed is called the *drawee*.
3. The person to whom the money is ordered to be paid is called the *payee*; and
4. Any person who purchases a bill of exchange is called the *buyer* or *remitter*.

280. Bills of exchange are the proper money of commerce. Suppose Mr. Isaac Wilson of the city of New York, ships 1000 bags of cotton, worth £96000, to Samuel Johns & Co. of Liverpool; and at about the same time William James of New York orders goods from Liverpool, of Ambrose Spooner, to the amount of eighty thousand pounds sterling. Now, Mr. Wilson draws a bill of exchange on Messrs. Johns & Co. in the following form: viz.,

Exchange for £80000. New York, July 30th, 1846.

Sixty days after sight of this my first Bill of Exchange (second and third of the same date and tenor unpaid*) pay to David C. Jones or order, eighty thousand pounds sterling, with or without further advice. ISAAC WILSON.

Messrs. Samuel Johns & Co.,
Merchants, Liverpool.

Let us now suppose that Mr. James purchases this bill of David C. Jones for the purpose of sending it to Ambrose

* Three bills are generally drawn for the same amount, called the first, second, and third, and together they form a set. One only is paid, and then the other two are of no value. This arrangement avoids the accidents and delays incident to transmitting the bills.

QUEST.—279. What is a bill of exchange? How many parties are there to a bill of exchange? Name them. 280. How do bills of exchange aid commerce? Name all the parties of the bill in this example.

Spooner of Liverpool, whom he owes. We shall then have all the parties to a bill of exchange; viz., Isaac Wilson, the *maker* or *drawer;* Messrs. Johns & Co., the *drawees;* David C. Jones, the *payee;* and William James, the *buyer* or *remitter.*

281. A bill of exchange is called an *inland bill,* when the drawer and drawee both reside in the same country; and when they reside in different countries, it is called a *foreign bill.* Thus, all bills in which the drawer and drawee reside in the United States, are inland bills; but if one of them resides in England or France, the bill is a foreign bill.

282. The time at which a bill is made payable varies, and is a matter of agreement between the drawer and buyer. They may either be drawn *at sight,* or at a certain number of days *after sight,* or at a certain number of days *after date.*

283. Days of Grace are a certain number of days granted to the person who pays the bill, after the time named in the bill has expired. In the United States and Great Britain three days are allowed.

284. In ascertaining the time when a bill payable so many days after sight, or after date, actually falls due, the day of presentment, or the day of the date, is not reckoned. When the time is expressed in months, *calendar months* are always understood.

If the month in which a bill falls due is shorter than the one in which it is dated, it is a rule not to go on into the next month. Thus a bill drawn on the 28th, 29th, 30th, or 31st of December, payable two months after date, would fall due

---

Quest.—281. What is an inland bill? What is a foreign bill? Are bills drawn between one state and another inland or foreign? 282. How is the time determined at which a bill is made payable? How are bills always drawn? 283. What are days of grace? How many days of grace are allowed in this country and in Great Britain? 284. In ascertaining the time when a bill is payable, what days are reckoned? When the time is expressed in months, what kind of months is understood? If the month in which the bill falls due is shorter than that in which it is drawn, what rule is observed?

on the last of February, except for the days of grace, and would be actually due on the third of March.

ENDORSING BILLS.

285. In examining the bill of exchange drawn by Isaac Wilson, it will be seen that Messrs. Johns & Co. are requested to pay the amount to David C. Jones or order; that is, either Mr. Jones or to any other person named by him. If Mr. Jones simply writes his name on the back of the bill, he is said to endorse it in *blank*, and the drawees must pay it to any rightful owner who presents it. Such rightful owner is called the *holder*, and Mr. Jones is called the *endorser*.

If Mr. Jones writes on the back of the bill, over his signature, "Pay to the order of William James," this is called *a special endorsement*, and William James is the *endorsee*, and he may either endorse in blank or write over his signature "Pay to the order of Ambrose Spooner," and the drawees, Messrs. Johns & Co., will then be bound to pay the amount to Mr. Spooner.

A bill drawn payable to bearer, may be transferred by mere delivery.

ACCEPTANCE.

286. When the bill drawn on Messrs. Johns & Co. is presented to them, they must inform the holder whether or not they will pay it at the expiration of the time named. Their agreement to pay it is signified by writing across the face of the bill, and over their signature the word "accepted," and they are then called the *acceptors*.

LIABILITIES OF THE PARTIES.

287. The drawee of a bill does not become responsible for its payment until after he has accepted. On the presenta-

QUEST.—285. What is an endorsement in blank? What is the person making it called? What is a special endorsement? What is the effect of an endorsement? How may a bill drawn to bearer be transferred? 286. What is an acceptance? How is it made? 287. When does the drawee of a bill become responsible for its payment?

tion of the bill, if the drawee does not accept, the holder should immediately take means to have the drawer and all the endorsers notified. Such notice is called a *protest*, and is given by a public officer called a *notary*, or *notary public*. If the parties are not notified in a reasonable time, they are not responsible for the payment of the bill.

If the drawer accepts the bill and fails to make the payment when it becomes due, the parties must be notified as before, and this is called *protesting the bill for non-payment*. If the endorsers are not notified in a reasonable time, they are not responsible for the amount of the bill.

PAR OF EXCHANGE—COURSE OF EXCHANGE.

288. The intrinsic *par of exchange*, is a term used to compare the coins of different countries with each other, with respect to their intrinsic values, that is, with reference to the amount of pure metal in each. Thus, the English sovereign, which represents the pound sterling, is intrinsically worth $4,861 in our gold, taken as a standard, as determined at the Mint in Philadelphia. This, therefore, is the value at which the sovereign must be reckoned, in estimating the par of exchange.

289. The *commercial par of exchange* is a comparison of the coins of different countries according to their market value. Thus, the market value of the English sovereign, varying from $4,83 to $4,85 (Art. **277**), the commercial par of exchange will fluctuate. It is, however, always determined when we know the value at which the foreign coin sells in our market.

QUEST.—If the drawee does not accept, what must the holder do? What is such notice called? By whom is it made? If the parties to the bill are not notified, what is the consequence? If the drawee accepts the bill and fails to make the payment, what must then be done? If the bill is not protested, what will be the consequence? 288. What do you understand by the intrinsic par of exchange? What is the intrinsic value of the English sovereign? 289. What is the commercial par of exchange? What is the commercial value of the English sovereign?

290. The *course of exchange* is the variable price which is paid at one place for bills of exchange drawn on another. The course of exchange differs from the intrinsic par of exchange, and also from the commercial par, in the same way that the market price of an article differs from its natural price. The commercial par of exchange would at all times determine the course of exchange, if there were no fluctuations in trade.

291. When the market price of a foreign bill is above the commercial par, the exchange is said to be at a premium, or in favor of the foreign place, because it indicates that the foreign place has sold more than it has bought, and that specie must be shipped to make up the difference. When the market price is below this par, exchange is said to be below par, or in favor of the place where the bill is drawn. Such place will then be a creditor, and the debt must be paid in specie or other property. It should be observed that a favorable state of exchange is advantageous to the buyer but not to the seller, whose interest, as dealer in exchange, is identified with that of the place on which the bill is drawn.

292. It was stated in Art. 277 that the exchange value of the pound sterling is $\$4,44\frac{4}{9} = 4,4444+$; that is, this value is the basis on which the bills of exchange are drawn. Now this value being below both the commercial and intrinsic value, the drawers of bills increase the course of exchange so as to make up this deficiency.

For example, if we add to the exchange value of the pound, 9 per cent, we shall have its commercial value, very nearly.

Thus, exchange value	- - - -	= $4,4444+
Nine per cent	- - - - - -	= ,3999+
which gives	- - - - - -	$4,8443

QUEST.—290. What do you understand by the course of exchange? How does it differ from the intrinsic par and the commercial par? What causes it to differ from the commercial par? 291. What is said when the price of a foreign bill is above the commercial par? When it is below it? To whom is a favorable state of exchange advantageous? To whom is it injurious? 292. What is the exchange value of the pound sterling?

and this is the average of the commercial value, very nearly. Therefore, when the course of exchange is at a premium of 9 per cent, it is at the commercial par, and as between England and this country it would stand near this point, but for the fluctuations of trade and other accidental circumstances.

INLAND BILLS.

293. We have seen that inland bills are those in which the drawer and drawee both reside in the same country (Art. 281).

EXAMPLES.

1. A merchant at New Orleans wishes to remit to New York \$8465, and exchange is $1\frac{1}{2}$ per cent premium. How much must he pay for such a bill?

2. A merchant in Boston wishes to pay in Philadelphia \$8746,50; exchange between Boston and Philadelphia is $1\frac{1}{4}$ per cent below par. What must he pay for a bill?

3. A merchant in Philadelphia wishes to pay \$9876,40 in Baltimore, and finds exchange to be 1 per cent below par: what must he pay for the bill?

ENGLAND.

294. It has already been stated that the exchanges between this country and England are made in pounds, shillings and pence, and that the exchange value of the pound sterling is \$4,44$\frac{4}{9}$, and that the premiums are all reckoned from this standard.

EXAMPLES.

1. A merchant in New York wishes to remit to Liverpool £1167 10*s*. 6*d*., exchange being at $8\frac{1}{2}$ per cent premium. How much must he pay for the bill in Federal money?

QUEST.—293. What are inland bills? 294. In what currency are the exchanges between this country and England made? What is the exchange value of the pound sterling?

First, £1167 10*s.* 6*d.*	- - -	= £1167.525
For $8\frac{1}{2}$ per cent multiply by	-	.085
the product is the premium	-	= 99.239625
this being added gives	- -	£1266.764625

which reduced to dollars and cents at the rate of $4,44$\frac{4}{9}$ to the pound, gives the amount which must be paid for the bill in dollars and cents.

2. A merchant has to remit £36794 8*s.* 9*d.* to London, how much must he pay for a bill in dollars and cents, ex change being $7\frac{3}{4}$ per cent premium?

3. A merchant in New York wishes to remit to London $67894,25, exchange being at a premium of 9 per cent. What will be the amount of his bill in pounds, shillings and pence?

NOTE.—Add the amount of the premium to the exchange value of the pound, viz. $4,44$\frac{4}{9}$, which in this case gives $4,84443, and then divide the amount in dollars by this sum, and the quotient will be the amount of the bill in pounds and the decimal of a pound.

4. A merchant in New York owes £1256 18*s.* 9*d.* in London; exchange at a nominal premium of $7\frac{1}{2}$ per cent: how much money in Federal currency will be necessary to purchase the bill?

5. I have $947,86 and wish to remit to London £364 18*s.* 8*d.*, exchange being at $8\frac{1}{4}$ per cent: how much additional money will be necessary?

FRANCE.

295. The accounts in France, and the exchange between France and other countries, are all kept in francs and centimes, which are hundredths of the franc. We see from the table that the value of the franc is 18.6 cents, which gives, very nearly, 5 francs and 38 centimes to the dollar. The rate of exchange is computed on the value 18.6 cents, but is often quoted by stating the value of the dollar in francs.

QUEST.—295. In what currency are the exchanges with France conducted? What is a centime? What is the value of a franc?

Thus, exchange on Paris is said to be 5 francs, 40 centimes, that is, one dollar will buy a bill on Paris of 5 francs and 40 hundredths of a franc.

EXAMPLES.

1. A merchant in New York wishes to remit 167556 francs to Paris, exchange being at a premium of $1\frac{1}{2}$ per cent. What will be the cost of his bill in dollars and cents?

Commercial value of the franc - -	18.6 cents
Add $1\frac{1}{2}$ per cent - - - - - -	279
Gives value for remitting - - -	18.879 cents;

then, $167556 \times 18.879 =$ $31632,89724,

which is the amount to be paid for the bill.

2. What amount in dollars and cents will purchase a bill on Paris for 86978 francs, exchange being at the rate of 5 francs and 2 centimes to the dollar?

First, $86978 \div 5.02 =$ $17326,274 + the amount.

Is this bill above or below par? What per cent?

3. How much money must be paid to purchase a bill of exchange on Paris for 68097 francs, exchange being 3 per cent below par?

4. A merchant in New York wishes to remit $16785,25 to Paris; exchange gives 5 francs 4 centimes to the dollar: how much can he remit in the currency of Paris?

HAMBURG.

296. Accounts and exchanges with Hamburg are generally made in the marc banco, valued, as we see in the table, at 35 cents.

EXAMPLES.

1. What amount in dollars and cents will purchase a bill of exchange on Hamburg for 18649 marcs banco, exchange being at 2 per cent premium?

QUEST.—What is meant when exchange on Paris is quoted at 5 francs 40 centimes? 296. In what are accounts kept at Hamburg? What is the value of the marc banco?

2. What amount will purchase a bill for 3678 marcs banco reckoning the exchange value of the marc banco at 34 cents? Will this be above or below the par of exchange?

ARBITRATION OF EXCHANGE.

297. Arbitration of exchange is the method by which the currency of one country is changed into that of another, through the medium of one or more intervening currencies, with which the first and last are compared.

98. When there is but one intervening currency it is called *simple arbitration;* and when there is more than one it is called *compound arbitration.* The method of performing this is called the *Chain Rule.*

299. The principle involved in arbitration of exchange is simply this: To pass from one system of values through several others, and find the true proportion or relation between the first and last. For example, suppose we wish to exchange 109150 pence into dollars by first changing them into shillings, then into pounds, and then into dollars. For this we have,

12 : 109150 : : 1*s*. : $109150 \times \frac{1}{12}$ = number of shillings.

20 : $109150 \times \frac{1}{12}$: : £1 : $109150 \times \frac{1}{12} \times \frac{1}{20}$ = No. of pounds.

£1 : \$4,444 : : $109150 \times \frac{1}{12} \times \frac{1}{20}$: $109150 \times \frac{1}{12} \times \frac{1}{20} \times \frac{4.444}{1}$:

hence the Chain Rule may be stated as follows:

Multiply the sum to be remitted by the following quotients, after having cancelled the common factors, viz., by a certain amount at the second place divided by its equivalent at the first; a certain amount at the third place by its equivalent at the second; a certain amount at the fourth place divided by its equivalent at the third, and so on to the last place.

QUEST.—297. What is arbitration of exchange? 298. When there is but one intervening currency, what is the exchange called? When there is more than one, what is it called? 299. What principle is involved in the arbitration of exchange? What is the Chain Rule? Give the rule.

NOTE.—In the above rule the amounts named are supposed to be expressed in the currency of the place from which the remittance is made. If in any case an amount is expressed in the currency of the place to which the remittance is made, the terms of the corresponding multiplier must be inverted. The example wrought above may be thus stated: Required to transmit 109150 pence to a second place where one piece of coin is worth 12 at the first place; thence to transmit it to a third where one piece is worth 20 at the second; thence to a fourth place where 4.444 pieces are equal to 1 at the third.

EXAMPLES.

1. A merchant wishes to remit $4888,40 from New York to London, and the exchange is 10 per cent. He finds that he can remit to Paris at 5 francs 15 centimes to the dollar, and to Hamburg at 35 cents per marc banco. Now, the exchange between Paris and London is 25 francs 80 centimes for £1 sterling, and between Hamburg and London $13\frac{3}{4}$ marcs banco for £1 sterling. How had he better remit?

1st. *To London direct.*

The amount to be remitted is $4888,40. The exchange value of £1 is $4,444, and since the exchange is at a premium of 10 per cent, the value of £1 is $4,444+,4444=$4,8884: hence,

$$\$4888,40 \times \frac{1}{4.8884} = £1000:$$

hence, if he remits direct he will obtain a bill for £1000.

2d. *Exchange through Paris.*

$$4888,40 \times \frac{\overset{1.03}{\cancel{5,15}}}{1} \times \frac{1}{\underset{5.16}{\cancel{25,80}}} = £975,7852 = £975\ 15s.\ 8\frac{1}{4}d.$$

Since 5,15 francs are equal to 1 dollar, the first multiplier will be this amount divided by $1; and since £1 is equal to 25.80 francs, the second multiplier will be £1 divided by this amount. Then by dividing by 5 and multiplying, we find that the amount remitted by the second method would be £975 15*s*. $8\frac{1}{4}$*d*.

3d. *Method through Hamburg.*

$$\$4888{,}40 \times \frac{1}{.35} \times \frac{1}{13.75} = 1015.771 = £1015\ 15s.\ 5d.$$

Since 1 marc banco is equal to 35 cents, it is 35 hundredths of a dollar: hence, the first multiplier is 1 marc banco divided by .35, and the second 1 divided by 13.75. The result shows that the best way to remit is through Hamburg, the next best direct, and the most unfavorable through Paris.

2. A merchant in London has sold goods in Amsterdam to the amount of 824 pounds Flemish, which could be remitted to London at the rate of 34*s.* 4*d.* Flemish per pound sterling. He orders it to be remitted circuitously at the following rates, viz., to France at the rate of 48*d.* Flemish per crown; thence to Vienna at 100 crowns for 60 ducats; thence to Hamburg at 100*d.* Flemish per ducat; thence to Lisbon at 50*d.* Flemish per crusado of 400 reas; and lastly, from Lisbon to England at 5*s.* 8*d.* per milrea: does he gain or lose by the circular exchange?

48*d.* Flemish	= 1 crown,
100 crowns	= 60 ducats,
1 ducat	= 100*d.* Flemish,
50*d.* Flemish	= 400 reas,
1 milrea or 1000 reas	= 68*d.* sterling.

$$824 \times \frac{1}{\overset{}{\cancel{48}}_{\,8}} \times \frac{\overset{10}{\cancel{60}}}{\cancel{100}} \times \frac{\cancel{100}}{1} \times \frac{\overset{8}{\cancel{400}}}{\cancel{50}} \times \frac{\overset{17}{\cancel{68}}}{\underset{\cancel{100}\ 25}{\cancel{1000}}} = \frac{824 \times 17}{25} = \frac{14008}{25}$$

$$= £560\ 6s.\ 4\tfrac{4}{5}d.$$

The direct exchange would give,

$$824 \times \frac{£1}{34s.\ 4d.\ \text{Flemish}} = 824 \times \tfrac{240}{412} = £480 \text{ sterling}.$$

Hence, the amount gained by circuitous exchange would be £80 6*s.* 4$\tfrac{4}{5}$*d.*

DUODECIMALS.

300. Duodecimals are denominate fractions in which 1 foot is the unit that is divided.

The unit 1 foot is first supposed to be divided into 12 equal parts, called inches or primes, and marked ′.

Each of these parts is supposed to be again divided into 12 equal parts, called seconds, and marked ″.

Each second is divided, in like manner, into 12 equal parts, called thirds, and marked ‴.

This division of the foot gives

1′ inch or prime - - - $= \frac{1}{12}$ of a foot.
1″ second is $= \frac{1}{12}$ of $\frac{1}{12}$ - $= \frac{1}{144}$ of a foot.
1‴ third is $= \frac{1}{12}$ of $\frac{1}{12}$ of $\frac{1}{12} = \frac{1}{1728}$ of a foot.

Hence, in duodecimals, the divisions of the foot increase from the lower denominations to the higher, according to the scale of twelves.

301. Duodecimals are added and subtracted like other denominate numbers, 12 of a lesser denomination making one of a greater, as in the following

TABLE.

12‴	make	1″	second.
12″	"	1′	inch or prime.
12′	"	1	foot.

EXAMPLES.

1. In 185′, how many feet? *Ans.* ——
2. In 250″, how many feet and inches? *Ans.* ——
3. In 4367‴, how many feet? *Ans.* ——
4. In 847″, how many feet? *Ans.* ——

Quest.—300. In Duodecimals, what is the unit that is divided? How is it divided? How are these parts again divided? What are the parts called? 301. How are duodecimals added and subtracted? How many of one denomination make 1 of the next greater?

EXAMPLES IN ADDITION AND SUBTRACTION.

1. What is the sum of 3*ft.* 6′ 3″ 2‴ and 2*ft.* 1′ 10″ 11‴?
2. What is the sum of 8*ft.* 9′ 7″ and 6*ft.* 7′ 3″ 4‴?
3. What is the difference between 9*ft.* 3′ 5″ 6‴ and 7*ft.* 3′ 6″ 7‴?
4. What is the difference between 40*ft.* 6′ 6″ and 19*ft.* 7‴?
5. What is the sum of 18*ft.* 9′ 11″ 5‴ and 17*ft.* 6′ 7″?
6. What is the difference between 27*ft.* 7‴ and 4*ft.* 9′ 10″ 9‴?

MULTIPLICATION OF DUODECIMALS.

302. It is known that feet multiplied by feet give square feet in the product. It is now required to show what fractions of the square foot will arise from multiplying feet by the divisions of the foot, and the divisions of the foot by each other.

EXAMPLES.

1. Multiply 6*ft.* 7′ 8″ by 2*ft.* 9′.

OPERATION.

	ft.		
	6	7′	8″
	2	9′	
2 × 8″ =		1′	4″
2 × 7′ =	1	2′	
2 × 6 =	12		
9′ × 8″ =			6″
9′ × 7′ =		5′	3″
9′ × 6 =	4	6′	
Prod.	18	3′	1″

Set down the multiplier under the multiplicand, so that feet shall fall under feet, and the corresponding divisions under each other. It is found most convenient to begin with the highest denomination of the multiplier, and multiply it by the lowest denomination of the multiplicand. Recollecting that 7′ expresses $\frac{7}{12}$ of a foot, and that 8″ expresses $\frac{8}{12}$ of $\frac{1}{12}$ of a foot, we see that 2 × 8″ will give 16-twelfths of twelfths of a square foot; that is, one-twelfth and four twelfths of one twelfth, or 4″. The 2 feet multiplied by 7′ give 14 twelfths of a square foot; that is, 1 square foot and two twelfths, or 2′. The feet multiplied by 6 give 12 square feet.

QUEST.—302. In multiplication how do you set down the multiplier? Where do you begin to multiply? How do you carry from one denomination to another?

Again, 9 inches or $\frac{9}{12}$ of a foot multiplied by 8 twelfths of $\frac{1}{12}$ of a foot, will give 72 twelfths of twelfths of twelfths of a square foot, which are equal to six twelfths of twelfths, or to 6″. Then 9′ × 7′ gives 63 twelfths of twelfths of a square foot, equal to 5′ and 3″: and 9′ × 6 gives 4 square feet and 6′.

303. Hence we see,

1st. *That feet multiplied by feet give square feet in the product.*

2d. *That feet multiplied by inches give twelfths of square feet in the product.*

3d. *That inches multiplied by inches give twelfths of twelfths of square feet in the product.*

4th. *That inches multiplied by seconds give twelfths of twelfths of twelfths of square feet in the product.*

2. Multiply 9*ft.* 4*in.* by 8*ft.* 3*in.*

OPERATION.

9	4′		
8	3′		
74	8′		
2	4′	0″	
77	0′	0″	*Ans.*

Beginning with the 8 feet, we say 8 times 4 are 32′, which is equal to 2 feet 8′: set down the 8′. Then say 8 times 9 are 72 and 2 to carry are 74 feet: then multiplying by 3′ we say, 3 times 4′ are 12″, equal to 1 inch: set down 0 in the second's place: then 3 times 9 are 27 and 1 to carry make 28′, equal to 2*ft.* 4′. Therefore the entire product is equal to 77*ft.*

3. How many solid feet in a stick of timber which is 25*ft.* 6*in.* long, 2*ft.* 7*in.* broad, and 3*ft.* 3*in.* thick?

4. Multiply 9*ft.* 2*in.* by 9*ft.* 6*in.* *Ans.* ——

5. Multiply 34*ft.* 10*in.* by 6*ft.* 8*in.* *Ans.* ——

6. Multiply 70*ft.* 9*in.* by 12*ft.* 3*in.* *Ans.* ——

7. How many cords and cord feet in a pile of wood 24 feet long, 4 feet wide, and 3*ft.* 6*in.* high?

8. Multiply 6*ft.* 9′ by 8*ft.* 6′. *Ans.* ——

QUEST.—303. Repeat the four principles.

9. How many cord feet in a pile of wood 25 feet long, 6 feet wide, and 5 feet high?

10. Multiply 16*ft.* 9′ by 11*ft.* 11″. *Ans.* ——

NOTE.—It must be recollected that 16 solid feet make 1 cord foot, (Art. 30).

INVOLUTION.

304. IF a number be multiplied by itself, the product is called the *second power*, or *square* of that number. Thus, $4 \times 4 = 16$: the number 16 is the 2d power or square of 4.

If a number be multiplied by itself, and the product arising be again multiplied by the number, the second product is called the 3d *power*, or *cube* of the number. Thus, $3 \times 3 \times 3 = 27$: the number 27 is the 3d *power*, or *cube* of 3.

The term *power* designates the product arising from multiplying a number by itself a certain number of times, and the number multiplied is called the *root.*

Thus, in the first example above, 4 is the root, and 16 the square or 2d power of 4.

In the 2d example, 3 is the root, and 27 the 3d power or cube of 3. The first power of a number is the number itself.

305. *Involution teaches the method of finding the powers of numbers.*

The number which designates the power to which the root is to be raised, is called the *index* or *exponent* of the power. It is generally written on the right, and a little above the root.

QUEST.—How many solid feet make a cord foot? 304. If a number be multiplied by itself once, what is the product called? If it be multiplied by itself twice, what is the product called? What does the term power mean? What is the root? What is the first power of a number? 305. What is Involution? What is the number called which designates the power? Where is it written?

Thus, 4^2 expresses the 2d power of 4, or that 4 is to be multiplied by itself once: hence,

$$4^2 = 4 \times 4 = 16.$$

For the same reason 3^3 denotes that 3 is to be raised to the 3d power, or cubed: hence,

$3^3 = 3 \times 3 \times 3 = 27$: we may therefore write

$4 =$	4	the 1st power of 4.
$4^2 = 4 \times 4 =$	16	the 2d power of 4.
$4^3 = 4 \times 4 \times 4 =$	64	the 3d power of 4.
$4^4 = 4 \times 4 \times 4 \times 4 =$	256	the 4th power of 4.
$4^5 = 4 \times 4 \times 4 \times 4 \times 4 =$	1024	the 5th power of 4.
&c.,	&c.,	&c.

Hence, to raise a number to any power,

Multiply the number continually by itself as many times less 1 as there are units in the exponent, and the last product will be the power sought.

EXAMPLES.

1. What is the 3d power of 125? *Ans.* ——
2. What is the cube of 7? *Ans.* ——
3. What is the square of 60? *Ans.* ——
4. What is the 4th power of 5? *Ans.* ——
5. What is the 5th power of 18? *Ans.* ——
6. What is the cube of 1? *Ans.* ——
7. What is the square of $\frac{1}{2}$? *Ans.* ——
8. What is the cube of .1? *Ans.* ——
9. What is the cube of $\frac{3}{4}$? *Ans.* ——
10. What is the square of .01? *Ans.* ——
11. What is the square of 6.12? *Ans.* ——
12. What is the 6th power of 10? *Ans.* ——
13. What is the cube of $3\frac{1}{4}$? *Ans.* ——
14. What is the 4th power of 36? *Ans.* ——
15. What is the cube of 8733? *Ans.* ——

QUEST.—What is the exponent of the square of a number? Of the cube? Of the fourth power? How do you raise a number to any power?

EVOLUTION.

306. We have seen that Involution teaches how to find the power when the root is given. Evolution is the reverse of Involution: it teaches how to find the root when the power is known. The root is that number which being multiplied by itself a certain number of times, will produce the given power.

The square root of a number is that number which being multiplied by itself once, will produce the given number.

The cube root of a number is that number which being multiplied by itself twice, will produce the given number.

For example, 6 is the square root of 36, because $6 \times 6 = 36$; and 3 is the cube root of 27, because $3 \times 3 \times 3 = 27$. The sign $\sqrt{}$ placed before a number denotes that its square root is to be extracted. Thus, $\sqrt{36} = 6$. The sign $\sqrt{}$ is called the radical sign, or the sign of the square root.

When we wish to express that the cube root is to be extracted, we place the figure 3 over the sign of the square root: thus, $\sqrt[3]{8} = 2$ and $\sqrt[3]{27} = 3$, and 3 is called the index of the root.

EXTRACTION OF THE SQUARE ROOT.

307. To extract the square root of a number, is to find a number which being multiplied by itself once, will produce the given number. Thus,

$$\sqrt{4} = 2; \text{ for } 2 \times 2 = 4;$$
$$\sqrt{9} = 3; \text{ for } 3 \times 3 = 9.$$

QUEST.—306. What is Evolution? What does it teach? What is the root of a number? What is the square root of a number? What is the cube root of a number? Make the sign denoting the square root. How do you denote the cube root? 307. What is required when we wish to extract the square root of a number?

Before proceeding to explain the rule for extracting the square root, let us first see how the squares of numbers are formed.

The first ten numbers are

1,	2,	3,	4,	5,	6,	7,	8,	9,	10	Roots.
1	4	9	16	25	36	49	64	81	100	Squares.

The numbers in the second line are the squares of those in the first; and the numbers in the first line are the *square roots* of the corresponding numbers of the second.

Now, it is evident that, *the square of a number expressed by a single figure will not contain any figure of a higher order than tens; and also, that if a number contains three figures, its root must contain tens and units*

The numbers 1, 4, 9, &c., of the second line, are called *perfect squares*, because they have exact roots.

Let us now see how the square of any number may be formed, say the number 36. This number is made up of 3 tens or 30, and 6 units

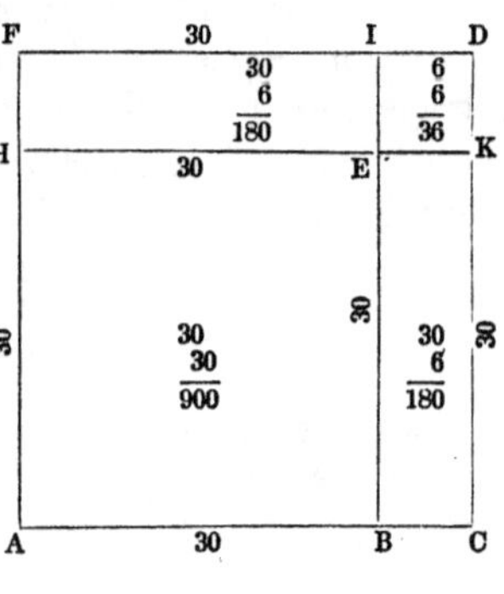

Let the line AB represent the 3 tens or 30, and BC the six units.

Let AD be a square on AC, and AE a square on the tens line AB.

Then ED will be a square on the unit line 6, and the rectangle EF will be the product of HE which is equal to the tens line, by IE which is equal to the unit line. Also, the rectangle BK will be the product of EB which is equal to the tens line, by the unit line BC. But the whole square

QUEST.—What is the greatest square of a single figure? What is the highest order of units that can be derived from the square of a single figure? How many perfect squares are there among the numbers that are less than one hundred?

on AC is made up of the square AE, the two rectangles FE and EC, and the square ED. Hence,

The square of two figures is equal to the square of the tens, plus twice the product of the tens by the units, plus the square of the units.

Let it now be required to extract the square root of 1296.

Since the number contains more than two places, its root will contain tens and units. But as the square of one ten is one hundred, it follows that the ten's place of the required root must be found in the figures on the left of 96. Hence, we point off the number into periods of two figures each.

```
  12̇ 9̇6(36
  9
66)396
   396
```

We next find the greatest square contained in 12, which is 3 tens or 30. We then square 3 tens which gives 9 hundred, and then place 9 under the hundred's place, and subtract.

This takes away the square AE and leaves the two rectangles FE and BK, together with the square ED on the unit line.

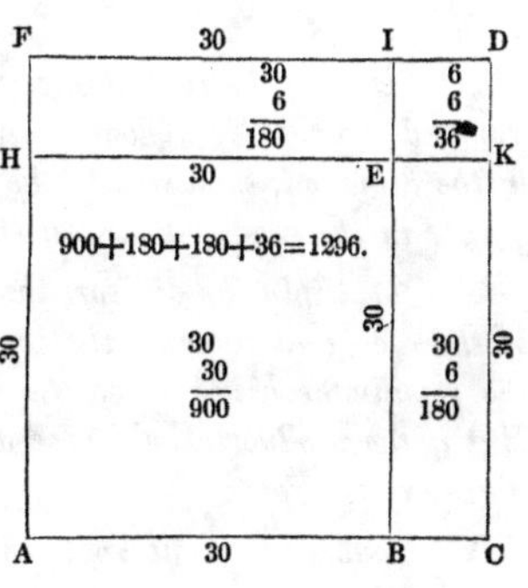

Now, since tens multiplied by units will give at least tens in the product, it follows that the area of the two rectangles FE and EC must be expressed by the figures of the given number at the left of the unit's place 6, which figures may also express a part of the square ED.

If, then, we divide the figures 39, at the left of 6, by twice the tens, that is, by twice AB or BE, the quotient will be BC or EK, the unit of the root.

QUEST.—What is the square of a number expressed by two figures equal to? In what places of figures will the square of the tens be found? In what places will the product of the tens by the units be found?

Then, placing BC or 6, in the root, and also in the divisor, and then multiplying the whole divisor 66 by 6, we obtain for a product the two rectangles FE and CE, together with the square ED.

Hence, the square root of 1296 is 36; or, in other words, 36 is the side of a square whose area is 1296.

CASE I.

308. To extract the square root of a whole number,

I. *Point off the given number into periods of two figures each, counted from the right, by setting a dot over the place of units, another over the place of hundreds, and so on.*

II. *Find the greatest square in the first period on the left, and place its root on the right after the manner of a quotient in division. Subtract the square of the root from the first period, and to the remainder bring down the second period for a dividend.*

III. *Double the root already found and place it on the left for a divisor. Seek how many times the divisor is contained in the dividend, exclusive of the right hand figure, and place the figure in the root and also at the right of the divisor.*

IV. *Multiply the divisor, thus augmented, by the last figure of the root, and subtract the product from the dividend, and to the remainder bring down the next period for a new dividend. But if the product should exceed the dividend, diminish the last figure of the root.*

V. *Double the whole root already found, for a new divisor, and continue the operation as before, until all the periods are brought down.*

EXAMPLES.

1. What is the square root of 263169?

QUEST.—308. What is the first step in extracting the square root of numbers? What the second? What the third? What the fourth? What the fifth? Give the entire rule.

We first place a dot over the 9, making the right hand period 69. We then put a dot over the 1 and also over the 6, making three periods.

OPERATION.

```
    2̇6 3̇1 6̇9(513
    25
101)131
    101
1023)3069
     3069
```

The greatest perfect square in 26, is 25, the root of which is 5. Placing 5 in the root, subtracting its square from 26, and bringing down the next period 31, we have 131 for a dividend, and by doubling the root we have 10 for a divisor. Now 10 is contained in 13, 1 time. Place 1 both in the root and in the divisor: then multiply 101 by 1; subtract the product and bring down the next period.

We must now double the whole root 51 for a new divisor, or we may take the first divisor after having doubled the last figure 1; then dividing we obtain 3, the third figure of the root.

309. Note 1.—There will be as many figures in the root as there are periods in the given number.

Note 2.—If the given number has not an exact root, there will be a remainder after all the periods are brought down, in which case ciphers may be annexed, forming new periods, each of which will give one decimal place in the root.

2. What is the square root of 36729?

In this example there are two places of decimals, which give two places of decimal in the root.

OPERATION.

```
      3̇ 6̇7 2̇9(191.64+.
      1
   29)267
      261
   381)629
       381
   3826)24800
        22956
   38324)184400
         153296
          31104 Rem.
```

Quest.—309. How many figures will there be in the root? If the given number has not an exact root, what may be done?

3. What is the square root of 213444 ? *Ans.* ——
4. What is the square root of 2268741 ? *Ans.* ——
5. What is the square root of 15193592 ? *Ans.* ——
6. What is the square root of 36372961 ? *Ans.* ——
7. What is the square root of 22071204 ? *Ans.* ——

CASE II.

310. To extract the square root of a decimal fraction,

I. *Annex one cipher, if necessary, so that the number of decimal places shall be even.*

II. *Point off the decimals into periods of two figures each, by putting a point over the place of hundredths, a second over the place of ten thousandths, &c.: then extract the root as in whole numbers, recollecting that the number of decimal places in the root will be equal to the number of periods in the given decimal.*

EXAMPLES.

1. What is the square root of .5 ?

We first annex one cipher to make even decimal places. We then extract the root of the first period, to which we annex ciphers, forming new periods.

OPERATION.

```
      .50(.707+
       49
  140)100
       000
 1407)10000
       9849
        151 Rem.
```

NOTE.—When there is a decimal and a whole number joined together the same rule will apply.

2. What is the square root of 3271.4207 ? *Ans.* ——
3. What is the square root of 4795.25731 ? *Ans.* ——
4. What is the square root of 4.372594 ? *Ans.* ——
5. What is the square root of .00032754 ? *Ans.* ——

QUEST.—310. How do you extract the square root of a decimal fraction? When there is a decimal and a whole number joined together, will the same rule apply?

6. What is the square root of .00103041? *Ans.* ——
7. What is the square root of 4.426816? *Ans.* ——
8. What is the square root of 47.692836? *Ans.* ——

CASE III.

311. To extract the square root of a vulgar fraction,

I. *Reduce mixed numbers to improper fractions, and compound fractions to simple ones, and then reduce the fraction to its lowest terms.*

II. *Extract the square root of the numerator and denominator separately, if they have exact roots; but when they have not, reduce the fraction to a decimal and extract the root as in Case II.*

EXAMPLES.

1. What is the square root of $\frac{23}{51}$ of $\frac{5}{37}$ of $4\frac{1}{2}$?
2. What is the square root of $\frac{2704}{4225}$? *Ans.* ——
3. What is the square root of $\frac{2216}{12544}$? *Ans.* ——
4. What is the square root of $\frac{275}{241}$? *Ans.* ——
5. What is the square root of $\frac{357}{476}$? *Ans.* ——
6. What is the square root of $\frac{478}{549}$? *Ans.* ——

EXTRACTION OF THE CUBE ROOT.

312. To extract the cube root of a number is to find a *second* number which being multiplied into itself twice, shall produce the given number.

Thus, 2 is the cube root of 8; for, $2 \times 2 \times 2 = 8$: and 3 is the cube root of 27; for, $3 \times 3 \times 3 = 27$.

Roots	1,	2,	3,	4,	5,	6,	7,	8,	9.
Cubes	1	8	27	64	125	216	343	512	729.

QUEST.—311. How do you extract the square root of a vulgar fraction?

From which we see, that the cube of units will not give a higher order than hundreds. We may also remark, that the cube of one ten or 10, is 1000: and the cube of 9 tens or 90, 729000; and hence, *the cube of tens will not give a lower denomination than thousands, nor a higher denomination than hundreds of thousands.* Hence also, if a number contains more than three figures its cube root will contain more than one; if the number contains more than six figures the root will contain more than two; and so on, every three figures from the right giving one additional place in the root, and the figures which remain at the left hand, although less than three, will also give one place in the root.

Let us now see how the cube of any number, as 16, is formed. Sixteen is composed of 1 ten and 6 units, and may be written $10 + 6$. Now to find the cube of 16 or of $10 + 6$, we must multiply the number by itself twice.

To do this we place the numbers thus			10 +	6
			10 +	6
Product by the units - - - - - - -			60 +	36
Product by the tens - - - - -		100 +	60	
Square of 16, - - - - -		100 +	120 +	36
Multiply again by 16 - - - - - - -			10 +	6
Product by the units - - - - -		600 +	720 +	216
Product by the tens - -	1000 +	1200 +	360	
Cube of 16 - - -	1000 +	1800 +	1080 +	216

1. By examining the composition of this number it will be found that the first part 1000 is the cube of the tens; that is,

$$10 \times 10 \times 10 = 1000.$$

2. The second part 1800 is equal to three times the square of the tens multiplied by the units; that is,

$$3 \times (10)^2 \times 6 = 3 \times 100 \times 6 = 1800.$$

3. The third part 1080 is equal to three times the square of the units multiplied by the tens; that is,

$$3 \times 6^2 \times 10 = 3 \times 36 \times 10 = 1080.$$

4. The fourth part is equal to the cube of the units; that is,

$$6^3 = 6 \times 6 \times 6 = 216.$$

Let it now be required to extract the cube root of the number 4096.

OPERATION.

$$\dot{4}\ 09\dot{6}(16$$
$$1$$
$$1^2 \times 3 = 3)3\ 0 \quad (9\text{–}8\text{–}7\text{–}6$$
$$\overline{16}^2 = 4\ 096.$$

Since the number contains more than three figures, we know that the root will contain at least units and tens.

Separating the three right hand figures from the 4, we know that the cube of the tens will be found in the 4. Now, 1 is the greatest cube in 4.

Hence, we place the root 1 on the right, and this is the tens of the required root. We then cube 1 and subtract the result from 4, and to the remainder we bring down the first figure 0 of the next period.

Now, we have seen that the second part of the cube of 16, viz., 1800, being three times the square of the tens multiplied by the units, will have no significant figure of a less denomination than hundreds, and consequently will make up a part of the 30 hundreds above. But this 30 hundreds also contains all the hundreds which come from the 3d and 4th parts of the cube of 16. If this were not the case, the 30 hundreds divided by three times the square of the tens would give the unit figure exactly.

Forming a divisor of three times the square of the tens, we find the quotient to be ten; but this we know to be too large. Placing 9 in the root and cubing 19, we find the result to be 6859. Then trying 8 we find the cube of 18 still too large; but when we take 6 we find the exact number. Hence, the cube root of 4096 is 16.

CASE I.

313. To extract the cube root of a whole number,

I. *Point off the given number into periods of three figures each, by placing a dot over the place of units, a second over the place of thousands, and so on to the left: the left hand period will often contain less than three places of figures.*

II. *Seek the greatest cube in the first period, and set its root on the right after the manner of a quotient in division. Subtract the cube of this figure from the first period, and to the remainder bring down the first figure of the next period, and call this number the dividend.*

III. *Take three times the square of the root just found for a divisor and see how often it is contained in the dividend, and place the quotient for a second figure of the root. Then cube the figures of the root thus found, and if their cube be greater than the first two periods of the given number, diminish the last figure, but if it be less, subtract it from the first two periods, and to the remainder bring down the first figure of the next period, for a new dividend.*

IV. *Take three times the square of the whole root for a new divisor, and seek how often it is contained in the new dividend: the quotient will be the third figure of the root. Cube the whole root and subtract the result from the first three periods of the given number, and proceed in a similar way for all the periods.*

EXAMPLES.

1. What is the cube root of 99252847?

$$\dot{9}9\ 25\dot{2}\ 84\dot{7}(463$$

$$4^3 = 64$$

$$4^2 \times 3 = 48)352 \quad \text{dividend}$$

First two periods - - - - $\dot{9}9\ 25\dot{2}$

$(46)^3 = 46 \times 46 \times 46 = \quad 97\ 336$

$3 \times (46)^2 = 6348\ )\ 19168$ 2d dividend

The first three periods - - $\dot{9}9\ 25\dot{2}\ 84\dot{7}$

$(463)^3 \quad = 99\ 252\ 847$

Ans. 463.

QUEST.—312. What is required when we are to extract the cube root of a number? 313 How do you extract the cube root of a whole number?

2. What is the cube root of 389017? *Ans.* ——
3. What is the cube root of 5735339? *Ans.* ——
4. What is the cube root of 32461759? *Ans.* ——
5. What is the cube root of 84604519? *Ans.* ——
6. What is the cube root of 259694072? *Ans.* ——
7. What is the cube root of 48228544? *Ans.* ——
8. What is the cube root of 27054036008? *Ans.* ——

CASE II.

314. To extract the cube root of a decimal fraction,

Annex ciphers to the decimals, if necessary, so that it shall consist of 3, 6, 9, *&c., places. Then put the first point over the place of thousandths, the second over the place of millionths, and so on over every third place to the right; after which extract the root as in whole numbers.*

Note 1.—There will be as many decimal places in the root as there are periods in the given number.

Note 2.—The same rule applies when the given number is composed of a whole number and a decimal.

Note 3.—If in extracting the root of a number there is a remainder after all the periods have been brought down, periods of ciphers may be annexed by considering them as decimals.

EXAMPLES.

1. What is the cube root of .127464? *Ans.* ——
2. What is the cube root of .870983875? *Ans.* ——
3. What is the cube root of 12.977875? *Ans.* ——
4. What is the cube root of 75.1089429? *Ans.* ——
5. What is the cube root of .353393243? *Ans.* ——
6. What is the cube root of 3.408862625? *Ans.* ——
7. What is the cube root of 27.708101576? *Ans.* ——

Quest.—314. How do you extract the cube root of a decimal fraction? How many decimal places will there be in the root? Will the same rule apply when there is a whole number and a decimal? In extracting the root if there is a remainder, what may be done?

CASE III.

315. To extract the cube root of a vulgar fraction,

I. *Reduce compound fractions to simple ones, mixed numbers to improper fractions, and then reduce the fraction to its lowest terms.*

II. *Then extract the cube root of the numerator and denominator separately, if they have exact roots; but if either of them has not an exact root, reduce the fraction to a decimal, and extract the root as in the last Case.*

EXAMPLES.

1. What is the cube root of $\frac{250}{686}$? *Ans.* ——
2. What is the cube root of $12\frac{19}{27}$? *Ans.* ——
3. What is the cube root of $31\frac{15}{343}$? *Ans.* ——
4. What is the cube root of $\frac{324}{1500}$? *Ans.* ——
5. What is the cube root of $\frac{4}{7}$? *Ans.* ——
6. What is the cube root of $\frac{5}{9}$? *Ans.* ——
7. What is the cube root of $\frac{2}{3}$? *Ans.* ——

ARITHMETICAL PROGRESSION.

316. If we take any number, as 2, we can, by the continued addition of any other number, as 3, form a *series* of numbers: thus,

2, 5, 8, 11, 14, 17, 20, 23, &c.,

in which each number is formed by the addition of 3 to the preceding number.

This series of numbers may also be formed by subtracting 3 continually from the larger number: thus,

23, 20, 17, 14, 11, 8, 5, 2.

A series of numbers formed in either way is called an *Arithmetical Series*, or an *Arithmetical Progression;* and the

---

Quest.—315. How do you extract the cube root of a vulgar fraction? 316. How do you form an Arithmetical Series?

number which is added or subtracted is called the *common difference.*

When the series is formed by the continued addition of the common difference, it is called an *ascending* series; and when it is formed by the subtraction of the common difference, it is called a *descending* series; thus,

2, 5, 8, 11, 14, 17, 20, 23, is an ascending series.
23, 20, 17, 14, 11, 8, 5, 2, is a descending series.

The several numbers are called *terms* of the progression: the first and last terms are called the *extremes*, and the intermediate terms are called the *means*.

317. In every arithmetical progression there are five things which are considered, any three of which being given or known, the remaining two can be determined. They are

1st, the first term;
2d, the last term;
3d, the common difference;
4th, the number of terms;
5th, the sum of all the terms.

318. By considering the manner in which the ascending progression is formed, we see that the 2d term is obtained by adding the common difference to the 1st term; the 3d, by adding the common difference to the 2d; the 4th, by adding the common difference to the 3d, and so on; *the number of additions being* 1 *less than the number of terms found.*

But instead of making the additions, we may multiply the common difference by the number of additions, that is, by 1 less than the number of terms, and add the first term to the product. Hence, we have

---

QUEST.—What is the common difference? What is an ascending series? What a descending series? What are the several numbers called? What are the first and last terms called? What are the intermediate terms called? 317. In every arithmetical progression how many things are considered? What are they? 318. How do you find the last term when the first term and common difference are known?

CASE I.

Having given the first term, the common difference, and the number of terms, to find the last term.

Multiply the common difference by 1 *less than the number of terms, and to the product add the first term.*

EXAMPLES.

1. The first term is 3, the common difference 2, and the number of terms 19: what is the last term?

We multiply the number of terms less 1, by the common difference 2, and then add the first term.

OPERATION.

18 number of terms less 1.
 2 common difference
36
 3 1st term.
39 last term.

2. A man bought 50 yards of cloth; he was to pay 6 cents for the first yard, 9 cents for the 2d, 12 cents for the 3d, and so on increasing by the common difference 3: how much did he pay for the last yard?

3. A man puts out $100 at simple interest, at 7 per cent; at the end of the first year it will have increased to $107, at the end of the 2d year to $114, and so on, increasing $7 each year: what will be the amount at the end of 16 years?

319. Since the last term of an arithmetical progression is equal to the first term added to the product of the common difference by 1 less than the number of terms, it follows, that the difference of the extremes will be equal to this product, and that the common difference will be equal to this product divided by 1 less than the number of terms. Hence, we have

CASE II.

Having given the two extremes and the number of terms of an arithmetical progression, to find the common difference.

Subtract the less extreme from the greater and divide the re-

QUEST.—319. How do you find the common difference, when you know the two extremes and number of terms?

mainder by 1 *less than the number of terms: the quotient will be the common difference.*

EXAMPLES.

1. The extremes are 4 and 104, and the number of terms 26: what is the common difference?

We subtract the less extreme from the greater and divide the difference by one less than the number of terms.

OPERATION.

$$\begin{array}{r} 104 \\ \underline{4} \\ 26 - 1 = 25)100(4 \\ \underline{100} \end{array}$$

2. A man has 8 sons, the youngest is 4 years old and the eldest 32, their ages increase in arithmetical progression: what is the common difference of their ages?

3. A man is to travel from New York to a certain place in 12 days; to go 3 miles the first day, increasing every day by the same number of miles; so that the last day's journey may be 58 miles: required the daily increase.

320. If we take any arithmetical series, as

3	5	7	9	11	13	15	17	19, &c.	
19	17	15	13	11	9	7	5	3	by reversing the order of the terms.
22	22	22	22	22	22	22	22	22	

Here we see that the sum of the terms of these two series is equal to 22, the sum of the extremes, multiplied by the number of terms; and consequently, the sum of either series is equal to the sum of the two extremes multiplied by half the number of terms; hence, we have

CASE III.

To find the sum of all the terms of an arithmetical progression,

Add the extremes together and multiply their sum by half the number of terms: the product will be sum of the series.

EXAMPLES.

1. The extremes are 2 and 100, and the number of terms 22: what is the sum of the series?

QUEST.—320. How do you find the sum of an arithmetical series?

We first add together the two extremes, and then multiply by half the number of terms.

OPERATION.

2	1st term
100	last term
102	sum of extremes
11	half the number of terms
1122	sum of series.

2. How many times does the hammer of a clock strike in 12 hours?

3. The first term of a series is 2, the common difference 4, and the number of terms 9: what is the last term and sum of the series?

4. If 100 eggs are placed in a right line, exactly one yard from each other, and the first one yard from a basket, what distance will a man travel who gathers them up singly, and places them in the basket?

GENERAL EXAMPLES.

1. What is the 18th term of an arithmetical progression of which the first term is 4 and the common difference 5?

2. The 18th term of an arithmetical progression is 89 and the common difference 5: what is the first term?

3. A flight of stairs has 18 steps; the first ascends but 12 inches in a vertical line, and each of the others 18: what is the entire ascent in a vertical line?

4. A debtor has 18 creditors; he owes to the largest creditor 89 dollars, and 5 dollars less to each of the others in succession: how much does he owe to the least?

5. A person travelled from Boston to a certain place in 8 days; he travelled 2 miles the first day, and every succeeding day he travelled farther than he did the preceding by an equal number of miles: the last day he travelled 23 miles: how much did he travel each day, and how much in all?

6. The number of terms is 22, the common difference 5, and the sum of the terms 1221: what is the least term?

7. A man is to receive $3000 in 12 payments, each succeeding payment to exceed the previous by $4: what will the last payment be?

GEOMETRICAL PROGRESSION

321. If we take any number, as 3, and multiply it continually by any other number, as 2, we form a series of numbers: thus,

3 6 12 24 48 96 192, &c.,

in which each number is formed by multiplying the number before it by 2.

This series may also be formed by dividing continually the largest number 192 by 2. Thus,

192 96 48 24 12 6 3.

A series formed in either way, is called a Geometrical Series, or a Geometrical Progression, and the number by which we continually multiply or divide, is called the *common ratio.*

When the series is formed by multiplying continually by the common ratio, it is called an *ascending series;* and when it is formed by dividing continually by the common ratio, it is called a *descending series.* Thus,

3 6 12 24 48 96 192 is an ascending series.
192 96 48 24 12 6 3 is a descending series.

The several numbers are called *terms* of the progression.

The first and last terms are called the *extremes*, and the intermediate terms are called the *means.*

322. In every Geometrical, as well as in every Arithmetical Progression, there are five things which are considered, any three of which being given or known, the remaining two can be determined. They are,

---

Quest.—321. How do you form a Geometrical Progression? What is the common ratio? What is an ascending series? What is a descending series? What are the several numbers called? What are the first and last terms called? What are the intermediate terms called? 322. In every geometrical progression, how many things are considered? What are they?

1st, the first term,
2d, the last term,
3d, the common ratio,
4th, the number of terms,
5th, the sum of all the terms.

By considering the manner in which the ascending progression is formed, we see that the second term is obtained by multiplying the first term by the common ratio; the 3d term by multiplying this product by the common ratio, and so on, the number of multiplications being one less than the number of terms. Thus,

$3 = 3$ 1st term,
$3 \times 2 = 6$ 2d term,
$3 \times 2 \times 2 = 12$ 3d term,
$3 \times 2 \times 2 \times 2 = 24$ 4th term, &c. for the other terms.

But $2 \times 2 = 2^2$, $2 \times 2 \times 2 = 2^3$, and $2 \times 2 \times 2 \times 2 = 2^4$.

Therefore, any term of the progression is equal to the first term multiplied by the ratio raised to a power 1 less than the number of the term.

CASE I.

Having given the first term, the common ratio, and the number of terms, to find the last term,

Raise the ratio to a power whose exponent is one less than the number of terms, and then multiply the power by the first term: the product will be the last term.

EXAMPLES.

1. The first term is 3 and the ratio 2: what is the 6th term?

$$2 \times 2 \times 2 \times 2 \times 2 = 2^5 = 32$$
3 1st term.
Ans. 96

QUEST.—How many must be known before the remaining ones can be found? What is any term equal to? How do you find the last term?

15

2. A man purchased 12 pears: he was to pay 1 farthing for the 1st, 2 farthings for the 2d, 4 for the 3d, and so on doubling each time: what did he pay for the last?

3. A gentleman dying left nine sons, and bequeathed his estate in the following manner: to his executors £50; his youngest son to have twice as much as the executors, and each son to have double the amount of the son next younger: what was the eldest son's portion?

4. A man bought 12 yards of cloth, giving 3 cents for the 1st yard, 6 for the 2d, 12 for the 3d, &c.: what did he pay for the last yard?

CASE II.

323. Having given the ratio and the two extremes to find the sum of the series.

Subtract the less extreme from the greater, divide the remainder by 1 less than the ratio, and to the quotient add the greater extreme: the sum will be the sum of the series.

EXAMPLES.

1. The first term is 3, the ratio 2, and last term 192; what is the sum of the series?

$192 - 3 = 189$ difference of the extremes,

$2 - 1 = 1$) 189 (189; then $189 + 192 = 381$ *Ans.*

2. A gentleman married his daughter on New Year's day, and gave her husband 1*s.* towards her portion, and was to double it on the first day of every month during the year: what was her portion?

3. A man bought 10 bushels of wheat on the condition that he should pay 1 cent for the 1st bushel, 3 for the 2d, 9 for the 3d, and so on to the last: what did he pay for the last bushel and for the 10 bushels?

4. A man has six children; to the 1st he gives $150; to the 2d $300, to the 3d $600, and so on, to each twice as much as the last: how much did the eldest receive, and what was the amount received by them all?

QUEST.—223. How do you find the sum of the series?

MENSURATION.

324. Mensuration is the process of determining the contents of geometrical figures, and is divided into two parts, the mensuration of surfaces and the mensuration of solids.

MENSURATION OF SURFACES.

325. Surfaces have length and breadth. They are measured by means of a square, which is called the *unit of surface.*

A square is the space included between four equal lines, drawn perpendicular to each other. Each line is called a side of the square. If each side be one foot, the figure is called a *square foot.*

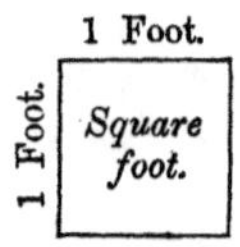

1 Foot.

1 Foot.

Square foot.

If the sides of a square be each four feet, the square will contain sixteen square feet. For, in the large square there are sixteen small squares, the sides of which are each one foot. Therefore, the square whose side is four feet, contains sixteen square feet.

The number of small squares that is contained in any large square is always equal to the product of two of the sides of the large square. As in the figure, $4\times4=16$ square feet. The number of square inches contained in a square foot is equal to $12\times12=144$.

326. A triangle is a figure bounded by three straight lines. Thus, BAC is a triangle.

QUEST.—324. What is mensuration? 325. What is a surface? What is a square? What is the number of small squares contained in a large square equal to? 326. What is a triangle?

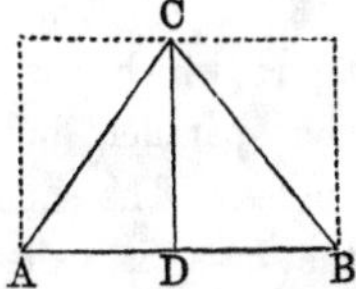

The three lines BA, AC, BC, are called *sides:* and the three corners, B, A, and C, are called *angles.* The side AB is called the *base.*

When a line like CD is drawn making the angle CDA equal to the angle CDB, then CD is said to be perpendicular to AB, and CD is called the *altitude* of the triangle. Each triangle CAD or CDB is called a right-angled triangle. The side BC, or the side AC, opposite the right angle, is called the *hypothenuse.*

The area or contents of a triangle is equal to half the product of its base by its altitude (Bk. IV. Prop. VI).*

EXAMPLES.

1. The base, AB, of a triangle is 50 yards, and the perpendicular, CD, 30 yards: what is the area?

We first multiply the base by the altitude, and the product is square yards, which we divide by 2 for the area.

OPERATION.

```
        50
        30
    2)1500
Ans.   750 square yards.
```

2. In a triangular field the base is 60 chains, and the perpendicular 12 chains: how much does it contain?

3. There is a triangular field, of which the base is 45 rods and the perpendicular 38 rods: what are its contents?

4. What are the contents of a triangle whose base is 75 chains and perpendicular 36 chains?

327. A rectangle is a four-sided figure like a square, in which the sides are perpendicular to each other, but the adjacent sides are not equal.

* All the references are to Davies' Legendre.

QUEST.—326. What is the base of a triangle? What the altitude? What is a right-angled triangle? Which side is the hypothenuse? What is the area of a triangle equal to? 327. What is a rectangle?

328. A parallelogram is a four-sided figure which has its opposite sides equal and parallel, but its angles not right-angles. The line DE, perpendicular to the base, is called the altitude.

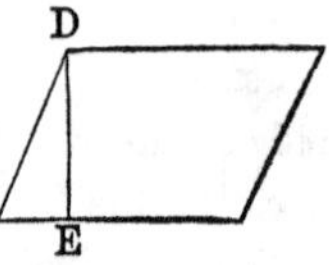

329. To find the area of a square, rectangle, or parallelogram,

Multiply the base by the perpendicular height, and the product will be the area (Bk. IV. Prop. V).

EXAMPLES.

1. What is the area of a square field of which the sides are each 66.16 chains?
2. What is the area of a square piece of land of which the sides are 54 chains?
3. What is the area of a square piece of land of which the sides are 75 rods each?
4. What are the contents of a rectangular field, the length of which is 80 rods and the breadth 40 rods?
5. What are the contents of a field 80 rods square?
6. What are the contents of a rectangular field 30 chains long and 5 chains broad?
7. What are the contents of a field 54 chains long and 18 rods broad?
8. The base of a parallelogram is 542 yards, and the perpendicular height 720 feet: what is the area?

330. A trapezoid is a four-sided figure ABCD, having two of its sides, AB, DC, parallel. The perpendicular EF is called the altitude.

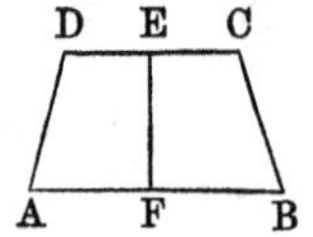

QUEST.—328. What is a parallelogram? 329. How do you find the area of a square, rectangle, or parallelogram? 330. What is a trapezoid?

331. To find the area of a trapezoid,

Multiply the sum of the two parallel sides by the altitude, and divide the product by 2, and the quotient will be the area (Bk. IV. Prop. VII).

EXAMPLES.

1. Required the area or contents of the trapezoid ABCD, having given AB=643.02 feet, DC=428.48 feet, and EF =342.32 feet.

We first find the sum of the sides, and then multiply it by the perpendicular height, after which, we divide the product by 2, for the area.

OPERATION.

$643.02 + 428.48 = 1071.50 =$ sum of parallel sides. Then, $1071.50 \times 342.32 = 366795.88$; and, $\frac{366795.88}{2} = 183397.94 =$ the area.

2. What is the area of a trapezoid, the parallel sides of which are 24.82 and 16.44 chains, and the perpendicular distance between them 10.30 chains?

3. Required the area of a trapezoid whose parallel sides are 51 feet, and 37 feet 6 inches, and the perpendicular distance between them 20 feet 10 inches.

4. Required the area of a trapezoid whose parallel sides are 41 and 24.5, and the perpendicular distance between them 21.5 yards.

5. What is the area of a trapezoid whose parallel sides are 15 chains, and 24.5 chains, and the perpendicular height 30.8 chains?

6. What are the contents when the parallel sides are 40 and 64 chains, and the perpendicular distance between them 52 chains?

QUEST.—331. How do you find the area of a trapezoid?

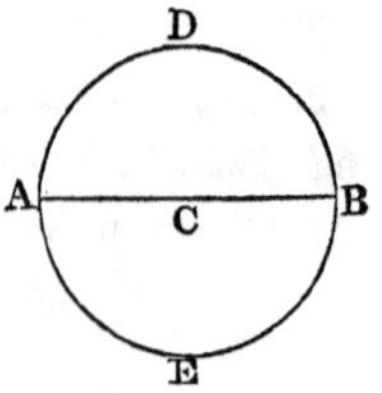

332. A circle is a portion of a plane bounded by a curved line, every part of which is equally distant from a certain point within, called the centre.

The curved line AEBD is called the *circumference;* the point C the *centre;* the line AB passing through the centre, a *diameter;* and CB the *radius.*

The circumference AEBD is 3.1416 times as great as the diameter AB. Hence, if the diameter is 1, the circumference will be 3.1416. Therefore, if the diameter is known, the circumference is found by multiplying 3.1416 by the diameter (Bk. V. Prop. XIV).

EXAMPLES.

1. The diameter of a circle is 8: what is the circumference?

The circumference is found by simply multiplying 3.1416 by the diameter.

OPERATION.

	3.1416
	8
Ans.	25.1328

2. The diameter of a circle is 186: what is the circumference?

3. The diameter of a circle is 40: what is the circumference?

4. What is the circumference of a circle whose diameter is 57?

333. Since the circumference of a circle is 3.1416 times as great as the diameter, it follows, that if the circumference is known, we may find the diameter by dividing it by 3.1416.

QUEST.—332. What is a circle? What is the centre? What is the circumference? What is the diameter? What the radius? How many times greater is the circumference than the diameter? How do you find the circumference when the diameter is known? 333. How do you find the diameter when the circumference is known?

EXAMPLES.

1. What is the diameter of a circle whose circumference is 157.08?

We divide the circumference by 3.1416, the quotient 50 is the diameter.

OPERATION:

3.1416)157.080(50
157.080

2. What is the diameter of a circle whose circumference is 23304.3888?

3. What is the diameter of a circle whose circumference is 13700? *Ans.* ——.

334. To find the area or contents of a circle,

Multiply the square of the diameter by the decimal .7854 (Bk. V. Prop. XII. Cor. 2).

EXAMPLES.

1. What is the area of a circle whose diameter is 12?

We first square the diameter, giving 144, which we then multiply by the decimal .7854: the product is the area of the circle.

OPERATION.

$\overline{12}^2 = 144$

$144 \times .7854 = 113.0976$

Ans. 113.0976

2. What is the area of a circle whose diameter is 5?

3. What is the area of a circle whose diameter is 14?

4. How many square yards in a circle whose diameter is $3\frac{1}{2}$ feet?

335. A sphere is a solid terminated by a curved surface, all the points of which are equally distant from a certain point within, called the centre. The line AD, passing through its centre C, is called the diameter of the sphere, and AC its radius.

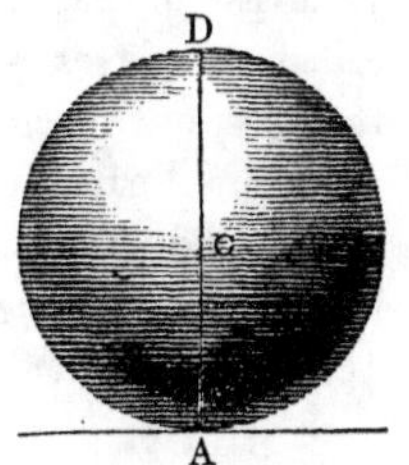

QUEST.—334. How do you find the area of a circle? 335. What is a sphere? What is a diameter? What is a radius?

336. To find the surface of a sphere,

Multiply the square of the diameter by 3.1416 (Bk. VIII. Prop. X. Cor.)

EXAMPLES.

1. What is the surface of a sphere whose diameter is 6?

We simply multiply the decimal 3.1416 by the square of the diameter: the product is the surface.

OPERATION.

$$\begin{array}{r} 3.1416 \\ 6^2 = \quad 36 \\ \hline \textit{Ans.} \quad 113.976 \end{array}$$

2. What is the surface of a sphere whose diameter is 14?

3. Required the number of square inches in the surface of a sphere whose diameter is 3 feet or 36 inches.

4. Required the area of the surface of the earth, its mean diameter being 7918.7 miles.

MENSURATION OF SOLIDS.

337. A cube is a body, or solid, having six equal faces, which are squares. If the sides of the cube be each one foot long, the solid is called a cubic or solid foot. But when the sides of the cube are one yard, as in the figure, the cube is called a cubic or solid yard. The base of the cube, which is the face on which it stands, contains $3 \times 3 = 9$ square feet. Therefore 9 cubes, of one foot each, can be placed on the base. If the solid were one foot high it would contain 9 cubic feet; if it were 2 feet high it would contain two tiers of cubes, or 18 cubic feet; and if it were 3 feet high, it would contain

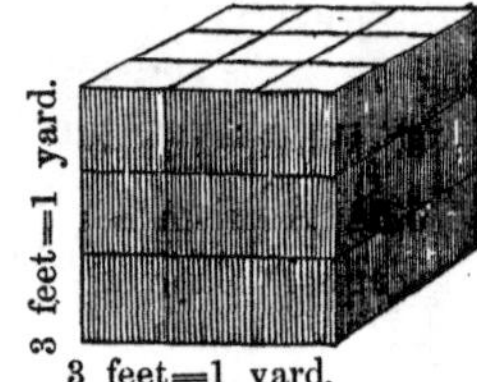

QUEST.—336. How do you find the surface of a sphere? 337. What is a cube? What is a cubic or solid foot? What is a cubic yard? How many cubic feet in a cubic yard?

three tiers, or 27 cubic feet. Hence, *the contents of a solid are equal to the product of its length, breadth, and height.*

338. To find the solidity of a sphere,

Multiply the surface by the diameter and divide the product by 6, *the quotient will be the solidity* (Bk. VIII. Prop. XIV. Sch. 3).

EXAMPLES.

1. What is the solidity of a sphere whose diameter is 12 ?

We first find the surface by multiplying the square of the diameter by 3.1416. We then multiply the surface by the diameter, and divide the product by 6.

OPERATION.

	$\overline{12}^2 = 144$
multiply by	3.1416
surface	$= 452.3904$
diameter	12
	6)5428.6848
solidity	$= 904.7808$

2. What is the solidity of a sphere whose diameter is 8 ?

3. What is the solidity of a sphere whose diameter is 16 inches ?

4. What is the solidity of the earth, its mean diameter being 7918.7 miles ?

5. What is the solidity of a sphere whose diameter is 12 feet ?

339. A prism is a solid whose ends are equal plane figures and whose faces are parallelograms.

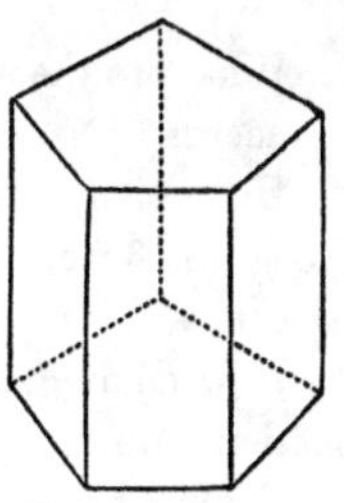

The sum of the sides which bound the base is called the perimeter of the base, and the sum of the parallelograms which bound the solid is called the convex surface.

340. To find the convex surface of a right prism,

QUEST.—What are the contents of a solid equal to ? 338. How do you find the solidity of a sphere ? 339. What is a prism ? What is the perimeter of the base ? What is the convex surface ?

Multiply the perimeter of the base by the perpendicular height, and the product will be the convex surface (Bk. VII. Prop. I).

EXAMPLES.

1. What is the convex surface of a prism whose base is bounded by five equal sides, each of which is 35 feet, the altitude being 52 feet?

2. What is the convex surface when there are eight equal sides, each 15 feet in length, and the altitude is 12 feet?

341. To find the solid contents of a prism,

Multiply the area of the base by the altitude, and the product will be the contents (Bk. VII. Prop. XIV).

EXAMPLES.

1. What are the contents of a square prism, each side of the square which forms the base being 16, and the altitude of the prism 30 feet?

We first find the area of the square which forms the base, and then multiply by the altitude.

OPERATION.

$$\overline{16}^2 = 256$$
$$30$$
$$\textit{Ans. } 7680$$

2. What are the solid contents of a cube, each side of which is 48 inches?

3. How many cubic feet in a block of marble, of which the length is 3 feet 2 inches, breadth 2 feet 8 inches, and height or thickness 5 feet?

4. How many gallons of water will a cistern contain, whose dimensions are the same as in the last example?

5. Required the solidity of a triangular prism, whose height is 20 feet, and area of the base 691.

QUEST.—340. How do you find the convex surface of a prism? 341. How do you find the solid contents of a prism?

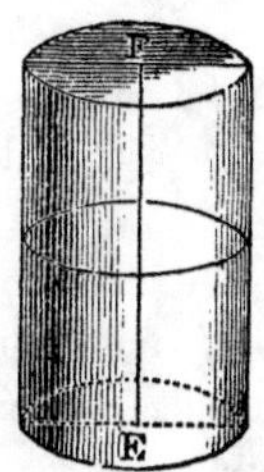

342. A cylinder is a round body with circular ends. The line EF is called the axis or altitude, and the circular surface the *convex surface* of the cylinder.

343. To find the convex surface of a cylinder,

Multiply the circumference of the base by the altitude, and the product will be the convex surface (Bk. VIII. Prop. I).

EXAMPLES.

1. What is the convex surface of a cylinder, the diameter of whose base is 20 and the altitude 40?

	OPERATION.
We first multiply 3.1416 by the diameter, which gives the circumference of the base. Then multiplying by the altitude, we obtain the convex surface.	3.1416 20 62.8320 40 *Ans.* 2513.2800

2. What is the convex surface of a cylinder whose altitude is 28 feet and the circumference of its base 8 feet 4 inches?

3. What is the convex surface of a cylinder, the diameter of whose base is 15 inches and altitude 5 feet?

4. What is the convex surface of a cylinder, the diameter of whose base is 40 and altitude 50 feet?

344. To find the solidity of a cylinder,

Multiply the area of the base by the altitude: the product will be the solid contents (Bk. VIII. Prop. II).

QUEST.—342. What is a cylinder? What is the axis or altitude? What is the convex surface? 343. How do you find the convex surface? 344. How do you find the solidity of a cylinder?

EXAMPLES.

1. Required the solidity of a cylinder of which the altitude is 11 feet, and the diameter of the base 16 feet.

We first find the area of the base, and then multiply by the altitude: the product is the solidity.

OPERATION.

$\overline{16}^2 = 256$
.7854
area base 201.0624
11
2111.6864

2. What is the solidity of a cylinder, the diameter of whose base is 40 and the altitude 29?

3. What is the solidity of a cylinder, the diameter of whose base is 24 and the altitude 30?

4. What is the solidity of a cylinder, the diameter of whose base is 32 and altitude 12?

5. What is the solidity of a cylinder, the diameter of whose base is 25 and altitude 15?

345. A pyramid is a solid formed by several triangular planes united at the same point S, and terminating in the different sides of a plane figure, as ABCDE. The altitude of the pyramid is the line SO, drawn perpendicular to the base.

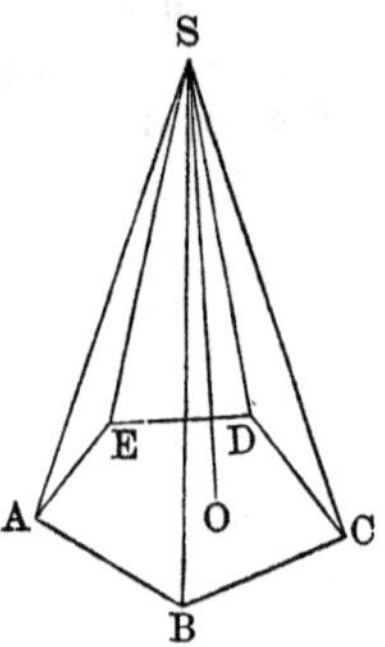

346. To find the solidity of a pyramid,

Multiply the area of the base by the altitude, and divide the product by 3 (Bk. VII. Prop. XVII).

QUEST.—345. What is a pyramid? What is the altitude of a pyramid? 346. How do you find the solidity of a pyramid?

EXAMPLES.

1. Required the solidity of a pyramid, of which the area of the base is 86 and the altitude 24.

We simply multiply the area of the base 86, by the altitude 24, and then divide the product by 3.

OPERATION.

```
       86
       24
      ---
      344
     172
    -----
   3)2064
Ans.  688
```

2. What is the solidity of a pyramid, the area of whose base is 365 and the altitude 36 ?

3. What is the solidity of a pyramid, the area of whose base is 207 and altitude 36 ?

4. What is the solidity of a pyramid, the area of whose base is 562 and altitude 30 ?

5. What are the solid contents of a pyramid, the area of whose base is 540 and altitude 32 ?

6. A pyramid has a rectangular base, the sides of which are 50 and 24; the altitude of the pyramid is 36: what are its solid contents ?

7. A pyramid with a square base, of which each side is 15, has an altitude of 24: what are its solid contents?

347. A cone is a round body with a circular base, and tapering to a point called the *vertex.* The point C is the vertex, and the line CB is called the axis or altitude.

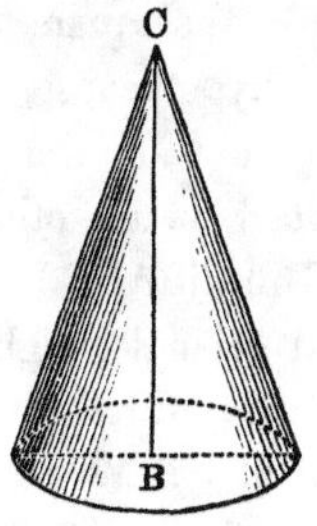

QUEST.—347. What is a cone? What is the vertex? What is the axis? 348. How do you find the solidity of a cone?

348. To find the solidity of a cone,

Multiply the area of the base by the altitude, and divide the product by 3; *or, multiply the area of the base by one-third of the altitude.* (Bk. VIII. Prop. V.)

EXAMPLES.

1. Required the solidity of a cone, the diameter of whose base is 6 and the altitude 11.

We first square the diameter, and multiply it by .7854, which gives the area of the base. We next multiply by the altitude, and then divide the product by 3.

OPERATION.

$6^2=36$

$36\times.7854=28.2744$

11

$3)311.0184$

Ans. 103.6728

2. What is the solidity of a cone, the diameter of whose base is 36 and the altitude 27?

3. What are the solid contents of a cone, the diameter of whose base is 35 and the altitude 27?

4. What is the solidity of a cone, whose altitude is 27 feet and the diameter of the base 20 feet?

RIGHT ANGLED TRIANGLE.

349. The properties of the right angled are so important as to be worthy of particular notice.

In every right angled triangle, the square described on the hypothenuse, is equal to the sum of the squares described on the other two sides.

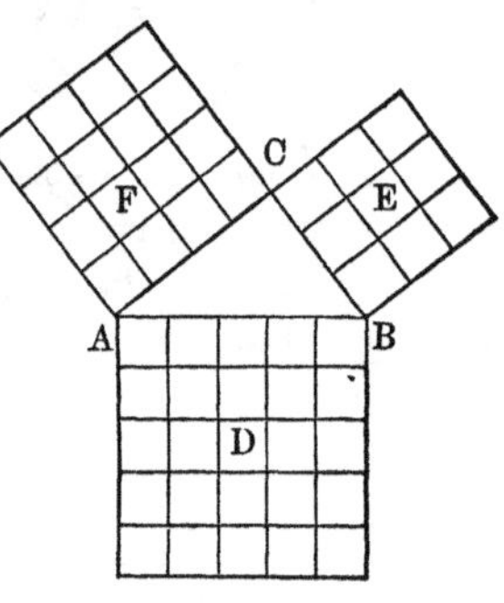

Thus, if ABC be a right angled triangle, right angled at C, then will the square D described on AB be equal to the sum of the squares E and F, described on the sides CB and AC. This is called the carpenter's theorem.

Hence, to find the hypothenuse when the base and perpendicular are known,

1st. *Square each side separately.* 2d. *Add the squares together.* 3d. *Extract the square root of the sum, and the result will be the hypothenuse of the triangle.*

EXAMPLES.

1. The wall of a building, on the brink of a river is 120 feet high, and the breadth of the river 70 yards: what is the length of a line which would reach from the top of the wall to the opposite edge of the river?

2. The side roofs of a house of which the eaves are of the same height, form a right angle with each other at the top. Now, the length of the rafters on one side is 10 feet, and on the other 14 feet: what is the breadth of the house?

3. What would be the width of the house, in the last example, if the rafters on each side were 10 feet?

350. When the hypothenuse and one side of a right angled triangle are known, to find the other side.

Square the hypothenuse and also the other given side, and take their difference: extract the square root of their difference, and the result will be the required side.

1. The height of a precipice on the brink of a river is 103 feet, and a line of 320 feet in length will just reach from the top of it to the opposite bank: required the breadth of the river.

2. The hypothenuse of a triangle is 53 yards, and the perpendicular 45 yards: what is the base?

3. A ladder 60 feet in length, will reach to a window 40 feet from the ground on one side of the street, and by turning it over to the other side, it will reach a window 50 feet from the ground: required the breadth of the street.

QUEST.—349. What is the property of a right angled triangle? When can you find the hypothenuse? How? 350. How do you find a side?

OF THE MECHANICAL POWERS.*

351. There are six simple machines, which are called *Mechanical powers.* They are, the *Lever*, the *Pulley*, the *Wheel* and *Axle*, the *Inclined Plane*, the *Wedge*, and the *Screw.*

352. To understand the power of a machine, four things must be considered.

1st. The power or force which acts. This consists in the efforts of men or horses, of weights, springs, steam, &c.

2d. The resistance which is to be overcome by the power. This generally is a weight to be moved.

3d. We are to consider the centre of motion, or *fulcrum*, which means a prop. The prop or fulcrum is the point about which all the parts of the machine move.

4th. We are to consider the respective velocities of the power and resistance.

353. A machine is said to be in equilibrium when the resistance exactly balances the power, in which case all the parts of the machine are at rest.

We shall first examine the lever.

354. The *Lever*, is a straight bar of wood or metal, which moves around a fixed point, called the fulcrum. There are three kinds of levers.

1st. When the fulcrum is between the weight and the power.

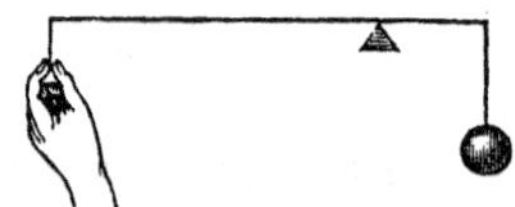

* This article is taken from a Practical Work for mechanics, entitled "Mensuration and Drawing."

QUEST.—351. How many simple machines are there? What are they called? 352. What things must be considered in order to understand the power of a machine? 353. When is a machine said to be in equilibrium? 354. What is a lever? How many kinds of levers are there? Describe the first kind.

2d. When the weight is between the power and the fulcrum.

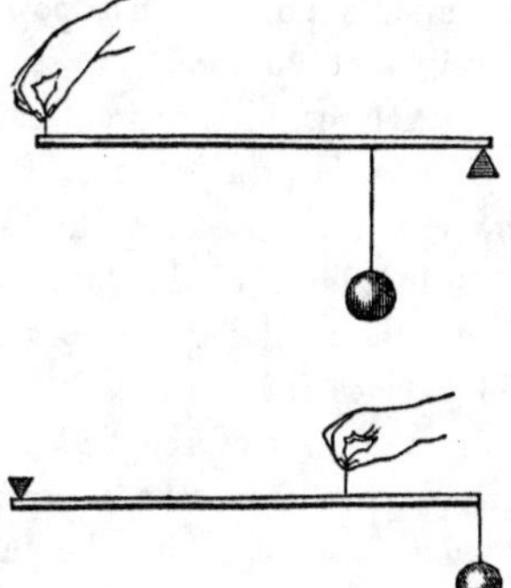

3d. When the power is between the fulcrum and the weight.

The parts of the lever from the fulcrum to the weight and power, are called the *arms* of the lever.

355. An equilibrium is produced in all the levers, when the weight multiplied by its distance from the fulcrum is equal to the product of the power multiplied by its distance from the fulcrum. That is,

The weight is to the power, as the distance from the power to the fulcrum, is to the distance from the weight to the fulcrum.

EXAMPLES.

1. In a lever of the first kind, the fulcrum is placed at the middle point: what power will be necessary to balance a weight of 40 pounds?

2. In a lever of the second kind, the weight is placed at the middle point: what power will be necessary to sustain a weight of 50 lbs.?

3. In a lever of the third kind, the power is placed at

QUEST.—Where is the weight placed in the second kind? Where is the power placed in the third kind? 355. When is an equilibrium produced in all the levers? What is then the proportion between the weight and power?

the middle point: what power will be necessary to sustain a weight of 25 lbs.?

4. A lever of the first kind is 8 feet long, and a weight of 60 lbs. is at a distance of 2 feet from the fulcrum: what power will be necessary to balance it?

5. In a lever of the first kind, that is 6 feet long, a weight of 200 lbs. is placed at 1 foot from the fulcrum: what power will balance it?

6. In a lever of the first kind, like the common steelyard, the distance from the weight to the fulcrum is one inch: at what distance from the fulcrum must the poise of 1 lb. be placed, to balance a weight of 1 lb.? A weight of $1\frac{1}{2}$ lbs.? Of 2 lbs.? Of 4 lbs.?

7. In a lever of the third kind, the distance from the fulcrum to the power is 5 feet, and from the fulcrum to the weight 8 feet: what power is necessary to sustain a weight of 40 lbs.?

8. In a lever of the third kind, the distance from the fulcrum to the weight is 12 feet, and to the power 8 feet: what power will be necessary to sustain a weight of 100 lbs.?

356. Remarks.—In determining the equilibrium of the lever, we have not considered its weight. In levers of the first kind, the weight of the lever generally adds to the power, but in the second and third kinds, the weight goes to diminish the effect of the power.

In the previous examples, we have stated the circumstances under which the power will exactly sustain the weight. In order that the power may overcome the resistance, it must of course be somewhat increased. The lever is a very important mechanical power, being much used, and entering indeed into all the other machines.

Quest.—356. Has the weight been considered in determining the equilibrium of the levers? In a lever of the first kind, will the weight increase or diminish the power? How will it be in the two other kinds?

OF THE PULLEY.

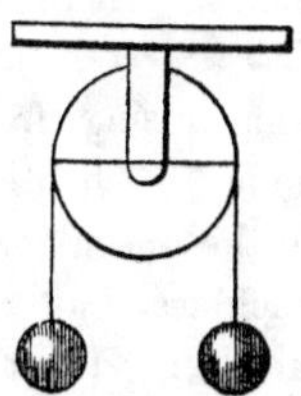

357. The pulley is a wheel, having a groove cut in its circumference, for the purpose of receiving a cord which passes over it. When motion is imparted to the cord, the pulley turns around its axis, which is generally supported by being attached to a beam above.

358. Pulleys are divided into two kinds, fixed pulleys and moveable pulleys. When the pulley is fixed, it does not increase the power which is applied to raise the weight, but merely changes the direction in which it acts.

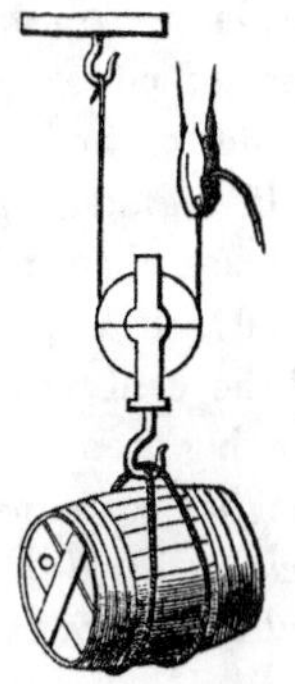

359. A moveable pulley gives a mechanical advantage. Thus, in the moveable pulley, the hand which sustains the cask does not actually support but one-half the weight of it; the other half is supported by the hook to which the other end of the cord is attached.

360. If we have several moveable pulleys, the advantage gained is still greater, and a very heavy weight may be raised by a small power. A longer time, however, will be required, than with the single pulley. It is indeed a general principle in machines, that *what is gained in power, is*

QUEST.—357. What is a pulley? 358. How many kinds of pulleys are there? Does a fixed pulley give any increase of power? 359. Does a moveable pulley give any mechanical advantage? In a single moveable pulley, how much less is the power than the weight? 360. Will an advantage be gained by several moveable pulleys?

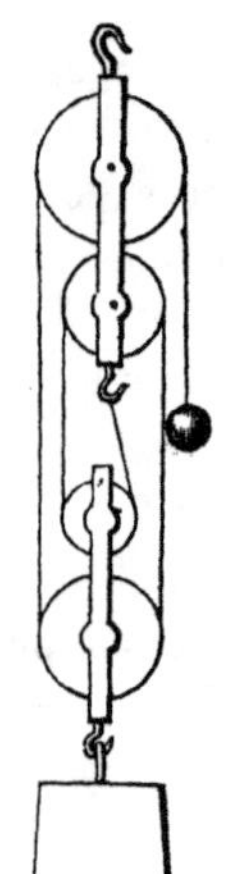

lost in time; and this is true for all machines. There is also an actual loss of power, viz. the resistance of the machine to motion, arising from the rubbing of the parts against each other, which is called the *friction* of the machine. This varies in the different machines, but must always be allowed for, in calculating the power necessary to do a given work. It would be wrong, however, to suppose that the loss was equivalent to the gain, and that no advantage is derived from the mechanical powers. We are unable to augment our strength, but, by the aid of science we so divide the resistance, that by a continued exertion of power, we accomplish that which it would be impossible to effect by a single effort.

If in attaining this result, we sacrifice time, we cannot but see that it is most advantageously exchanged for power.

361. It is plain, that in the moveable pulley, all the parts of the cord will be equally stretched, and hence, each cord running from pulley to pulley, will bear an equal part of the weight; consequently the *power will always be equal to the weight, divided by the number of cords which reach from pulley to pulley.*

EXAMPLES.

1. In a single immoveable pulley, what power will support a weight of 60 lbs.?

2. In a single moveable pulley, what power will support a weight of 80 lbs.?

3. In two moveable pulleys with 5 cords, (see last fig.,) what power will support a weight of 100 lbs.?

QUEST.—State the general principle in machines. What does the actual loss of power arise from? What is this rubbing called? Does this vary in different machines? 361. In the moveable pulley, what proportion exists between the cord and the weight?

WHEEL AND AXLE.

362. This machine is composed of a wheel or crank—firmly attached to a cylindrical axle. The axle is supported at its ends by two pivots, which are of less diameter than the axle around which the rope is coiled, and which turn freely about the points of support. In order to balance the weight, we must have

The power to the weight, as the radius of the axle, to the length of the crank, or radius of the wheel.

EXAMPLES.

1. What must be the length of a crank or radius of a wheel, in order that a power of 40 lbs. may balance a weight of 600 lbs. suspended from an axle of 6 inches radius?

2. What must be the diameter of an axle that a power of 100 lbs. applied at the circumference of a wheel of 6 feet diameter may balance 400 lbs.!

INCLINED PLANE.

363. The inclined plane is nothing more than a slope or declivity, which is used for the purpose of raising weights. It is not difficult to see that a weight can be forced up an inclined plane, more easily than it can be raised in a vertical line. But in this, as in the other machines, the advantage is obtained by a partial loss of power.

QUEST.—362. Of what is the machine called the wheel and axle, composed? How is the axle supported? Give the proportion between the power and the weight. 363. What is an inclined plane?

Thus, if a weight W, be supported on the inclined plane ABC, by a cord passing over a pulley at F, and the cord from the pulley to the weight be parallel to the length of the plane AB, the power P, will balance the weight W, when

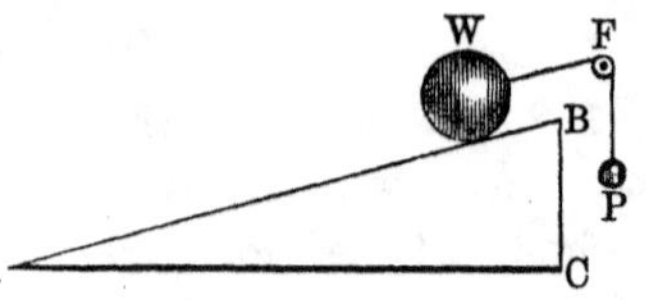

P : W : : height BC : length AB.

It is evident that the power ought to be less than the weight, since a part of the weight is supported by the plane.

EXAMPLES.

1. The length of a plane is 30 feet, and its height 6 feet: what power will be necessary to balance a weight of 200 lbs.?

2. The height of a plane is 10 feet, and the length 20 feet: what weight will a power of 50 lbs. support?

3. The height of a plane is 15 feet, and length 45 feet: what power will sustain a weight of 180 lbs.?

THE WEDGE.

364. The wedge is composed of two inclined planes, united together along their bases, and forming a solid ACB. It is used to cleave masses of wood or stone. The resistance which it overcomes is the attraction of cohesion of the body which it is employed to separate. The wedge acts principally by being struck with a hammer, or mallet, on its head, and very little effect can be produced with it, by mere pressure.

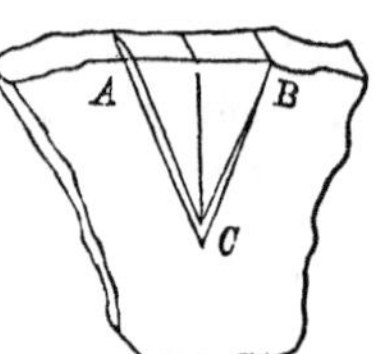

All cutting instruments are constructed on the principle

QUEST.—What proportion exists between the power and weight when they are in equilibrium? 364. What is the wedge? What is it used for? What resistance is it used to overcome?

of the inclined plane or wedge. Such as have but one sloping edge, like the chisel, may be referred to the inclined plane, and such as have two, like the axe and the knife, to the wedge.

THE SCREW.

365. The screw is composed of two parts—the screw S, and the nut N.

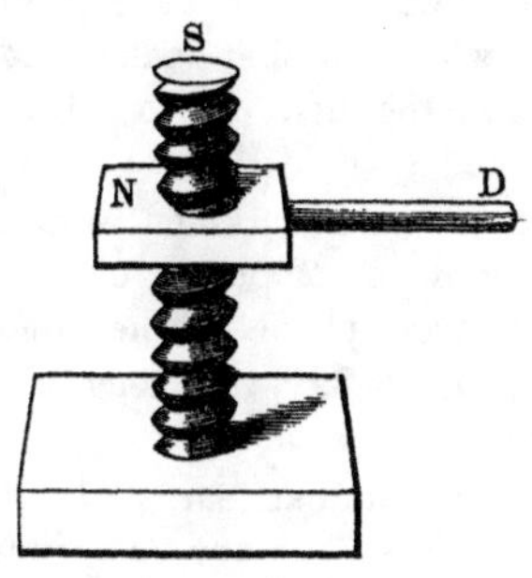

The screw S, is a cylinder with a spiral projection winding around it, called the *thread*. The nut N is perforated to admit the screw, and within it is a groove into which the thread of the screw fits closely.

The handle D, which projects from the nut, is a lever which works the nut upon the screw. The power of the screw depends on the distance between the threads. The closer the threads of the screw, the greater will be the power; but then the number of revolutions made by the handle D, will also be proportionably increased; so that we return to the general principle—what is gained in power is lost in time. The power of the screw may also be increased by lengthening the lever attached to the nut.

The screw is used for compression, and to raise heavy weights. It is used in cider and wine-presses, in coining, and for a variety of other purposes.

QUEST.—365. Of how many parts is the screw composed? Describe the screw. What is the thread? What the nut? What is the handle used for? To what uses is the screw applied?

PROMISCUOUS QUESTIONS.

1. Two persons have put in trade each a certain sum; that which the first contributed is to that of the second as 11 to 15: the first put in $1359: what did the second contribute?

2. Twelve workmen working 12 hours a day have made in 12 days 12 pieces of cloth, each piece 75 yards long. How many pieces of the same stuff would have been made, each piece 25 yards long, if there had been 7 more workmen?

3. A workman earns $18,50 by working 12 days in 14, during these 14 days he spends 50 cents a day for his board and gives 4 cents a day to the poor; on Sunday he triples the alms. How long will it take him at this rate to pay his rent, which is $56, and a debt of $11,50?

4. How much time would it require to receive $80 of interest with a capital of $400, knowing that $600 placed at the same rate would produce an interest of $90 every three years?

5. If $100 at interest gains $3 every nine months, what capital would be necessary to gain $800 every two years?

6. Four partners have gained $21175; the first is to have $4250 more than the second; the second $1700 more than the third; the third $1175 more than the fourth: what is the share of each?

7. The sum of two numbers is 5330, their difference 1999: what are the two numbers?

8. A person was born on the 1st of October, 1792, at 6 o'clock in the morning; what was his age on the 21st of September, 1839, at half past 4 in the afternoon?

9. A merchant bought 80 yards of cloth, then sold 140 yards, after which there remained to him one half the quan-

tity he had in the store before his last purchase: what was this quantity?

10. Sound travels about 1142 feet in a second If then the flash of a cannon be seen at the moment it is fired, and the report heard 45 seconds after, what distance would the observer be from the gun?

11. A person having a certain sum borrowed $65,50, and then paid a debt of $94,90; he received $56,75 which was due him, and found that he had $49,30 after having expended $9,30. How much had he at first?

12. A house which was sold a second time for $7180, would have given a profit of $420 if the second proprietor had purchased it $130 cheaper than he did: at what price did he purchase it?

13. A person purchased 78000 quills, for half of which he gave $4,50 per thousand, and for the rest $87\frac{1}{2}$ cents per hundred; he sells them at $1\frac{1}{4}$ cents each: what is his profit supposing he takes 265 for his own use?

14. In order to take a boat through a lock from a certain river into a canal, as well as to descend from the canal into the river, a body of water is necessary $46\frac{1}{2}$ yards long, 8 yards wide, and $2\frac{3}{5}$ yards deep. How many cubic yards of water will this canal throw into the river in a year, if 40 boats ascend and 40 descend each day except Sundays and eight holidays?

15. How many scholars are there in a class, to which if 11 be added the number will be augmented one-sixteenth?

16. A person being asked the time, said, the time past noon is equal to $\frac{1}{5}$ of the time past midnight: what was the hour?

17. What number is that which being augmented by 85, and this sum divided by 9, will give 25 for the quotient?

18. Three travellers have 1377 miles to go before they reach the end of their journey; the first goes 30 miles a day, the second 27, and the third 24: how many days should one set out after another that they may arrive together?

19. A company numbering sixty-six shareholders have constructed a bridge which cost $200000: what will be the gain of each partner at the end of 22 years, supposing that 6400 persons pass each day, and that each pays one cent toll, the expense for repairs, &c., being $5 per year for each shareholder?

20. The entire length of the walls of a fort is 495 yards, their height $8\frac{1}{2}$ yards, and their thickness 3 yards: how many years has it taken to construct them, each cubic yard having cost 16 francs, and the expenses having been 20086 francs per year; and what will this sum amount to in dollars and cents, at the custom house value?

21. One-fifth of an army was killed in battle, $\frac{1}{6}$ part was taken prisoners, and $\frac{1}{10}$ died by sickness: if 4000 men were left, how many men did the army at first consist of?

22. A person delivered to another a sum of money to receive interest for the same at 4 per cent per annum. At the end of three years he received for principal and interest £176 8*s*. What was the sum lent?

23. A snail in getting up a pole 20 feet high, was observed to climb up 8 feet every day, but to descend 4 feet every night: in what time did he reach the top of the pole?

24. Four merchants A, B, C, and D, trade together; A clears £76 4*s*. in 6 months, B £57 10*s*. in 5 months, C 100 guineas in 12 months, and D, with a stock of 200 guineas, clears £78 15*s*. in 9 months. Required each man's stock.

25. Three merchants traded together as follows: A put in $2500 for 3 months, B $1750 for 5 months, and C $2000 for 2 months: C's gain was $147,50. What must A and B receive for their respective shares, and what was the whole gain?

26. Three different kinds of wine were mixed together in such a way that for every 3 gallons of one kind there were 4 of another, and 7 of a third: what quantity of each kind was there in a mixture of 292 gallons?

27. Divide £500 among four persons, so that when A has £$\frac{1}{2}$, B shall have $\frac{1}{3}$, C $\frac{1}{4}$, and D $\frac{1}{5}$.

28. Two partners have invested in trade $1600, by which they have gained $300; the gain and stock of the second amount to $1140. What is the stock and gain of each?

29. How many planks 15 feet long and 15 inches wide will floor a barn $60\frac{1}{2}$ feet long and $33\frac{1}{2}$ feet wide?

30. A merchant bought a quantity of wine for $430. He sold 55 quarts of it for $24,50, and gained 5 cents a quart: how much wine had he at first?

31. Twenty-five workmen have agreed to labor 12 hours a day for 24 days, to pay an advance made to them of $900; but having lost each an hour per day, five of them engage to fulfil the agreement by working 12 days: how many hours per day must these labor?

32. If a person receives $1 for $\frac{4}{7}$ of a day's work, how much is that a day?

33. If $14\frac{3}{10}$ pieces of ribbon cost $26,50, how much is that a piece?

34. What number is that of which $\frac{1}{2}$, $\frac{1}{3}$, and $\frac{1}{4}$ added together, will make 48?

35. A landlord being asked how much he received for the rent of his property, answered, after deducting 9 cents from each dollar for taxes and repairs, there remains $3014,30. What was the amount of his rents?

36. A person traded 360 yards of linen for cloth worth $1,62 per yard: how many yards of cloth has he received, and for how much has he sold the linen per yard, knowing that the price of a yard of cloth is equal to that of $2\frac{3}{4}$ yards of linen?

37. If 165 pounds of soap cost $16,40, for how much will it be necessary to sell 390 pounds, in order to gain the price of 36 pounds?

38. What is the height of a wall which is $14\frac{1}{2}$ yards in length, and $\frac{7}{10}$ of a yard in thickness, and which has cost $406, it having been paid for at the rate of $10 per cubic yard?

39. If the tare of a quantity of merchandise is 54*lb.* 7*oz.*, what is the gross weight, the tare being 4*lb.* in 100?

40. At what rate per cent will \$1720,75 amount to \$2325,86 in 7 years?

41. In what time will \$2377,50 amount to \$2852,42, at 4 per cent per annum?

42. What principal put at interest for 7 years, at 5 per cent per annum, will amount to \$2327,89?

43. What difference is there between the interest of \$2500 for $4\frac{3}{4}$ years, at 6 per cent, and half that sum for twice the time, at half the same rate per cent?

44. If, when I sell cloth at 8*s*. 9*d*. per yard I gain 12 per cent, what will be the gain per cent when it is sold for 10*s*. 6*d*. per yard?

45. A tea-dealer purchased 120*lb*. of tea, $\frac{2}{5}$ of which he sold at 10*s*. 6*d*. per. *lb*.; but the rest being damaged, he sold it at a loss of £3 12*s*., after which he found he had neither gained nor lost. What did it cost him per *lb*., and what was the damaged tea sold for?

46. A piece of cloth containing 5000 ells Flemish was sold for \$21250, by which the gain upon every yard was equal to $\frac{1}{8}$ of the prime cost of an English ell. What was the first cost of the whole piece?

47. A person lent a certain sum at 4 per cent per annum; had this remained at interest 3 years, he would have received for principal and interest \$9676,80. What was the principal?

48. Three persons purchased a house for \$9202; the first gave a certain sum; the second three times as much; and the third one and a half times as much as the two others together: what did each pay?

49. A piece of land of 165 acres was cleared by two companies of workmen; the first numbered 25 men and the second 22; how many acres did each company clear, and what did the clearing cost per acre, knowing that the first company received \$86 more than the second?

50. The greatest of two numbers is 15 and the sum of their squares is 346: what are the two numbers?

51. A water tub holds 147 gallons; the pipe usually brings in 14 gallons in 9 minutes: the tap discharges, at a medium, 40 gallons in 31 minutes. Now, supposing these to be left open, and the water to be turned on at 2 o'clock in the morning; a servant at 5 shuts the tap, and is solicitous to know in what time the tub will be filled in case the water continues to flow.

52. A thief is escaping from an officer. He has 40 miles the start, and travels at the rate of 5 miles an hour; the officer in pursuit travels at the rate of 7 miles in an hour: how far must he travel before he overtakes the thief?

53. Five merchants were in partnership for four years; the first put in $60, then, 5 months after, $800, and at length $1500, 4 months before the end of the partnership; the second put in at first $600, and 6 months after $1800; the third put in $400, and every six months after he added $500; the fourth did not contribute till 8 months after the commencement of the partnership; he then put in $900, and repeated this sum every 6 months; the fifth put in no capital, but kept the accounts, for which the others agreed to pay him $1,25 a day. What is each one's share of the gain, which was $20000?

54. A traveller leaves New Haven at 8 o'clock on Monday morning, and walks towards Albany at the rate of 3 miles an hour; another traveller sets out from Albany at 4 o'clock on the same evening, and walks towards New Haven at the rate of 4 miles an hour: now supposing the distance to be 130 miles, where on the road will they meet?

55. An employer has 45 workmen, by each of whom he gains 15 cents a day: how long a time would it require for them to gain him $468,93, and what must he pay them during this time, he paying each $1,25 a day?

56. When it is 12 o'clock at New York, what is the hour at London, New York being 75° of longitude west of London?

Since the circumference of the earth is supposed to be divided into 360 degrees (Art. 40), and since the sun apparently passes through these 360° every twenty-four hours, it follows that in a single hour it will pass through one twenty-fourth of 360°, or 15°. Hence, there are

15° of motion in 1 hour of time,
1° of motion in 4 minutes
1′ of motion in 4 seconds.

If two places, therefore, have different longitudes, they will have different times, and the difference of time will be one hour for every 15° of longitude, or 4 minutes for each degree, and 4 seconds for each minute. It must be observed that the place which is most *easterly* will have the time first, because the sun travels from east to west.

To return then to our question. The difference of longitude between London and New York being 75°, the difference of time will be found in minutes by multiplying 75° by 4, giving 300 minutes, or 5 hours. Now since New York is west of London, the time will be *later* in London; that is, when it is twelve o'clock at New York, it will be 5, P. M. in London; or when it is 12 at London, it will be 7, A. M. at New York.

OPERATION.

$$\begin{array}{r} 75^\circ \\ 4 \\ \hline 60)300 \\ \hline \textit{Ans.}\ \ 5\ \textit{hours.} \end{array}$$

57. Boston is 6° 40′ east longitude from the city of Washington: when it is 6 o'clock P. M. at Washington, what is the hour at Boston?

The 6 degrees being multiplied by 4 give 24 minutes of time, and the 40 minutes being multiplied by 4 give 160 seconds, or 2 minutes 40 seconds. The sum is 26*m.* 40*sec.*, and since Boston is east of Washington the time is later at Boston.

OPERATION.

$$\begin{array}{r} 6 \times 4 = 24m. \\ 40 \times 4 = 160'' = \ 2m.\ 40sec. \\ \hline 26m.\ 40sec. \\ \hline \textit{Ans.}\ \ 26m.\ 40sec.\ \textit{past}\ 6. \end{array}$$

58. The difference of longitude of two places is 85° 20′: what is the difference of time?

ANSWERS.

REDUCTION.

Ex.	*Ans.*	*Ex.*	*Ans.*
2.	1155*s.* 13860*d.* 55440*far.*	15.	2410*cr.* £602 10*s.*
3.	52405*far.*	16.	25920*s.* 5184*cr.* £1296.
4.	37245*hf. d.*	17.	——
5.	——	18.	14*oz.*
6.	5726*two d.*	19.	184800*gr.* 13200*gr.*
7.	2767*tr. d.* 8301*d.*	20.	26215*gr.*
8.	4343*six d.*	21.	122*lb.* 2*oz.* 18*pwt.* 9*gr.*
9.	216*cr.* 432*hf. cr.* 1080*s.* 2160*six d.* 12960*d.* 51840*far.*	22.	——
		23.	300℥ 2400ʒ 7200℈ 144000*gr.*
10.	1493*s.* 5975*tr. d.* 71700*far.*	24.	157℔ 7℥ 4ʒ.
		25.	86962*gr.*
11.	2880*d.* 240*s.* £12.	26.	30℔ 4℥ 3ʒ 2℈ 7*gr.*
		27.	——
		28.	26880*lb.*
12.	——	29.	14*T.*
13.	99*gu.* 4*s.* 4*d.*	30.	42292*lb.*
14.	£105.	31.	20*T.* 13*cwt.* 1*qr.* 22*lb.* 4*oz.*

Ex.	*Ans.*	*Ex.*	*Ans.*
32.	——	48.	93*A*. 2*R*. 16*P*.
33.	403*T*. 19*cwt*. 1*qr*. 24*lb*. 6*oz*. 231654496*dr*.	49.	818*M*. 162*A*. 3*R*. 23*P*.
		50.	——
34.	5024*na*.	51.	19840*S*. *ft*.
35.	864*yds*.	52.	3760128*S*. *in*.
36.	78*E*. *E*. 1*qr*.	53.	440*cords*.
37.	1197*E*. *E*. 23940*na*.	54.	43742*cords* 32*S*. *ft*.
		55.	——
38.	9996*yd*. 7996*E*. *E*. 1*yd*. 6664*E*. *Fr*.	56.	13*tuns*
		57.	37800*pt*.
39.	3768*fur*. 150720*rd*.	58.	970$\frac{1}{5}$*hf*. *ank*.
		59.	85248*gi*.
40.	——	60.	——
41.	88000*yd*. 264000*ft*. 3168000*in*. 9504000*bar*.	61.	32832*pt*.
		62.	297216*hf*. *pt*.
		63.	1800*gal*.
42.	200613*ft*. 6*in*.	64.	1408*pk*.
43.	4755801600*bar*. 1172*qrs*. 4*bu*. 1*gal*. 3184*bar*. over.	65.	——
		66.	1240*sacks*.
		67.	31557600*sec*.
44.	12374*P*.	68.	189733554*sec*.
45.	——	69.	240*yr*. 10*da*. 4*hr*. 28*m*. 38*sec*.
46.	2214262*sq*. *ft*. 72*sq*. *in*.		
47.	2800*P*.	70.	——

ADDITION.

4.	787676921.	9.	6001001250561.
5.	10570011.	10.	6000037684799.
6.	15371781930.	11.	128738075326
7.	45105211.	12.	21890459447.
8.	——	13.	——

Ex.	Ans.
14.	1819857171437.
15.	$108,892.
16.	$1057,87.
17.	$800,076.
18.	——
19.	$5498,043.
20.	$67476,840.
21.	£684 5*s.* 7*d.*
22.	£205 3*s.* 10*d.*
23.	——
24.	£240 6*s.* 8¾*d.*
25.	2382*lb.* 1*oz.* 16*pwt.*
26.	344*lb.* 1*oz.* 19*pwt.* 20*gr.*
27.	462*lb.* 9*oz.* 14*pwt.*
28.	——
29.	511℔ 11℥ 3ʒ.
30.	294℔ 0℥ 7ʒ.
31.	36℔ 5℥ 6ʒ 1℈ 18*gr.*
32.	464℔ 0℥ 5ʒ.
33.	——
34.	3030*cwt.* 1*qr.* 27*lb.*
35.	92*cwt.* 2*qr.* 15*lb.* 10*oz.*
36.	347*lb.* 7*oz.* 6*dr.*
37.	471*yd.* 2*qr.* 1*na.*
38.	——
39.	3821*E. Fr.*
40.	4768*E. Fl.* 0*qr.* 2*na.*
41.	489*L.* 1*mi.* 6*fur.*
42.	4487*fur.* 35*rd.* 5*yd.*
43.	——
44.	644*ft.* 0*in.* 1*bar.*
45.	509*A.* 2*R.* 18*P.*
46.	4797*A.* 2*R.* 11*P.*

Ex.	Ans.
47.	403*A.* 1*R.* 1*P.*
48.	——
49.	3209*tun.* 0*hhd.* 27*gal.*
50.	5422*pun.* 57*gal.* 2*qt.*
51.	460*tier.* 29*gal.* 1*qt.*
52.	297*gal.* 2*qt.*
53.	——
54.	323*bar.* 1*fir.* 4*gal.*
55.	3150*hhd.* 16*gal.* 1*qt.*
56.	5220*hhd.* 4*gal.* 2*qt.*
57.	528*L. ch.* 13*bu.* 2*pk.*
58.	——
59.	3842*qr.* 6*bu.* 2*pk.*
60.	409*scows* 12*L. ch.* 19*bu.*
61.	4299*yr.* $7\frac{87}{112}$*mo.* 2*wk.*
62.	525*mo.* 0*wk.* 4*da.*
63.	——
64.	4444*hr.* 23*m.* 50*sec.*

APPLICATIONS.

Ex.	Ans.
1.	1605260*acres.*
2.	1st 3 *yrs.* 42390529*A.* *last* " 4530902*A.*
3.	$2051423,77.
4.	——
5.	15995942 *coins.* $5668663=*value.*
6.	*Imports*, $303955539. *Exports*, $287820350.
7.	1287462.
8.	3617900.
9.	——

Ex.	*Ans.*	*Ex.*	*Ans.*
10.	29884 *to Br. N. Amer.* 66770 *to U. S.* 96654 *entire number.*	17.	1104087.
11.	138500*tons.*	18.	In 1790, 3924829. 1800, 5305941. 1810, 7265579. 1820, 9638191. 1830, 12861192. 1840, 17063350.
12.	231771*men.*		
13.	$70560.		
14.	681 *No. of vessels.* 403 *sail vessels.* 144 *steam vessels.*	19.	In 1790, 607897 1800, 893041. 1810, 1191359. 1820, 1627428. 1830, 1998318. 1840, 2487355.
15.	$ 977911 *of gold.* 1567420 *of silver.* 2545331 *entire sum.*		
16.	——		

SUBTRACTION.

1.	81328.	19.	79℔ 10℥ 6ʒ.
2.	7559.	20.	——
3.	£7 18*s.* 9¾*d.*	21.	13ʒ 0℈ 15*gr.*
4.	338918990202409993.	22.	8℔ 10℥ 7ʒ.
5.	——	23.	12*T.* 17*cwt.* 3*qr.*
6.	499972609093220149.	24.	2*cwt.* 2*qr.* 26*lb.*
7.	149299788316514071.	25.	——
8.	$179,577.	26.	134*lb.* 14*oz.* 13*dr.*
9.	$79,324.	27.	134*yds.* 2*qr.* 3*na.*
10.	——	28.	124*E. E.* 3*qr.* 3*na.*
11.	$999,955.	29.	96*E. Fr.* 2*qr.* 1*na.*
12.	$107,576.	30.	——
13.	$506,034.	31.	17*L.* 2*mi.* 6*fur.*
14.	$985,997.	32.	34*rd.* 4*yd.*
15.	——	33.	4*rd.* 3¾*yd.* 2*ft.*
16.	9*oz.* 17*pwt.* 20*gr.*	34.	3*ft.* 0*in.* 1*bar.*
17.	15*lb.* 3*oz.* 16*pwt.*	35.	——
18.	2*oz.* 18*pwt.* 21*gr.*	36.	37*A.* 2*R.* 34*P.*

Ex.	*Ans.*
37.	1*A.* 1*R.* 26*P.*
38.	4*A.* 2*R.* 39*P.*
39.	7*tun* 2*hhd.* 55*gal.*
40.	——
41.	1*tier.* 1*gal.* 3*qt.*
42.	7*gal.* 2*qt.* 1*pt.*
43.	1*bar.* 3*fir.* 7*gal.*
44.	107*bar.* 1*fir.* 4*gal.*
45.	——
46.	63*hhd.* 2*gal.* 3*qt.*
47.	27*L. ch.* 0*bu.* 1*pk.*
48.	2*weys* 4*qr.* 2*bu.*
49.	52*qr.* 6*bu.* 3*pk.*
50.	——

Ex.	*Ans.*
51.	2*yr.* $11\frac{5}{112}$*mo.* 3*wk.*
52.	127*mo.* 3*wk.* 6*da.*
53.	147*da.* 21*hr.* 56*min.*
54.	52*hr.* 50*min.* 54*sec.*

PROMISCUOUS EXAMPLES.

Ex.	*Ans.*
55.	£3 9*s.*
56.	£121 17*s.* $0\frac{1}{4}$*d.*
57.	£980 2*s.* 1*d.*
58.	180*T.* 11*cwt.* 12*lb.*
59.	5*yr.* $11\frac{10}{112}$*mo.* 2*wk.* 6*da.* 9*hr.* 22*min.*
60.	34° 35′ = *dif. of lat.* 165° 18′ = " " *long*

61. Newton's age was 84*yr.* 2*mo.* 26*da.*
Euler's " " 76*yr.* 4*mo.* 22*da.*
Lagrange's " 77*yr.* 2*mo.* 11*da.*
Laplace's " " 78*yr.* 4*da.*
From Newton's death to Jan. 1st, 1846, was - - -. 118*yr.* 9*mo.* 12*da.*
" Euler's " " " 62*yr.* 3*mo.* 24*da.*
" Lagrange's " " " 32*yr.* 8*mo.* 21*da.*
" Laplace's " " " 18*yr.* 9*mo.* 5*da.*

Ex.	*Ans.*
62.	3279*hhd.*
63.	937215*diff.* 2428921 *pop. of state.*
64.	13277872 *diff.* 18535786 *whole pop.*
65.	——
66.	$92449341,16.
67.	£5742078.
68.	$52315291.
69.	$2458211.
70.	——
71.	$49282,03.
72.	$3466051,78.

Ex.		Ans.
73.	From the founding of St. Augustine,	280*yr.* 6*mo.* 9*da*
	" " " Jamestown,	238*yr.* 10*mo.* 4*da.*
	" " Battle of Princeton,	69*yr.* 2*mo.* 14*da.*
	" " Surrender of Cornwallis,	64*yr.* 4*mo.* 29*da.*
	" Washington's Inauguration,	56*yr.* 10*mo.* 17*da.*
	" Washington's Death,	46*yr.* 3*mo.* 3*da.*
	" the French Berlin Decree,	39*yr.* 3*mo.* 26*da.*
	" " Orders in Council,	38*yr.* 4*mo.* 6*da.*
	" " Declaration of War,	33*yr.* 8*mo.* 29*da.*
	" " Capture of the Guerriere,	33*yr.* 6*mo.* 29*da.*
	" " " " " Macedonian,	33*yr.* 4*mo.* 23*da.*
	" " " " York,	32*yr.* 10*mo.* 20*da.*
	" " " " Fort George,	32*yr.* 9*mo.* 21*da.*
	" " Defeat at Sackett's Harbor,	32*yr.* 9*mo.* 20*da.*
	" " Battle of Lake Erie,	32*yr.* 6*mo.* 7*da.*
	" " " of Chippewa,	31*yr.* 8*mo.* 12*da.*
	" " " of Niagara,	31*yr.* 7*mo.* 23*da.*
	" " Sortie from Fort Erie,	31*yr.* 6*mo.*
	" " Battle of New Orleans,	31*yr.* 2*mo.* 9*da.*
	" " Death of Adams and Jefferson,	19*yr.* 8*mo.* 13*da.*
	" " Compromise Bill,	13*yr.* 1*mo.* 5*da.*
	" " Death of Lafayette,	12*yr.* 9*mo.* 28*da.*
	" " Removal of the Cherokees,	7*yr.* 9*mo.* 22*da.*

MULTIPLICATION.

1.	6776368.	9.	2324684880333.
2.	68653214.	10.	71109696492112.
3.	4563272.	11.	——
4.	1301922.	19.	129359360000.
6.	——	20.	13729103000000.
7.	556321146764.	21.	664763206000000.
8.	1747125213301.	22.	8799238229600000.

23. 2526426017908695000000.

24. 1093689368445084378777040.

25. ——

26. 8371562339213807802080112

Ex.			*Ans.*
27.			72058988008174745973090826
28.			95666032459647278072171264.
29.	4896.	23.	£2 0*s*. 6*d*.
30.	234048.	24.	£2 1*s*. 8½*d*.
31.	——	25.	——
32.	314986464.	26.	£2 19*s*. 6*d*.
		27.	£8 3*s*. 9*d*
	Art. 67.	28.	£16 8*s*. 7½*d*.
		29.	£22 13*s*.
1.	{ $70,840. 85,008.	30.	——
	99,176.	31.	11*lb*. 6*oz*. 9*pwt*. 12*gr*.
2.	$7834,14.	32.	197*yd*. 1*qr*. 1*na*.
3.	$12517,68.	33.	£3 19*s*. 4½*d*.
4.	$77079,456.	34.	£66 19*s*. 6*d*.
5.	——	35.	——
6.	$341,25.	36.	£65 19*s*. 9½*d*.
7.	$98,94.	37.	£208 13*s*. 9*d*.
8.	$813,020.	38.	£154 12*s*. 3*d*.
9.	$5869,75.	39.	£42 1*s*. 6*d*.
10.	——	40.	——
11.	$2426,15.	41.	£819 6*s*.
12.	$15169,50.	42.	690*lb*. 8*oz*. 18*pwt*. 16*gr*.
13.	$162,25.	43.	75*A*. 3*R*. 39*P*.
14.	$21935,214.	45.	£19 10*s*. 8¾*d*.
15.	——	46.	——
16.	$963,66.	47.	£33 3*s*. 1¾*d*.
17.	$18844,01.	48.	£83 2*s*. 8*d*.
18.	£81 6*s*. 10*d*.	49.	£137 7*s*. 3*d*.
19.	24*T*. 7*cwt*. 3*qr*. 27*lb*. 8*oz*.	50.	£698 2*s*.
20.	——	51.	——
21.	£1 10*s*. 2*d*.	52.	222*cwt*. 18*lb*.
22.	£1 9*s*. 2*d*.	53.	15*cwt*. 27*lb*. 11*oz*. 7*dr*.

Ex	*Ans.*	*Ex.*	*Ans.*
55.	£2687 18*s*. 3*d*.	65.	£566 5*s*. $3\frac{1}{5}$*d*.
56.	£351 2*s*. $7\frac{1}{2}$*d*.	66.	£17038 10*s*. 11*d*. $2\frac{1}{2}$*far*.
57.	——	67.	£12422 2*s*. 7*d*. $2\frac{6}{7}$*far*.
58.	£62 1*s*. $7\frac{1}{2}$*d*.	68.	£2875 0*s*. $7\frac{1}{2}$*d*.
59.	£15299 18*s*. 4*d*.	69.	——
60.	£51 7*s*. $2\frac{1}{2}$*d*.	70.	£658 0*s*. 11*d*. $1\frac{3}{4}$*far*.
61.	£344 0*s*. 6*d*.	71.	£50 12*s*. 9*d*.
62	——	72.	£2 10*s*. 2*d*. $2\frac{3}{4}$*far*

DIVISION

2.	$407294\frac{1080}{1754}$.	9.	$1318096555\frac{104990}{374567}$
3.	$13195133\frac{1842}{3574}$.	10.	——
4.	$1251392011\frac{13010}{45705}$.	11.	$9948157977\frac{81605}{678957}$.
		12.	$59085714\frac{84}{121}$.
5.	——	13.	$1258127\frac{15785}{57149}$.
6.	$14243757748\frac{35411}{47143}$.	14.	123456789.
7.	$15395919\frac{12214}{37149}$.	15.	——
8.	$30001000\frac{6347}{57143}$.	16.	$119191753\frac{90107}{123456}$.
17.	$900184442401827462422 6\frac{548424}{987675}$.		

	ART. 79.		ART. 81.
1.	132.	3.	$17085\frac{29}{56}$.
2.	4871000.	4.	$67639\frac{21}{72}$.
3.	——	5.	$6129\frac{11}{110}$.
4.	7128368.	6.	$3095\frac{87}{144}$.
5.	918546.	8.-	——
	ART. 80.	9.	$5203802\frac{61}{144}$.
2.	387.	10.	$11805558\frac{53}{55}$.
3.	133.	11.	$39096821\frac{64}{121}$.
4.	201.	12.	$34297219\frac{62}{108}$.

Ex.	*Ans.*
13.	——
14.	$10823637\frac{12}{99}$.
15	$650\frac{248}{720}$.
16.	$206190192477\frac{443604690}{479001600}$.

ART. 82.

Ex.	*Ans.*
2.	$15655794\frac{9075}{36500}$.
3.	——
4.	$16871651\frac{2944}{47150}$.
5.	$47462565\frac{9649}{31500}$.
6.	$8905748\frac{722934}{754000}$.
7.	$1421076222\frac{42715}{754000}$.
8.	——
9.	$4087692937\frac{381715}{1749000}$.
10.	$7943859\frac{39}{90}$.
11.	$119092\frac{12348}{14400}$.
12.	71400714374.
13.	——
14.	$1924530\frac{652547}{3714900}$.

EXAMPLES IN DENOMINATE NUMBERS.

Ex.	*Ans.*
2.	£15 19*s.* 9*d.* $1\frac{1}{3}$*far.*
3.	£1 11*s.* 5*d.* $1\frac{2}{3}$*far.*
4.	15*s.* 1*d.* $3\frac{15}{66}$*far.*
5.	——
6.	£9 18*s.* 3*d.* $3\frac{92}{179}$*far.*
7.	9*yd.* 2*qr.* $1\frac{1}{5}$*na.*
8.	4*A.* 32*P.* 20*sq. yd.* 1*sq. ft.* 72*sq. in.*
9.	8*lb.* 7*pwt.* $15\frac{45}{80}$*gr.*
10.	——
11.	2*E. E.* 3*qr.* $\frac{107}{142}$*na.*
12.	£1 17*s.* 6*d.*
13.	6*d.* $1\frac{16}{112}$*far.*
14.	17*s.* $11\frac{3}{4}$*d.*
15.	——
16.	£40 10*s.* 6*d.*
17.	1*s.* 9*d*
18.	$10\frac{3}{4}$*d.*
19.	4*s.* $9\frac{36}{252}$*d.*
20.	——
21.	10*oz.* 15*pwt.* $14\frac{88}{125}$*gr.*
22.	17*cwt.* 2*qr.* 11*lb.* 14*oz.* $13\frac{1}{27}$*dr.*
23.	39*E. Fl.* 1*qr.* 3*na.* $1\frac{1}{2}$*in.*

APPLICATIONS.

Ex.	*Ans.*
1.	21*T.* 10*cwt.* 2*qr.* $6\frac{2}{9}$*lb.*
2.	——
3.	$73,296+.
4.	$0,0899+.
5.	$1 283+.
6.	20–37868 *rem.*
7.	——
8.	$1,255+. $1,244+
9.	$25141072,267+.
10.	$0,06+.
11.	$166743,259+.
12.	——
13.	37–2588 *rem.*
14.	38–190 *rem*
15.	$334477,744+.

PROPERTIES OF THE 9's.

ART. 85.

Ex.	*Ans.*
1.	2. 7.
2.	——
3.	1. 7.

ART. 86.

Ex.	*Ans.*
2	140487982–7.
3.	177234105–3.

ART. 87.

Ex.	*Ans.*
2.	503602–7. *Excess in minuend*, 7.
3.	45991735–7. *Excess in minuend*, 8.

VULGAR FRACTIONS.

ART. 98.

Ex.	Ans.
1.	$\frac{35}{8}$.
2.	$\frac{21}{11}$.
3.	$\frac{162}{13}$.
4.	$\frac{132}{17}$.
5.	——
6.	$\frac{108}{17}$.
7.	$\frac{126}{11}$.
8.	$\frac{195}{27}$.

ART. 99.

Ex.	Ans.
1.	$\frac{4}{17}$.
2.	——
3.	$\frac{22}{741}$.
4.	$\frac{21}{456}$.
5.	$\frac{15}{126}$.
6.	$\frac{30}{147}$.
7.	——
8.	$\frac{9}{492}$.

ART. 100.

Ex.	Ans.
1.	$\frac{67}{282}$
2.	$\frac{34}{783}$.
3.	$\frac{127}{4188}$.
4.	——
5.	$\frac{327}{5684}$.
6.	$\frac{1249}{4815}$.
7.	$\frac{327}{1825}$.
8.	$\frac{94}{3896}$.

ART. 101.

Ex.	Ans.
1.	——
2.	$\frac{16}{8}$, $\frac{16}{4}$, $\frac{16}{2}$.
3.	$\frac{34}{24}$, $\frac{34}{16}$, $\frac{34}{8}$.
4.	$\frac{147}{7}$, $\frac{147}{2}$.
5.	$\frac{127}{7}$.
6.	——
7.	$\frac{246}{42}$, $\frac{246}{21}$.
8.	$\frac{449}{12}$, $\frac{449}{4}$.

ART. 102.

Ex.	Ans.
2.	$\frac{51}{57}$, $\frac{68}{76}$, $\frac{102}{114}$, $\frac{119}{133}$, $\frac{153}{171}$, $\frac{255}{285}$, $\frac{289}{323}$.
3.	$\frac{117}{153}$, $\frac{156}{204}$, $\frac{208}{272}$, $\frac{91}{119}$, $\frac{65}{85}$, $\frac{143}{187}$.

ART. 103.

Ex.	*Ans.*
3.	$\frac{16}{64}, \frac{8}{32}, \frac{4}{16}, \frac{2}{8}, \frac{1}{4}$
4.	$\frac{30}{90}, \frac{20}{60}, \frac{15}{45}, \frac{12}{36}, \frac{10}{30}, \frac{6}{18}, \frac{5}{15}, \frac{4}{12}, \frac{3}{9}, \frac{2}{6}, \frac{1}{3}$.

GREATEST COMMON DIVISOR.

1.	4.		2D METHOD.
2.	45.		ART. 105.
3.	630.	2.	43, 3, 5.
4.	267.	3.	3, 3, 2, 2, 2, 5.
5.	——		
6.	12.		ART. 106.
7.	8.	2.	3.
8.	4.	3.	25.
9.	3.	4.	——
10.	——	5.	15

LEAST COMMON MULTIPLE.

3.	840.	11.	——
4.	147.		2D METHOD.
5.	840.	3.	1260.
6.	——	4.	7200.
7.	78.	5.	2520.
8.	84.	6.	1008.
9.	1008.	7.	——
10.	156.	8.	10800.

REDUCTION OF VULGAR FRACTIONS.

	ART. 110.	6.	$\left\{ 2\frac{77}{125}, 24, 7\frac{1700}{6941}, 13\frac{47712}{72301}. \right.$
2.	$12\frac{3}{8}$.	7.	$219\frac{2}{4}$.
3.	$2\frac{5}{7}$*yd.*	8.	$191\frac{15}{25}$.
4.	$5\frac{6}{9}$*bu.*	9.	14.
5.	——	10.	——

Ex.	Ans.
11.	$10731\frac{55}{349}$.

ART. 111.

Ex.	Ans.
3.	$\frac{287}{6}$.
4.	$\frac{34513}{51}$, $\frac{7899}{9}$, $\frac{69047}{100}$, $\frac{38177}{104}$.
5.	$\frac{148261}{175}$, $\frac{91772}{104}$, $\frac{59267822}{879}$.
6.	——
7.	$\frac{28278}{151}$.
8.	$\frac{1346}{9}$.
9.	$\frac{37219}{99}$.
10.	$\frac{1749383049}{99999}$.
11.	——
12.	$\frac{16106}{9}$.
13.	$77\frac{7}{7} = 78$.

ART. 112.

Ex.	Ans.
2.	$\frac{1}{3}$.
3.	$\frac{1}{8}$.
4.	——
5.	$\frac{117}{265}$.
6.	$\frac{2}{13}$.
7.	$\frac{7}{9}$.
8.	$\frac{7}{9}$.
9.	——
10.	$\frac{4}{7}$.
11.	$\frac{187}{515}$.
12.	$\frac{41}{51}$.
13.	$\frac{69}{349}$.
14.	——
15.	$\frac{183}{2381}$.

ART. 113.

Ex.	Ans.
2.	$\frac{135}{9}$.
3.	$\frac{24325}{175}$.
4.	$\frac{332497}{181}$
5.	——
6.	$\frac{14863}{89}$.
7.	$\frac{52258}{17}$.

ART. 114.

Ex.	Ans.
2.	$\frac{819}{8}$.
3.	$\frac{15}{7}$.
4.	——

ART. 115.

Ex.	Ans.
3.	$\frac{375}{10504}$.
4.	$\frac{1681}{175560}$.
5.	$\frac{3045}{344} = 8\frac{293}{344}$.

ART. 116.

Ex.	Ans.
2.	$\frac{381}{760}$.
3.	——
4.	$\frac{99}{332}$.
5.	$345\frac{4}{5}$.
6.	$\frac{7}{42936}$.
7.	$\frac{28098520}{636899967}$.

ART. 117.

Ex.	Ans.
4.	——
5.	$\frac{525}{600}$, $\frac{1080}{600}$, $\frac{22200}{600}$.
6.	$\frac{200}{50}$, $\frac{62}{50}$, $\frac{1550}{50}$.
7.	$\frac{1080}{144}$, $\frac{248}{144}$, $\frac{900}{144}$.
8.	$\frac{518}{126}$, $\frac{1026}{126}$, $\frac{1575}{126}$.
9.	——
10.	$\frac{5}{28}$, $\frac{9}{28}$.
11.	$\frac{20664}{4032}$, $\frac{14976}{4032}$, $\frac{16576}{4032}$, $\frac{26712}{4032}$.
12.	$\frac{126}{210}$, $\frac{140}{210}$, $\frac{30}{210}$, $\frac{105}{210}$.
13.	$\frac{2304}{4032}$, $\frac{2240}{4032}$, $\frac{1512}{4032}$, $\frac{3528}{4032}$, $\frac{76608}{4032}$.

Ex.	*Ans.*
14.	——

ART. 118.

Ex.	*Ans.*
3.	$\frac{2}{12}, \frac{1}{12}, \frac{9}{12}$.
4.	$\frac{42}{84}, \frac{24}{84}, \frac{24}{84}$.
5.	$\frac{15}{24}, \frac{92}{24}, \frac{18}{24}$.
6.	$\frac{154}{24}, \frac{228}{24}, \frac{127}{24}$.
7.	——

ART. 119.

Ex.	*Ans.*
2.	$\frac{36}{45}, \frac{40}{45}, \frac{9}{45}$.
3.	$\frac{122}{8}, \frac{51}{8}, \frac{44}{8}$.
4.	$\frac{72}{360}, \frac{60}{360}, \frac{320}{360}$.
5.	$\frac{67}{120}, \frac{18}{120}, \frac{300}{120}$.
6.	——
7.	$\frac{450}{144}, \frac{624}{144}, \frac{1200}{144}, \frac{2079}{144}$.
8.	$\frac{6}{12}, \frac{8}{12}, \frac{9}{12}, \frac{10}{12}$.
9.	$\frac{36}{90}, \frac{60}{90}, \frac{50}{90}, \frac{63}{90}$.
10.	$\frac{16}{48}, \frac{36}{48}, \frac{40}{48}, \frac{42}{48}, \frac{33}{48}, \frac{34}{48}$.

REDUCTION OF DENOMINATE FRACTIONS.

ART. 121.

Ex.	*Ans.*
3.	——
4.	$\frac{5}{4536}$*hhd.*
5.	£$\frac{4}{300}$.
6.	£$\frac{1}{2880}$.
7.	$\frac{3}{504}$*hhd.*
8.	——
9.	$\frac{187}{182880}$*da.*
10.	$\frac{3}{448}$*cwt.*
11.	$\frac{1}{71680}$*T.*
12.	£$\frac{3}{7}$.
13.	——
14.	$\frac{1}{288}$*lb.*
15.	$\frac{1}{140}$*lb.*
16	$\frac{11}{819}$*hhd.*

ART. 122.

Ex.	*Ans.*
2.	64*lb.*
3.	32*d.*
4.	480*m.*
5.	$\frac{480}{9}$*P.*
6.	——
7.	$\frac{3}{8}$*gal.*
8.	$\frac{128}{5}$*pt.*
9.	$\frac{259200}{7}$*sec.*
10.	5040*gi.*
11.	——
12.	$\frac{7}{4}$*d.*
13.	$\frac{4}{5}$*pwt.*
14.	$\frac{4}{7}$*lb.*
15.	$\frac{4233600}{71}$*sec.*
16.	——
17.	$\frac{224}{17}$*na.*
18.	$\frac{174240}{23}$*ft.*
19.	$\frac{2448}{11}$℈.

ART. 123.

Ex.	*Ans.*
2.	9*oz.* 12*pwt.*
3.	——
4.	2*R.* 20*P.*
5.	3*s.* 4*d.*
6.	52*gal.* 2*qt.*

Ex.	Ans.
7.	24*gal.* 2*qt.*
8.	——
9.	7*oz.* 4*pwt.*
10.	1*pi.* 1*hhd.* 31*gal.* 2*qt.*
11.	7*oz.* 4*pwt.*
12.	1*mi.* 6*fur.* 16*rd.*
13.	——
14.	8*yd.* 1*qr.* $1\frac{1}{3}$*na.*
15.	1*pi.* 1*hhd.* 7*gal.*
16.	29*gal.* 1*qt.* $1\frac{7}{11}$*pt.*
17.	98*da.* 8*hr.* 4*m.* $36\frac{12}{13}$*sec.*
18.	——
19.	5*s.* 4*d.*
20.	6*cwt.* 3*qr.* 6*lb.*
21.	10*S. ft.* 216*S. in.*
22.	30*gal.* $2\frac{6286}{138637}$*qt.*

Art. 124.

Ex.	Ans.
3.	——
4.	$\frac{1}{18}$*hhd.*
5.	$\frac{5}{16}$*cwt.*
6.	$\frac{1}{72}$*hhd.*
7.	$\frac{79}{63360}$*mi.*
8.	——

Ex.	Ans.
9.	$\frac{3661}{2592000}$*da.*
10.	$\frac{61}{480}$*da.*
11.	$\frac{61}{960}$ of 2*da.* $\frac{61}{1440}$ of 3*da.* $\frac{61}{1920}$ of 4*da.* $\frac{61}{4800}$ of 10*da.* $\frac{61}{12000}$ of 25*da.*
12.	——
13.	$\frac{23}{54}$*s.*
14.	$\frac{14}{9}$*cwt.*
15.	$\frac{751}{1800}$*lb.*
16.	$\frac{34}{45}$*E. E.*
17.	——
18.	£$\frac{379}{480}$.
19.	$\frac{1}{4}$*groat.*
20.	$\frac{4}{7}$*quarter.*
21.	$\frac{7}{16}$*yd.*
22.	——
23.	$\frac{51}{5760}$℔.
24.	$\frac{5}{336}$*hhd.*
25.	$\frac{23}{5832}$*S. yd.*
26.	$\frac{71}{144}$*L. ch.*
27.	——
28.	$\frac{223}{432}$*hhd.*

ADDITION OF VULGAR FRACTIONS.

Art. 126.

Ex.	Ans.
2.	£$2\frac{1}{7}$.
3.	$\frac{14}{3}$.
4.	2.
5.	——
6.	$\frac{60}{11}$.

Art. 127.

Ex.	Ans.
2.	£$\frac{453}{378}$.
3.	$\frac{10194}{945}$
4.	$\frac{1195}{1008}$.

Ex.	*Ans.*
5.	——
6.	$\frac{1309}{60}$.
7.	$\frac{46353}{3080}$.

ART. 128.

Ex.	*Ans.*
2.	$84\frac{193}{280}$.
3.	$31\frac{101}{105}$.
4.	——
5.	$61\frac{23}{105}$.
6.	$170\frac{53}{140}$.
7.	$8\frac{489}{616}$.
8.	$6\frac{783}{880}$.
9.	——

ART. 129.

Ex.	*Ans.*
2.	$14\frac{1}{18}$*in.* = $\frac{253}{648}$*yd.*
3.	2*da.* $14\frac{1}{2}$*hr.*
4.	2*qr.* 17*lb.* 1*oz.* $3\frac{1}{15}$*d.*
5.	1*mi.* 3*fur.* 18*rd.*
7.	——
8.	1*cwt.* 1*qr.* 27*lb.* 13*oz.*
9.	11*s.* $6\frac{23}{42}$*d.*
10.	£3 12*s.*
11.	2*oz.* 10*pwt.* 12*gr.*
12.	——
13.	2*E. E.* 4*qr.* $0\frac{2}{3}$*na.*
14.	3*fur.* 25*rd.* 3*yd.* $1\frac{19}{35}$*in.*
15.	2*R.* 20*P.* 11*Sq.ft.* $58\frac{1}{35}$*Sq. in.*
16.	3*hhd.* 37*gal.* $3\frac{1}{5}$*qt.*
17.	——
18.	2*da.* 2*hr.* 12*m.*
19.	55*da.* 10*hr.* $8\frac{1}{4}$*m.*

SUBTRACTION OF VULGAR FRACTIONS.

ART. 131.

Ex.	*Ans.*
2.	$\frac{166}{105}$.
3.	$\frac{3281}{9765}$.
4.	——

ART. 132.

Ex.	*Ans.*
2.	$\frac{1}{12}$.
3.	$2\frac{1}{4}$.
4.	$\frac{19}{68}$.
5.	$\frac{1}{32}$.
6.	——
7.	$\frac{3}{80}$.
8.	$\frac{77}{24}$.
9.	$\frac{1717137}{8242920}$.
10.	$75\frac{3}{4}$.
11.	——
12.	$364\frac{4}{15}$.
13.	$47\frac{3}{5}$.
14.	$\frac{1}{40}$.
15.	$9\frac{14}{15}$.
16.	——

ART. 133.

Ex.	*Ans.*
2.	11*hr.* 59*m.* $59\frac{1}{3}$*sec.*
3.	1*lb.* 8*oz.* 16*pwt.* 16*gr.*
4.	16*gal.* 2*qt.* 1*pt.* $3\frac{7}{95}$*gi.*
5.	9*s.* 3*d.*
6.	——

Ex.	Ans.
7.	4*cwt.* 1*qr.* 15*lb.* 1*oz.* $9\frac{3}{5}$*dr.*
8.	14*s.* $3\frac{3}{7}$*d.*
9.	9*oz.* 7*pwt.* 12*gr.*
10.	7*cwt.* 1*qr.* 27*lb.* 8*oz.*
11.	——
12.	1*mi.* 1*fur.* 16*rd.*
13.	202*da.* 21*hr.* 45*m.* $32\frac{1}{7}$*sec.*
14.	Troy oz. 1*pwt.* $18\frac{1}{2}$*gr.* greater.

MULTIPLICATION OF VULGAR FRACTIONS.

ART. 135.

Ex.	Ans.
2.	$3\frac{1}{12}$.
3.	——
4.	$33\frac{24}{47}$.
5.	$42\frac{1}{3}$.
6.	$124\frac{101}{145}$.
7.	$493\frac{119}{196}$.
8.	——
9.	$119\frac{69629}{87146}$.

ART. 136.

Ex.	Ans.
3.	$\frac{7}{8}$.
4.	$8\frac{1}{10}$.
5.	$\frac{53}{56}$.
6.	——
7.	20.
8.	$\frac{23}{84}$.
9.	$14\frac{124}{231}$.
10.	$2\frac{8}{21}$.
11.	——
12.	540.
13.	$\frac{11753}{4087964}$.

ART. 137.

Ex.	Ans.
3.	$608\frac{7}{12}$.
4.	$5987\frac{2}{9}$.
5.	——
6.	$10622\frac{2}{3}$.

GENERAL EXAMPLES.

Ex.	Ans.
1.	$1\frac{157}{308}$.
2.	$\frac{303}{700}$.
3.	$\frac{140}{8181}$.
4.	——
5.	\$$5\frac{1}{4}$.
6.	\$36
7.	\$$7\frac{13}{16}$.
8.	£2 3*s.* 9*d.*
9.	——
10.	\$$8\frac{7}{16}$.
11.	\$$267\frac{3}{8}$.

DIVISION OF VULGAR FRACTIONS.

ART. 139.

Ex.	Ans.
2.	$\frac{2}{37}$.
3.	$1\frac{121}{285}$.
4.	——
5.	$\frac{379}{19005}$.
6.	$\frac{37}{152}$.

Ex.	*Ans.*	*Ex.*	*Ans.*
7.	$\frac{61}{777}$.	14.	$58\frac{31758}{33341}$.
Art. 140.		15.	$1\frac{7803}{59356}$.
1.	$\frac{7}{8}$.	16.	21.
2.	$29\frac{1}{4}$.	17.	$\frac{5}{1296}$.
3.	——	18.	——
4.	$68\frac{5877}{7029}$.	19.	$8\frac{8}{21}$*cts.*
5.	153801.	20.	$4\frac{4}{49}$*cts.*
6.	$30\frac{10942}{25553}$.	21.	$5,25.
7.	$\frac{2}{3}$.	22.	\$2$\frac{94}{125}$.
8.	——	23.	——
9.	$71\frac{91}{182}$.	24.	$88\frac{512}{1701}$*cts.*
10.	$20\frac{20}{39}$.	25.	$95\frac{5}{16}$*cts.*
11.	$\frac{63}{64}$.	26.	$19\frac{23}{51}$*lb.*
12.	$9\frac{8}{13}$.	27.	\$2$\frac{454}{3535}$.
13.	——	28.	——

DECIMAL FRACTIONS.

Art. 143.		15.	——
1.	41.3.	16.	.000003.
2.	16.000003.	17.	.00039.
3.	5.09.		
4.	65.015.	**Art. 144.**	
5.	——		
6.	2.00000003.	1.	$17.389.
7.	.492.	2.	$92.895.
8.	3000.0021.	3.	——
9.	47.0021.	4.	$47.25.
10.	——	5.	$39.397.
11.	39.640.	6.	$12.003.
12.	300.000840.	7.	$147.04.
13.	.650.	8.	——
14	50000.04.	9.	$4.006.

Ex.	*Ans.*	*Ex.*	*Ans.*
10.	$14.039.	13.	——
11.	$149.332.	14.	$0.058.
12.	$1328.005.	15.	$3856.02.

ADDITION OF DECIMAL FRACTIONS.

Ex.	Ans.	Ex.	Ans.
1.	1306.1805.	11.	31.02464.
2.	528.697893.	12.	1.110129.
3.	——	13.	——
4.	1.5415.	14.	$641.249.
5.	446.0924.	15.	.111.
6.	27.2087.	16.	4.0006.
7.	88.76257.	17.	2.413009.
8.	——	18.	——
9.	1835.599.	19.	$1132.365.
10.	397.547.		

SUBTRACTION OF DECIMAL FRACTIONS.

Ex.	Ans.	Ex.	Ans.
2.	3277.9121.	13.	17.949.
3.	249.60401.	14.	.699993.
4.	9.888899.	15.	——
5.	——	16.	.999.
6.	2.7696.	17.	6373.9.
7.	1571.85.	18.	565.007497.
8.	.6946.	19.	20.9942.
9.	.89575.	20.	——
10.	——	21.	10.030181.
11.	1379.25922.	22.	2.0294.
12.	99.706.		

MULTIPLICATION OF DECIMAL FRACTIONS.

Ex.	Ans.	Ex.	Ans.
2.	329.307391.	4.	——
3.	742.036196.	5.	26.99178

Ex.	*Ans.*	*Ex.*	*Ans.*
6.	10376.283913.	14.	——
7.	275539.5065.	15.	933.8253150762.
8.	.020621125.	16.	.25.
9.	——	17.	.0025.
10.	175.26788356.	18.	.00715248.
11.	.00043204577.	19.	——
12.	215.67436625.	20.	.02860992.
13.	.000000000294.	21.	2.435141056.

CONTRACTION IN MULTIPLICATION.

2.	258.13005.	4.	——
3.	162.525.	5.	3566163.

DIVISION OF DECIMAL FRACTIONS.

ART. 152.

2.	2.22.	10.	254.7347748. 25473.47748. 254734.7748. 2547347.748. 25473477.48. 254734774.8.
3.	8.522.		
4.	33.331.		
5.	——		
6.	12420.5.	11.	——
7.	25.05068. 250.5068. 2505.068. 25050.68. 250506.8.	12.	1918+.
		13.	.00473+.
		14.	1.74412.
		15.	69.7125.
8.	48.65961. 4865.961. 48659.61. 486596.1. 4865961.	16.	——
		17.	12976+.
		18.	.0049589+.

ART. 154.

9.	41.622. 416.22. 4162.2. 41622. 416220. 4162200.	2.	10970.
		3.	60200.
		4.	——
		5.	100.

Ex.	Ans.	Ex.	Ans.
6.	10. 100. 1000. 30. 20. 2000. 12. 1200. 500000.	4.	$11055.2925
		5.	——
		6.	$140.625.
		7.	$20.87.
		8.	$3731.123.
		9.	224.58*S. yd.*
		10.	——
		11.	$365.61525.
	ART. 155.	12.	$1.35
3.	8.3111+.	13.	269 *acres* = *area.* $13573.204 = *cost.* $50.458+ = *average price.*
4.	1.563+.		
5.	——		
	ART. 156.	14.	$7631.8855 = *eldest son's share.* $5723.914125 = *share of each of the other sons.*
1.	339.513001*yd.*		
2.	155.1011*lb.*		
3.	$88.141.		

CONTRACTION IN DIVISION.

2.	——	4.	11.5834036625–6 *rem.*
3.	35.2843–3 *rem.*	5.	3202.8870–1 *rem.*

REDUCTION OF VULGAR FRACTIONS TO DECIMALS.

2.	.125. .0159+.	11.	.000000488+.
3.	——	12.	.8571+.
4.	.5. .0028+.	13.	——
5.	1.496+.	14.	.2571+.
6.	1.333+. .1629+. .792+. 4.666+.	15.	.8947+.
7.	.02343+.	16.	.008033+.
8.	——	17.	.23903+.
9.	.0003.	18.	——
10.	.2224+.	19.	1.5555+.

Ex.	Ans.	Ex.	Ans.
20.	.15909+.	23.	——
21.	$100.8.	24.	1.25.
22.	$17.85.	25.	3.0339+.

REDUCTION OF DENOMINATE DECIMALS.

Art. 160.

1.	.0546875*lb.*
2.	£.325.
3.	——
4.	.029166*da.*+.
5.	3.9375*pk.*
6.	.375*da.*
7.	71.151*mi.*+.
8.	——
9.	.00396*bar.*+.
10.	1.5*yd.*
11.	.6625*lb.*
12.	.7282*yr.*+.
13.	——
14.	£25.977+.
15.	.9375*cwt.*
16.	.7391*mi.*+.
17.	.2325*T.*
18.	——
19.	.7129*da.*+.

Art. 161.

1.	£19.8635+.
2.	£2.325.
3.	.625*s.*
4.	——
5.	2.5*yd.*
6.	1.046875*lb.*
7.	10.16666+.
8.	£.3729+.
9.	——
10.	£.2325757+.
11.	.12968*lb.*+.
12.	.05*oz.*
13.	.398763*lb.*+ *troy.* (See Arith., Art. 20, p. 23.)
14.	——
15.	.633928*cwt.*+.
16.	.042965*cwt.*+. (See Arith., Art. 20, p. 23.)
17.	.3125*yd.*
18.	.55*E. E.*
19.	——
20.	.48125*A.*
21.	.00992*hhd.*+.
22.	.104166*ch.*+.
23.	.07472*yr.*+.
24.	——
25.	.26175*A.*
26.	.1005113*mi.*+.

Art. 162.

1.	2*qr.* 14*lb.*
2.	2*qt.* 1*pt.*
3.	——
4.	20*gal.* 1*qt.*
5.	136*da.* 21*hr.*

Ex.	*Ans.*	*Ex.*	*Ans.*
6.	1*s.* 8¼*d.*+.	14.	12*dr.*
7.	1*qr.* 14*oz.* 5*dr.*+.	15.	208*da.* 3*hr.* 23*m.* 33*sec.*+.
8.	——	16.	£2 1*s.* 10*d.*+
9.	2*mi.* 24*rd.* 5*yd.* 10*in.*+.	17.	£5 12*s.* 9½*d.*+.
10.	6*s.* 9*d.*	18.	——
11.	6*cwt.* 3*qr.*	19.	1*coom* 1*strike* 2*pk.* 3*qt.* 1*pt.*+.
12.	8*P.*		
13.	——	20.	1℥ 13+

CIRCULATING OR REPEATING DECIMALS.

Art. 164.

3. Factors of den. 5×5×2. decimal value .06.
4. 37×5×2. .0878+.
5. ——
6. 2×2×2×2×2×2×2×2×5. .01328125.

Art. 165.

7. 5×5×5×2. .028.
8. 5×5×5×5. .0176.
9. 2×2×2×2×2×2×2. .1328125.

REDUCTION OF CIRCULATING DECIMALS.

Art. 175.

3. ——
4. $\frac{85}{143}$, $\frac{4}{11}$, $\frac{1}{7}$.

Art. 176.

4. $\frac{5}{36}$, $7\frac{269}{495}$, $\frac{29}{666}$, $37\frac{49}{90}$, $\frac{223}{330}$, $\frac{75434}{99999}$.
5. $\frac{34}{45}$, $\frac{217}{495}$, $\frac{21}{225}$, $4\frac{1256}{1665}$, $\frac{163}{16500}$, $\frac{41}{90}$.

Art. 177. *Sec.* 2.

2. 2.4`181818′+. .5`925925′+. .008`497133′+.

Sec. 4.

1. 165.16`416416′+. .04`040404′+. .03`777777′+.
2. .5`333333′+. .4`757575′+. 1.7`577577′+.

ART. 178.

Ex.	*Ans.*
2.	.1875.
3.	.0 0344827586206896551724137931′+.
4.	——

ADDITION OF CIRCULATING DECIMALS.

2.	95.2ˋ829647′+.	5.	47.4ˋ754481′+.
3.	69.74ˋ203112′+.	6.	——
4.	55.6ˋ209780437503′+.		

SUBTRACTION OF CIRCULATING DECIMALS.

2.	45.7ˋ755′+.	6.	——
3.	2.9ˋ957′+.	7.	4.619ˋ525′+.
4.	5.09.	8.	1.0923ˋ7+.
5.	.65ˋ370016280906′+.	9.	1.3462ˋ937′+.

MULTIPLICATION OF CIRCULATING DECIMALS.

2.	——	7.	——
3.	1.093ˋ086′+.	8.	11.ˋ068735402′+.
4.	1.6411ˋ7+.	9.	.81654ˋ168350′+.
5.	1.7183ˋ39′+.	10.	189.301ˋ977′+
6.	1.4710ˋ037′+.		

DIVISION OF CIRCULATING DECIMALS.

2.	13.570413ˋ961038′+.	6.	3.1ˋ45′+.
3.	——	7.	3.ˋ8235294117647058′+.
4.	7.719ˋ54′+.	8.	——
5.	26.7837ˋ428571′+.	9.	15.48ˋ423′+.

RATIO AND PROPORTION OF NUMBERS.

ART. 183.

Ex.	Ans.
1.	2.
2.	4.
3.	4.
4.	——
5.	$\frac{1}{5}$.
6.	$\frac{3}{10}$.
7.	$\frac{1}{2}$.
8.	$\frac{1}{3}$.
9.	——
10.	$\frac{1}{10}$.
11.	$\frac{1}{25}$.

ART. 185.

Ex.	Ans.
1.	3. 5. $\frac{1}{2}$. $\frac{1}{9}$. $\frac{15}{16}$.

ART. 187.

Ex.	Ans.
1.	9 : 8 : : 18 : 16.
2.	——
3.	16 : 9 : : 48 : 27.
4.	13 : 19 : : 52 : 76.
5.	16 : 21 : : 80 : 105.
6.	35 : 42 : : 210 : 252.
7.	——
8.	23 : 45 : : 207 : 405.

ART. 188.

Ex.	Ans.
1.	6.
2.	$\frac{1}{3}$.
3.	$\frac{3}{10}$.

OF CANCELLING.

ART. 190.

Ex.	Ans.
3.	——
4.	11.
5.	9.
6.	$2\frac{1}{2}$.
7.	8.
8.	——
9.	12.
10.	6.

ART. 191.

Ex.	Ans.
1.	$\frac{63}{512}$.
2.	27.
3.	——
4.	323.
5.	$15\frac{369}{385}$.

ART. 192.

Ex.	Ans.
1.	2 : 8 : : 1 : 4.
2.	2 : 7 : : 6 : 21.
3.	——

RULE OF THREE.

APPLICATIONS.

Ex.	Ans.
1.	\$330.
2.	£9 6*s*. 8*d*.
3.	\$2762,50.
4.	3300*lb*.
5.	——

Ex	*Ans.*	*Ex.*	*Ans.*
6.	32 *men.*	17.	£1913 6*s.* 8*d.*
7.	3*hr.* 2*m.* $49\frac{43}{53}$*sec.*	18.	$6\frac{2}{3}$*oz.*
8.	$16432\frac{1}{2}$ *miles.*	19.	£1270 1*s.* $9\frac{11}{49}$*d.*
9.	$121,875.	20.	——
10.	——	21.	£39637 10*s.*
11.	20*days.*	22.	120 *yards.*
12.	4200*bu.*	23.	190 *guineas.*
13.	£253 10*s.* $\frac{3}{4}$*d.*	24.	$19\frac{43}{95}$*da.*
14.	6*oz.* $15\frac{9}{25}$*dr.*	25.	——
15.	——	26.	108000.
16.	£8 16*s.* $2\frac{38}{365}$*d.*		

27. Rate 3*mi.* 1*fur.* $18\frac{2}{11}$*rd.* per hour. Time 20*hr.* 21*m.* $25\frac{5}{7}$*sec.*

Ex	*Ans.*	*Ex.*	*Ans.*
28.	17 *times round.*	36.	——
29.	$4\frac{12}{17}$ *days.*	37.	8*s.* 2*d.* $3\frac{24545}{29117}$*far.*
30.	——		
31.	13*oz.* per day.	38.	$112,86.
32.	588000*lb.* total weight. 42000*lb.* spoiled.	39.	whole weight, 588000*lb.* they received, 546000*lb.*
33.	2018*E. Fl.* 2*qr.*		
34.	£9 3*s.* 9*d.*	40.	whole weight 9408000*oz.* 14*oz* per day.
35.	$35\frac{1}{4}$*yd.* baize. cost 11*s.* 2*d.* $1\frac{3}{141}$*far.* per yard.	41.	——

RULE OF THREE BY ANALYSIS.

Ex	*Ans.*	*Ex.*	*Ans.*
2.	87 *miles.*	11.	105 *days.*
4.	$78.	12.	——
5.	504 *miles.*	13.	106*yd.* 2*ft.*
6.	$2,08.	14.	$5\frac{1}{3}$ *days.*
7.	——	15.	27 *days.*
8.	$380.	16.	1*lb.* 5*oz.* $9\frac{3}{5}$*dr.*
9.	$0,429+.	17.	——
10.	30 *days.*	18.	64 *bottles.*

Ex.	*Ans.*	*Ex.*	*Ans.*
19.	25*yr.* 202*da.* 2*hr.*	21.	$67\frac{8}{11}$*gal.*
20.	He gained $246,75.	22.	——

RULE OF THREE BY CANCELLING

Ex.	Ans.	Ex.	Ans.
2.	8*s.* 8*d.*	13.	54 *days.*
3.	$400.	14.	6 *days.*
4.	160 *days.*	15.	12 *days.*
5.	2700*lb.*	16.	——
6.	——	17.	3600.
7.	$12.	18.	$8\frac{1}{6}$ *yards.*
8.	60 *days.*	19.	16 *months.*
9.	23352*bu.*	20.	121 *yards.*
10.	£7 4*s.*	21.	——
11.	——	22.	£41, *D's part.* £61 10*s. paid by each of the others.*
12.	20 *days.*		

EXAMPLES INVOLVING FRACTIONS.

Ex.	Ans.	Ex.	Ans.
2.	£1 18*s.* 6*d.*	13.	$52.50.
3.	£682 18*s.* 9*d.*	14.	$638.08+
4.	£112 12*s.* $9\frac{57}{150}$*d.*	15.	——
5.	——	16.	$21 10*s.* $1\frac{1}{2}$*d.*
6.	£102 7*s.* 7*d.*+.	17.	62.734375 *days.*
7.	1.431\frac{9}{11}$.	18.	£77 3*s.* $7\frac{3}{4}$*d.*
8.	14.58\frac{1}{3}$.	19.	$5.625.
9.	14 *days.*	20.	——
10.	——	21.	2700.
11.	.005172 *guineas.*+.	22.	293.28*gals.*
12.	.71428*cwt.*+.	23.	$114\frac{30}{31}$*hrs.*

24. At the equator, $1038\frac{336}{359}$ *miles.*
" Madras, &c. $1010\frac{344}{359}$ *miles.*
" Madrid, " $794\frac{344}{359}$ *miles.*
" Petersburg " $519\frac{141}{359}$ *miles.*

DOUBLE RULE OF THREE.

Ex.	*Ans.*	*Ex.*	*Ans.*
5	——	13.	14*oz.*
6.	11676*bu.*	14.	540*S. yd.* 36*yd.* long.
7.	36 *days.*	15.	——
8.	2808*qrs.*	16.	27 *men.*
9.	168.	17.	£11 $2\frac{2}{39}$*d.*
10.	——	18.	4*da.* 11*hr.* $54\frac{393}{1568}$*m.*
11.	10 *days.*	19.	$8571\frac{3}{7}$.
12.	9600 *men.*	20.	——

PRACTICE.

13.	£91 1*s.* $\frac{99}{112}$*d.*	32.	——
14.	£44 9*s.* $8\frac{13}{14}$*d.*	33.	£348 14*s.* $1\frac{17}{40}$*d.*
15.	£38 3*s.* $3\frac{6}{7}$*d.*	34.	$135,375.
16.	£19 1*s.* $3\frac{15}{56}$*d.*	35.	$15.
17.	——	36.	$1854.
18.	$1,575.	37.	——
19.	$547,50.	38.	£199 1*s.* $10\frac{7}{8}$*d.*
20.	$5.	39.	$1095.
21.	$108.	40.	$7.
22.	——	41.	$8,10.
23.	£10 15*s.* $\frac{233}{256}$*d.*	42.	——
24.	£18 7*s.* $6\frac{7}{16}$*d.*	43.	£550 11*s.* $10\frac{1}{2}$*d.*
25.	£3 7*s.* $5\frac{59}{80}$*d.*	44.	£661 17*s.* $3\frac{3}{40}$*d.*
26.	£2051 13*s.* $10\frac{59}{256}$*d.*	45.	£445 5*s.* $4\frac{73}{128}$*d.*
27.	——	46.	$65,90625.
28.	$1617.	47.	——
29.	$643,75.	48.	$148,2890625.
30.	$519,75.	49.	$48,91796875.
31.	£1877 10*s.* $9\frac{39}{160}$*d.*		

TARE AND TRET.

Ex.	Ans.	Ex.	Ans.
1.	65*cwt.* 1*qr.* 19*lb.*	6.	$808,71.
2.	4*cwt.* 1*qr.* 15*lb.* 8*oz.*	7.	£912 14*s.* $5\frac{4}{7}$*d.*
3.	——	8.	——.
4.	$265,16.	9.	3*T.* 8*cwt.* 3*qr.* 5*lb.*
5.	£59 4*s.* $3\frac{9}{56}$*d.*	10.	6*T.* 12*cwt.* 3*qr.* $3\frac{1}{2}$*lb.* Value, $306,724685.

Ex.	Ans.
11.	In leaf, $26,0586+ per *cwt.* In rolls, $29,5504+ " "
12.	Net weight, 161*cwt.* 1*qr.* 8.961*lb.*+. Value, $1177,709+.
13.	——
14.	Net weight, 27*cwt.* 2*qr.* $6\frac{1}{8}$*lb.* Value, $232.837+.
15.	Net weight, 10*cwt.* $27\frac{1}{2}$*lb.* Value, $4,0417+.
16.	Net weight, 20*cwt.* 2*qr.* $13\frac{1}{8}$*lb.* 358.312*gal.*+ Freight, $444,306+.
17.	Net weight, 298*cwt.* 2*qr.* 23,454*lb.*+ Freight, $355,464+.

PERCENTAGE.

Art. 204.

Ex.	Ans.
1.	——
2.	432*bar.*
3.	42*hhd.*
4.	$10,80.
5.	$24,25.
6.	——
7.	205 *boxes.*
8.	742*gal.* 3*qt.* $3\frac{1}{5}$*gi.*

Art. 205.

Ex.	Ans.
1.	$15\frac{1}{22}$ *per cent.*
2.	55 " "
3.	——
4.	20 *per cent*
5.	$16\frac{2}{3}$ " "

SIMPLE INTEREST.

ART. 207.

Ex.	Ans.
4.	——
5.	$803,25.
6.	$450,32760.
7.	$4853,844.
8.	$643,83375.
9.	——
10.	$235,764.
11.	$1205,9208.
12.	$1375,8144.
13.	$12959,584.
14.	——

ART. 208.

Ex.	Ans.
3.	$1225,511.
4.	$44,20845.
5.	$6015,4272.
6.	$357,9165.
7.	——
8.	$270,5175.
9.	$2953,22685.
10.	$7765,4504.
11.	$1578,6625.
12.	——

ART. 210.

Ex.	Ans.
3.	{ $3,2727875. $145,1545.
4.	$67,278.
5.	$5,56529+.
6.	$0,4314+.
7.	——
8.	$1,13559.
9.	$13,68.

Ex.	Ans.
10.	$31,928125.
11.	$121,77275
12.	——
13.	$609,45776.

ART. 212.

Ex.	Ans.
2.	$16474,3855+.
3.	$1449,70998.
4.	$371,86875.
5.	——
6.	$266,277.
7.	$14096,5+.
8.	$4479,618.
9.	$1149,552.
10.	——
11.	$40900,9335.
12.	$6418,50195.

ART. 213.

Ex.	Ans.
3.	$511,5357.
4.	$167,30.
5.	——
6.	$555,465.
7.	$4200.
8.	$2643,8386.
9.	$7,00.
10.	——
11.	$587,6311+.
12.	$3974,5187+.
13.	$3329,1468+.
14.	$1137,0592+.
15.	——
16.	$678,2599+.
17.	$4824,2366.

Art. 214.

Ex.	*Ans.*	*Ex.*	*Ans.*
3.	$43,2049+.	4.	$8,1855.

5. Interest at 4 per cent, $45,837.
" 5 " " $57,29625.
" $5\frac{1}{2}$ " " $63,025875.
" 6 " " $68,7555.
" 7 " " $80,21475.
" $7\frac{1}{2}$ " " $85,944375.
" 8 " " $91,674.
" $8\frac{1}{2}$ " " $97,403625.
" 9 " " $103,13325.

6. ——
7. $380,28952.
8. $669,7096875.
9. $25571,2473+.

Art. 215.

2. £45 8*s.* $1\frac{3}{4}$*d.*
3. ——
4. £45 10*s.* 2*d.*+
5. £662 3*s.*+
6. £216 1*s.* $10\frac{1}{4}$*d.*+
7. £219 18*s.* $0\frac{1}{4}$*d.*+
8. ——
9. £68 5*s.* 10*d.*+

Art. 216.

1. $394,3256+.
2. $697,986.
3. $3339,6+.
4. ——
5. $4640,5326.
6. $1976,6305+.

Art. 217.

2. $5359,363+.
3. $8921,618+.
4. ——

Art. 220.

2. $3976,848+.
3. $575,569+.
4. $1424,849+.
5. ——

Art. 221.

2. 5 *per cent.*

Art. 222.

2. 2*yr.* 6*mo.*

REDUCTION OF CURRENCIES.

3. £1073 18*s.* $1\frac{1}{2}$*d.*
4. $1967,892+.
5. ——
6. $2551,733+.

COMPOUND INTEREST.

Art. 223.		Art. 224.	
2.	$57,3048+.	2.	$578,740+.
3.	$73,015+.	3.	$8611,128+.
4.	$41,216+.	4.	——
5.	——	5.	$7058,617+.
		6.	$2647,996+.
6.	$48165,938+.	7.	£11 18*s.* 11*d.*+.
7.	$14523,553+.	8.	$9974,685+.

LOSS AND GAIN.

Art. 225.			
		6.	——
3.	Loss of $0,75.	7.	$337,50.
4.	$0,966+.	8.	$217,50.
		9.	$4,108.
Art. 227.		10.	25 *per cent.*
1.	——	11.	——
2.	12 *per cent.*	12.	Whole gain, $13,00. Gain, 20 *per cent.*
3.	$1,25.		
4.	$1,20.	13.	$2,05.
5.	$6,33⅓.	14.	$1,031¼.

COMMISSION AND BROKERAGE.

2.	$49725.	11.	$46260.
4.	——	12.	$23700.
5.	$7235,752+.	13.	$25420,195.
6.	$7814,516+.	14.	——
7.	15 *tons.*	15.	$965,30.
8.	$213532,50.	16.	$1995.
		17.	$13573,56.
9.	——	18.	149*T.* 18*cwt.* 2*qr.* 12*lb.* 6*oz.* $1\frac{1}{95}$*dr.*
10.	$59110.		

BANK DISCOUNT.

Art. 239. Ex.	Ans.	Art. 240. Ex.	Ans.
1.	——	2.	$344,59+.
2.	$15240,54.	3.	$5734,32+.
3.	$5,840+.	4.	$695,64+.
4.	$3393,504.	5.	$118,85+.
5.	$29,0097+.	6.	——
6.	——	7.	$1740,61+.
		8.	$1057,51+

DISCOUNT.

Ex.	Ans.	Ex.	Ans.
2.	$1551,918+.	9.	——
3.	£33 17*s.* $7\frac{3}{4}$*d.*+.	10.	$3869,407+.
4.	——	11.	1*bu.* $2\frac{2}{15}$*qts.*
5.	£223 5*s.* 8*d.*+.	12.	$2109,236+.
6.	$5620,175+.	13.	$2763,694+.
7.	$702,485+.	14.	——
8.	£804 19*s.* 5*d.*+.	15.	He lost $6,473+.

INSURANCE.

Ex.	Ans.	Ex.	Ans.
2.	$5168,59.	5.	——
3.	$237,60. $158,40.	6. 7.	$504. $39,375.
4.	$252. $126. $84. $56. $42.	8. 9. 10. 11.	$306,25. $450. —— $18,75.

ASSESSING TAXES.

Ex.	Ans.	Ex.	Ans.
2.	.4685+ *per cent.*	3.	$37901125.

EQUATION OF PAYMENTS.

Ex.	*Ans.*	*Ex.*	*Ans.*
2.	12 *months.*	9.	——
3.	——	10.	6*mo.* 6*days.*
4.	9 *months.*	11.	7*mo.* 3*da.*
5.	$68\frac{22}{31}$ *days.*	12.	*Jan.* 25*th.*
6.	$8\frac{2}{3}$ *months.*	13.	8*mo.* 10*da.*
7.	$67\frac{6}{19}$ *days.*	14.	——

PARTNERSHIP OR FELLOWSHIP.

2. A's share, \$1714,285+.
   B's share, \$285,714+.

3. A's share, £4030.
   B's " £3980.
   C's " £3980.
   D's " £4010.

4. A's share, \$5000.
   B's " \$2500.
   C's " \$3333,33+.
   D's " \$2500.
   E's " \$6666,67+.

DOUBLE FELLOWSHIP.

2. A's share, £9 12*s.*
   B's " £14 8*s.*

3. ——

GENERAL EXAMPLES IN FELLOWSHIP.

1. *75 cents on the dollar.*
   A's part, \$375,27$\frac{3}{4}$.
   B's " \$171.
   C's " \$968,42$\frac{1}{4}$.
   D's " \$532,05.

2. A paid \$3000.
   B " \$3000.
   C " \$9000.
   A's gain, \$250.
   B's " \$250.
   C's " \$750.

3. 1st son's share, \$3333,33$\frac{1}{3}$.
   2d " \$3000.
   3d " \$3000.
   4th " \$2666,66$\frac{2}{3}$.

4. \$750 *widow's gain.*
   \$375 *younger son's gain.*
   \$3000 *widow's share.*
   \$1500 *younger son's do.*

5. ——

6. A's loss, \$46,526+.
   B's " \$130,273+.
   C's " \$238,213+.
   D's " \$334,988+.

7. A's share, \$36.
   B's " \$28.

ALLIGATION MEDIAL.

Ex.	Ans.	Ex.	Ans.
2.	\$0,84375.	4.	——
3.	$\$0,28\frac{36}{43}$.	5.	73°.

ALLIGATION ALTERNATE.

ART. 252.

2. 2*lb.* at 8*cts.*; 2*lb.* at 10*cts.*; 6*lb.* at 14*cts.*

3. 3 parts of 16 *carats.*; 2 " of 18 "; 3 " of 23 "; 5 " of 24 "

4. 6*gal.* at 10*s.*; 3*gal.* at 14*s.*; 4*gal.* at 21*s.*; 8*gal.* at 24*s.*

ART. 253.

2. ——

3. 12*bu.* of oats.; 12*bu.* of barley.; 12*bu.* of rye.; 96*bu.* of wheat.

4. 32*gal.* of spirits.; 32*gal.* of Eng. brandy.; 40*gal.* of French do.

ART. 254.

2. 29*lb.* at 5*s.*; $14\frac{1}{2}$ at 6*s.*; $14\frac{1}{2}$ at 8*s.*; 29 at 9*s.*

3. 45*gal.* at 4*s.*; 5*gal.* at 6*s.*; 5*gal.* at 8*s.*; 5*gal.* at 10*s.*

4. ——

CUSTOM HOUSE BUSINESS.

1.	\$2812,5.	4.	\$1442,875.
2.	\$418,068.	5.	——
3.	\$251,45+.		

TONNAGE OF VESSELS.

1.	$225\frac{9}{19}$ *tons.*	4.	300.14 *tons*+.
2.	438.59 *tons*+.	5.	
3.	$729\frac{97}{855}$ *tons.*		

GAUGING.

Art. 267.

Ex.	Ans.
2.	32.4938*in.*
3.	28.1010*in.*

Art. 268.

Ex.	Ans.
1.	197.459 *wine gal.*+.
2.	162.613 *beer gal.*+.
3.	——
4.	147.384 *wine gal.*+.

LIFE INSURANCE.

2.	$144.
3.	$189,55.

ENDOWMENTS AND ANNUITIES.

1.	$228,11+.
2.	——

EXCHANGE.

Art. 293.

1.	$8591,975.
2.	$8637,168+.
3.	$9777,636.

Art. 294.

1.	$5630,065.
2.	——
3.	£14014 18*s.* 2*d*+.
4.	$6005,368+.
5.	$807,873+.

Art. 295.

2.	7 *per cent** *above par.*
3.	——
4.	84597 *francs* 66 *centimes.*

Art. 296.

1.	$6657,693.
2.	$1250,52, 3 *per cent** *nearly below par.*

DUODECIMALS.

Art. 301.

1.	15*ft.* 5′.
2.	——
3.	2*ft.* 6′ 3″ 11‴.
4.	5*ft*, 10′ 7″.

EXAMPLES IN ADDITION AND SUBTRACTION.

1.	5*ft.* 8′ 2″ 1‴.
2.	15*ft.* 4′ 10″ 4‴.
3.	——
4.	21*ft.* 6′ 5″ 5‴.

* The percentage is estimated upon the custom house value.

Ex.	Ans.
5.	36*ft.* 4′ 6″ 5‴.
6.	22*ft.* 2′ 1″ 10‴.

ART. 303.

Ex.	Ans.
3.	214*ft.* 1′ 1″ 6‴.
4.	——

Ex.	Ans.
5.	232*ft.* 2′ 8″
6.	866*ft.* 8′ 3″.
7.	2 *cords* 5 *cord feet.*
8.	57*ft.* 4′ 6″.
9.	——
10.	185*ft.* 6′ 4″ 3‴.

INVOLUTION.

Ex.	Ans.
1.	1953125.
2.	343.
3.	3600.
4.	——
5.	1889568.
6.	1.
7.	$\frac{1}{4}$.
8.	.001.
9.	——
10.	.0001.
11.	37.4544.
12.	1000000.
13.	$34\frac{21}{64}$.
14.	——
15.	666024768837.

EXTRACTION OF THE SQUARE ROOT.

ART. 309.

Ex.	Ans.
3.	462.
4.	1506.23+.
5.	3897.89+.
6.	——
7.	4698.

ART. 310.

Ex.	Ans.
2.	57.19+.
3.	69.247+.
4.	2.091+.
5.	——
6.	.0321.
7.	2.104.
8.	6.906.

ART. 311.

Ex.	Ans.
1.	.5236+.
2.	——
3.	.4203+.
4.	1.0682+
5.	.86602+.
6.	.93309+.

EXTRACTION OF THE CUBE ROOT.

ART. 313.

Ex.	Ans.
2	——
3.	179.
4.	319.
5.	439
6.	638.

Ex.	Ans.
7.	——
8.	3002.

Art. 314.

Ex.	Ans.
1.	.5032+.
2.	.955.
3.	2.35.
4.	——
5.	.707.
6.	1.505.

Ex.	Ans.
7.	3.026.

Art. 315.

Ex.	Ans.
1.	$\frac{5}{7}$.
2.	——
3.	$3\frac{1}{3}$.
4.	$\frac{3}{5}$.
5.	.829+.
6.	.822+.
7.	——

ARITHMETICAL PROGRESSION.

Art. 318.

2. $1,53.
3. $205.

Art. 319.

2. 4 *years.*
3. 5 *miles.*

Art. 320.

2. ——
3. Last term 34. Sum 162.

4. 5 *miles* 1300 *yards.*

GENERAL EXAMPLES.

1. 89.
2. 4.
3. ——
4. $4.
5. 3 *miles each day.* 100 " *in all.*
6. 3.
7. $272.

GEOMETRICAL PROGRESSION.

Art. 322.

2. ——
3. £25600.
4. $61,44.

Art. 323.

2. £204 15*s.*
3. $196,83. $295,24.
4. ——

MENSURATION.

Art. 326.

2. 36 *acres.*
3. 5*A.* 1*R.* 15*P.*
4. 135*A.*

Ex.	Ans.
	ART. 329.
1.	437*A*. 2*R*. 34*P*.+.
2.	291*A*. 2*R*. 16*P*.
3.	35*A*. 0*R*. 25*P*.
4.	20*A*.
5.	40*A*.
6.	15*A*.
7.	24*A*. 1*R*. 8*P*.
8.	130080*sq. yds.*
	ART. 331.
2.	21*A*. 0*R*. 39.824*P*.
3.	921*sq. ft.* 10′ 6″.
4.	704.125*sq. yds.*
5.	60*A*. 3*R*. 12.8*P*.
6.	270*A*. 1*R*. 24*P*.
	ART. 332.
2.	584.3376.
3.	125.664.
4.	——
	ART. 333.
2.	7418.
3.	4360.835+.
	ART. 334.
2.	19.635.
3.	153.9384.
4.	1.069016+.
	ART. 336.
2.	615.7536.
3.	4071.5136.
4.	196996571.722104*sq. ms.*
	ART. 338.
2.	268.0832.
3.	2144.6656.
4.	259992792079.869+.
5.	904.7808*sq. ft.*
	ART. 340.
1.	9100*sq. ft.*
2.	1440 *sq. ft.*
	ART. 341.
2.	110592*solid in.*
3.	$42\frac{2}{9}$*solid ft.*
4.	$315\frac{65}{77}$*gal.*
5.	13820*solid ft.*
	ART. 343.
2.	233.333+*sq. ft.*
3.	2827.44*sq. in.*
4.	6283.2*sq. ft.*
	ART. 344.
2.	36442.56.
3.	13571.712.
4.	9650.9952.
5.	7363.125.
	ART. 346.
2.	4380.
3.	2484.
4.	5620.

Ex.	Ans.
5.	5760.
6.	14400.
7.	1800.

ART. 348.

Ex.	Ans.
2.	9160.9056.
3.	8659.035.
4.	2827.44.

ART. 349.

Ex.	Ans.
1.	241.86*ft.*
2.	17.204*ft.*
3.	14.142*ft.*

ART. 350.

Ex.	Ans.
1.	302.9702*ft.*
2.	28*yd.*
3.	77.8875*ft.*

MECHANICAL POWERS.

ART. 355.

Ex.	Ans.
1.	40*lb.*
2.	25*lb.*
3.	50*lb.*
4.	20*lb.*
5.	40*lb.*
6.	1*in.*, $1\frac{1}{2}$*in.*, 2*in.*, 4*in.*, &c.
7.	64*lb.*
8.	150*lb.*

ART. 361.

Ex.	Ans.
1.	60*lb.*
2.	40*lb.*
3.	20*lb.*

ART. 362.

Ex.	Ans.
1.	$7\frac{1}{2}$*ft.*
2.	$1\frac{1}{2}$*ft.*

ART. 363.

Ex.	Ans.
1.	40*lb.*
2.	100*lb.*
2.	60*lb.*

PROMISCUOUS QUESTIONS.

Ex.	Ans.
1.	\$1853,131+.
2.	57 *pieces.*
3.	12*wk.* $3\frac{5}{7}\frac{1}{7}$*da.*
4.	4 *years.*
5.	——
6.	4th *partner's share* \$2500. 3d " " \$3675. 2d " " \$5375. 1st " " \$9625.
7.	*Greater number*, $3664\frac{1}{2}$. *Less* " $1665\frac{1}{2}$.
8.	46*yr.* 11*mo.* 20*da.* $10\frac{1}{2}$*hr.*
9.	120 *yards.*
10.	——
11	\$31,25.
12.	\$6760 *price he would have paid.* \$6890 *price actually paid.*

Ex.			*Ans.*
13.	$454.9375.	16.	3 *o'clock.*
14.	23599680 *cubic yards.*	17.	140.
15.	——		

18. The second $6\frac{3}{8}$ days after the third.
The first $5\frac{1}{10}$ days after the second.

19.	$4646,36+.	27.	A's share, £194 16*s.* $1\frac{19}{77}$*d.* B's " £129 17*s.* $4\frac{64}{77}$*d.* C's " £97 8*s.* $0\frac{48}{77}$*d.* D's " £77 18*s.* $5\frac{23}{77}$*d.*
20.	——		
21.	7500 *men.*		
22.	£157 10*s.*	28.	Gain of 1st, $120. " 2d, $180.
23.	4 *days.*		
24.	A's stock, £304 16*s.* B's " £276 C's " £210	29.	$108\frac{7}{75}$.
		30.	——
		31.	10 *hours.*
25.	——	32.	$1,75.
26.	$62\frac{4}{7}$ *gal.* of the 1st kind. $83\frac{3}{7}$ " " 2d " 146 " " 3d "	33.	$1,853+.
		34.	$44\frac{4}{13}$.
		35.	——

36. $130\frac{10}{11}$ *yards of cloth.*
Price of linen per yard, \$0,58$\frac{10}{11}$.

37.	\$42,34$\frac{2}{11}$.	42.	$1724,363+.
38.	4 *yards.*	43.	356,25.
39.	12*cwt.* 16*lb.* 15*oz.*	44.	$34\frac{2}{5}$ *per cent.*
40.	——	45.	——
41.	4*yr.* 11*mo.* $27\frac{765}{951}$*da.*		

46. Cost per yard, \$4,90$\frac{10}{111}$.
Entire cost, \$18378,37$\frac{31}{37}$.

47. $8640.

48. $920,20 = what the 1st gave.
$2760,60 = " 2d "
$5521,20 = " 3d "

49. \$28$\frac{2}{3}$ amount paid each workman.
1st company cleared $87\frac{36}{47}$ *acres.*
2d " " $77\frac{11}{47}$ "
Cost of clearing, \$8$\frac{82}{495}$ *per acre.*

Ex.	*Ans.*
50.	——
51.	6 *o'clock* 3*m.* $48\frac{114}{217}$*sec.*
52.	140 *miles.*
53.	Share of the 1st, $2019,651+. " " 2d, $4871,803+. " " 3d, $4815,805+. " " 4th, $6467,739+. " " 5th, $1825.
54.	$69\frac{3}{7}$ miles from New-Haven.
55.	——
58.	5*hr.* 41*m.* 20*sec.*

THE END.

www.ingramcontent.com/pod-product-compliance
Lightning Source LLC
LaVergne TN
LVHW020107110826
845151LV00001B/89

* 9 7 8 1 4 2 5 5 4 4 2 4 9 *